Children's
SCIENCE
ENCYCLOPEDIA

Children's
SCIENCE
ENCYCLOPEDIA

Author & Illustrator
A.H. Hashmi

Editor
Rajiv Garg
M.Sc., M.Tech

V&S PUBLISHERS

Published by:

V&S PUBLISHERS

F-2/16, Ansari road, Daryaganj, New Delhi-110002
23240026, 23240027 • *Fax:* 011-23240028
Email: info@vspublishers.com • *Website:* www.vspublishers.com

Regional Offi ce : Hyderabad
5-1-707/1, Brij Bhawan (Beside Central Bank of India Lane)
Bank Street, Koti, Hyderabad - 500 095
040-24737290
E-mail: vspublishershyd@gmail.com

Branch Offi ce : Mumbai
Jaywant Industrial Estate, 2nd Floor–222, Tardeo Road
Opposite Sobo Central Mall, Mumbai – 400 034
022-23510736
E-mail: vspublishersmum@gmail.com

Follow us on:

All books available at **www.vspublishers.com**

© Copyright: V&S Publishers
ISBN 978-93-813849-5-4
Edition 2016

Printed at : Param Offseters, Okhla, New Delhi-110020

PREFACE

Children are inquisitive by nature. They are forever curious about the world around them and want to know more about it. Though they have textbooks for their reference, still they can never quench their thirst for knowledge completely. Encyclopedias are specifically designed to bridge this gap. They help to nurture the hidden Albert Einstein or Marie Curie in the children.

Children's Science Encyclopedia in 17 volumes is one such unparalleled effort in this direction. All the essential subject areas, from Universe and Environment, Plants and Animals, to Human Body, are covered in this comprehensive encyclopedia. Readable entries integrated with informative illustrations make the text lively and interesting, thus enabling better understanding of the topics covered.

Matchless in approach and presentation, *Children's Science Encyclopedia* is certain to find appreciation in the eyes of students and also general readers, for being a one-stop comprehensive guide of the various aspects of Science. A must buy for school libraries and also individual households!

— Publishers

Contents

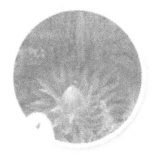

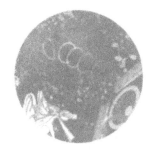

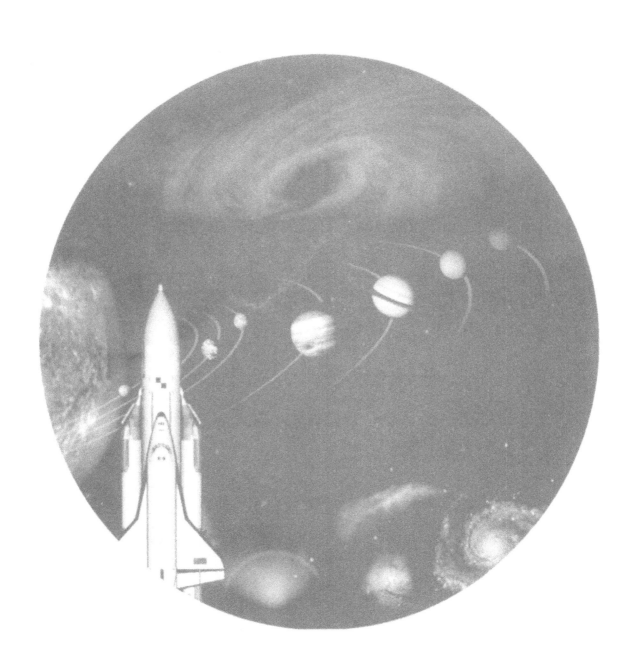

SPACE

Space is the vast region with no known boundaries beyond the Earth's atmosphere. It is the empty area extending in all directions to infinity. The solar system, the stars, galactic dust and galaxies, regions between the galaxies, all form parts of space. Space has no atmosphere. It appears black during the day as well as at night. It does not inhabit any living organism.

Man has been trying to collect information about space with the help of powerful radio telescopes, rockets, satellites, space-crafts and space probes. These space explorations have revealed many mysteries and enhanced man's knowledge about it.

Space probe Mariner 10 gave valuable information about Venus and Mercury in 1974

Space extends to infinity in all directions

STARS

Stars are huge spheres of glowing gases. They vary considerably in size, colour, temperature, brightness and distance from the Earth. The Sun is also a medium-sized star. It is comparatively nearer to the Earth, hence it looks bigger in size.

Stars appear to twinkle because they are seen through the Earth's atmosphere. Stars also seem to move from east to west across the sky. This 'movement' is actually caused by the spinning of the Earth. Stars are often classified according to their size. The four major sizes are super giant, giant, medium and dwarf.

The colour of a star indicates its surface temperature. The hottest stars are blue. Red stars are somewhat cooler, having a temperature of about 2800°C. The Sun and other yellow stars have surface temperature of about 5500°C. Rigel, the blue-white giant star, has a diameter 80 times larger than that of the Sun and is 60,000 times brighter than it.

A spacecraft takes three days to reach the Moon from the Earth. It will take several months to reach the Sun and may take thousands of years to reach the nearest star. Such long distances covered in months and years are very difficult to be measured in kilometres. Therefore, scientists use units like Light Year and Parsec (PC) for measuring the distances of stars. A Light Year is the distance travelled by light in one year at a speed of 3,00,000 km. per second; it is equal to 9.4607×10^{12} km. One Parsec is equal to 3.26 light years or 30.857×10^{12} km.

It takes 1.3 seconds for the moonlight to come to us. Light from the Sun reaches the Earth in 8 minutes 18 seconds. It should take 4.2 light years for the light to reach the Earth from Proxima Centauri, the nearest star next to the Sun. The farthest star in our galaxy lies at a distance of about 63,000 light years (19.325 Parsec).

A star as it appears to the naked eye

A star seen with the help of a telescope

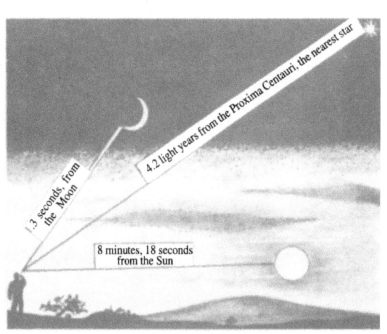

4.2 light years from the Proxima Centauri, the nearest star

1.3 seconds, from the Moon

8 minutes, 18 seconds from the Sun

Light from stars must take a long time to reach the Earth

12

PULSARS, BLACK HOLES AND QUASARS

Pulsars

Pulsars are rapidly spinning stars that emit pulses of radio waves at regular intervals. The word 'pulsar' stands for 'Pulsating Radio Star'.

When a big star explodes, its outer shell scatters to form a nebula while its core contracts into a denser star called the neutron star. Its neutrons are tightly compressed and it has extremely high density. Neutron stars are very small and dull. The average diameter of a neutron star is 10 km. These neutron stars are called pulsars.

Radio pulses coming from pulsars produce a 'tick' like sound on radio telescope. Neutron stars are like lighthouses in space. Ordinary pulsars flash at an interval of every one or half a second. The most rapid flashing pulsar, NP 0532 lies in the Crab Nebula and emits 30 pulses per second. The oldest and the least frequency pulsar is NP 0527 which emits pulses every 3.7 seconds. All pulsars emit pulses at a rate of one pulse per 0.03 second to one pulse per 4 seconds.

Most of the pulsars are not visible through optical telescopes. They may be observed with the help of radio telescopes. Only two pulsars – NP 0532 in the Crab Nebula and PSR 0833-35 in Gum Nebula – can be seen with the help of optical telescope. Scientists have so far discovered more than 100 pulsars.

Black Holes

When very big stars, three times bigger than the Sun, collapse by the force of their own gravity, some black regions are created in the space. These are called black holes. They have gravitational force so enormously high that no matter, not even light can escape from it.

The first black hole was detected in 1972. It was in a binary star of Cygnusx-1 which is a powerful source of X-rays. It is a small member of the binary star, which is completely black. It is not a neutron star and, therefore, it is called black hole. Normally black holes emit X-rays and infrared radiation which help in their detection in space. The mass of black holes may be equal to that of 100 million Suns.

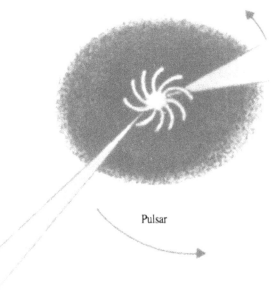

Pulsar

Radio beam

Study of a pulsar through a telescope

13

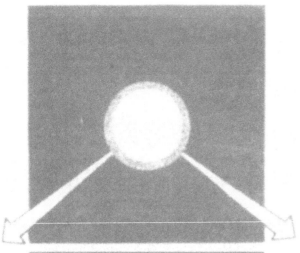

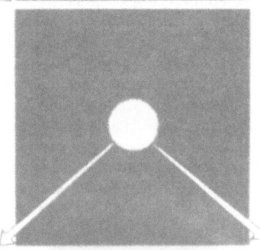

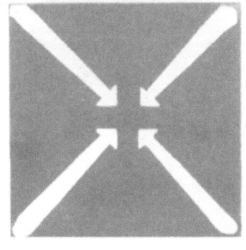

When a star has consumed its hydrogen, its outer shell swells and it turns red. It is considered the stage of collapse of the star. After explosion the star disappears, leaving behind a black hole

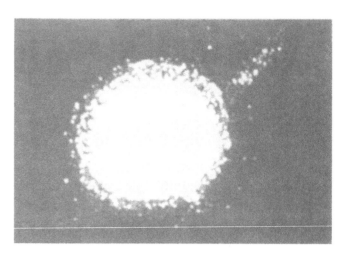

Quasars are a source of radio waves

Quasars

Quasar is an abbreviated form of Quasi Stellar Radio sources. Quasars appear like a star. Through an optical telescope they look like an ordinary dull star, but radio telescope observations have indicated that they are the source of radio waves. Quasar 3C-273 was discovered by Maarten Schmidt in 1962. The value of its redshift was 0.158. Redshift is an effect of change in frequency and is seen in moving objects. If the object is approaching, its light will shift towards the violet end of the spectrum, and if the object is moving away, its light will shift to the red end of the spectrum. Redshift indicates that the source of light was receding away.

Quasars emit radio waves and X-rays with light. The size of a quasar is 1/1,00,000 of our galaxy but its brightness is 100-200 times more. So far 12000 quasars have been discovered.

14

GALAXIES

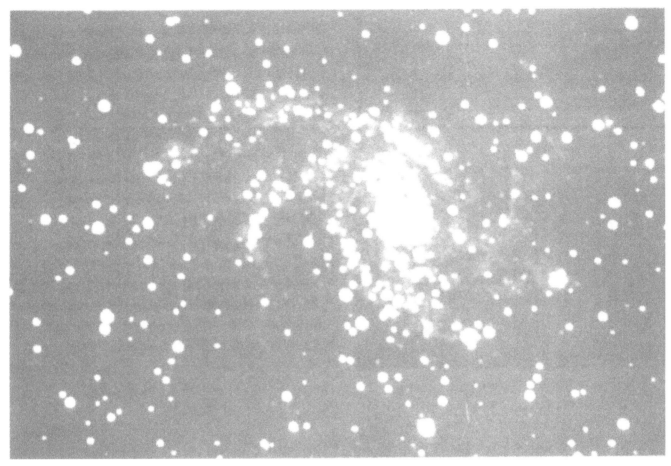

Spiral galaxy, Andromeda

At night, a milky white band of light is seen in the sky. It is called the Milky Way. Italian astronomer Galileo was the first to observe it with the help of his telescope and declared that it was a giant cluster of millions of stars. It is a galaxy. It comprises numerous solar systems and our solar system is one of its members.

Galaxies are giant clusters of stars held together by their mutual power of gravitation. Because of their vastness they are called 'peninsulas of the universe'. Millions of galaxies may be seen with the help of a powerful telescope. They lie at a distance of 1,000 to 1,00,00,000 light years. Most of the galaxies appear scattered in the sky.

It is assumed that huge masses of gases or proto-galaxies were formed as a result of the

The Milky Way

15

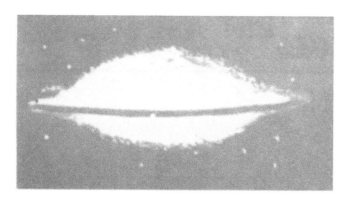

A spiral galaxy

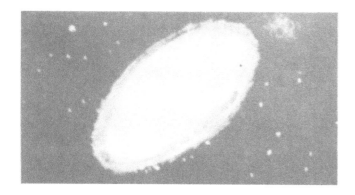

An elliptical galaxy

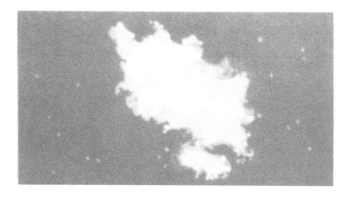

An irregular galaxy

primordial explosion in the universe and they started rotating at their own speed. Galaxies owe their different sizes and forms to their varying speed. Galaxies so far known are spiral, elliptical and irregular in shape.

Ours is a spiral galaxy. Its spiral arms are widespread. Our solar system lies near the edge of the galaxy. Our galaxy measures about 1,00,000 light years (30,600 pc). Its centre is covered under galactic dust particles and lies at about 32,000 light years (9,800 pc) from the Sun.

Our galaxy is estimated to be 12,000 to14,000 million years old and contains around 100,000 million stars.

Our galaxy rotates on its axis – faster at the centre than at the edges. The middle part completes one rotation on its axis in about 50,000 years. The Sun and its neighbouring stars revolve round the centre of the galaxy in their orbit at an average speed of 250 km. per second.

THE SUN

The Sun is one of the stars in the Milky Way. It looks bigger than other stars because it is the closest star to the Earth. However, it is smaller than some other stars. Betelgeuse, a red giant star, is 800 times bigger than the Sun.

The Sun lies about 150 million km. from the Earth. Its diameter is about 1,400,000 km., i.e., 109 times the Earth's diameter. Its gravitation is 28 times more than the gravitation of the Earth.

The Sun lies at a distance of about 32,000 light years from the centre of the galaxy. It takes the Sun about 225 million years to complete one revolution around the centre of the galaxy with a speed of 250 km. per second. This period is called cosmic year. The Sun, like the Earth, rotates on its axis as well. It is a gaseous mass and, therefore, can rotate at varying speeds at different latitudes. It rotates once on its axis in 24-25 days at the poles and in 34-37 days at the equator.

The Sun is composed of about 75% hydrogen and almost 25% helium. It may be called a big hydrogen bomb because it releases huge amount of heat and light as a result of nuclear fusion. The Sun is directly responsible for all life on Earth. It provides the Earth with all of its light, heat and energy.

The glowing surface of the Sun which we see is called photosphere. It has a temperature of about 6000° Celsius while the temperature of the core is 15,000,000° Celsius.

The glowing flames constantly arising from the photosphere are called Solar prominences which rise up to a height of 1,000,000 km.

The dark spots noticed on the surface of the Sun are called Sunspots. They are cooler than the surrounding area. The life span of sunspots varies from a few hours to many weeks. Larger sunspots may have temperatures up to 4000°-5000° Celsius. Some of them are made of many folded layers larger than our Earth's size.

When sunspots persist for longer periods, they cause solar flares and high solar prominences which create upheavals in the ionosphere resulting with disturbances in our radio communications.

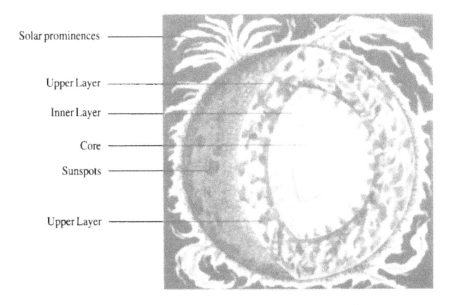

Solar prominences

Upper Layer

Inner Layer

Core

Sunspots

Upper Layer

The Sun is a huge mass of gases

THE SOLAR SYSTEM

The Solar System is located at a distance of about 30,000 to 33,000 light years from the centre of the Milky Way galaxy. The Sun is a star like other stars with the only difference that it is nearer to us as compared to other stars. It has a family of its own. The Sun and its family put together is called the Solar System. The solar system consists of nine planets, their satellites, asteroids, comets, meteoroids and other debris.

Planets are heavenly bodies which, like our Earth, revolve round the Sun. They do not have their own light. They look bright as the Sun's light falls on them and is reflected from their surfaces.

The nine planets are—Mercury, Venus, Earth, Mars, Jupiter, Saturn, Uranus, Neptune and Pluto.

Mercury, Venus, Jupiter and Saturn may be seen from the Earth with naked eyes. These were known to the astronomers thousands of years ago. Three planets, namely Uranus, Neptune and Pluto were discovered later after the invention of telescope. Uranus was discovered in 1781, Neptune in 1846 and Pluto in 1930.

In the recent years, astronomers have explored the possibility of existence of a tenth planet. There is also a fair chance of discovering more moons, especially around Jupiter and Saturn.

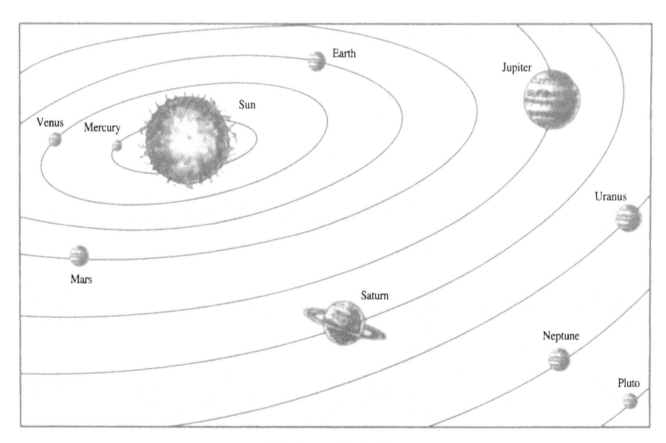

The Sun and its family

MERCURY

Mercury is the planet closest to the Sun. It is smaller than some of the small satellites of the planets. Because of its nearness to the Sun and its small size, it is difficult to see it without a telescope but it can be seen easily immediately after sunset or before sunrise.

It takes Mercury 58.7 Earth days to make a complete rotation on its axis and 88 days to make a complete orbit around the Sun. It is the fastest revolving planet.

The distance of Mercury from the Sun does not remain constant because it has a long narrow orbit resembling a lemon in shape. It moves very slowly. Its one day is equal to about 59 days of our Earth.

One part of this planet is always in the sunlight for a longer period to raise the day's temperature to over 350° Celsius, enough to melt tin and lead. The other part of the planet where it is night, the temperature is as low as –170° Celsius.

The pictures of the planet relayed by the space probe *Mariner-10* showed that it is like the Moon. It has rocks, cliffs and craters. It has no satellite and no atmosphere.

Facts About Mercury

- Mean distance from the Sun: 0.39 au@
- Minimum distance from the Earth: 0.54 au@
- Planet's day: 58 days and 15 hours of Earth
- Its sideral period (year): 88 days of Earth
- Diameter: 4,880 km.
- Mass: 0.06 times of the Earth's mass
- Surface temperature: 350° Celsius during the day and –170° Celsius during the night.
- Gravitation: 0.38
- Density (water = 1): 5.5

Mercury, the planet nearest to the Sun

@ *It is an astronomical unit representing mean distance between the Earth and the Sun. It is equal to 149,598,500 km.*

A view of Mercury's surface

VENUS

You must have seen a very bright star in the evening. It is visible in the morning also. Often it is called the 'Morning Star' or the 'Evening Star'. However, it is not a star but a planet called Venus, the one nearest to the Earth. It is the second planet from the Sun. Through a telescope it looks like our Moon. We get its full view when it is very far away from us. It is the brightest of all the planets.

Many new facts about Venus have come to light by space probes. Venus is the hottest planet in the solar system. The temperature rises up to 480° Celsius at its equator. Lead, tin and zinc melt at this temperature. Venus has very thick clouds which reach up to a height of 55 km. Thick clouds make studies of the planet's surface nearly impossible. The temperature of clouds at the higher level drops down to –35° Celsius. This red hot planet is shrouded by icy clouds.

The atmosphere on this planet contains 90 to 95 per cent of carbon dioxide. Some hydrogen and water vapour are also present. Its surface atmospheric pressure is 100 times more than that on the Earth. No astronaut can survive in the atmosphere of Venus, nor can he tolerate its excessive heat.

Radio waves have indicated that there are mountains and valleys on this planet. The planet has no satellites. The Sun rises in the west and sets in the east on this planet.

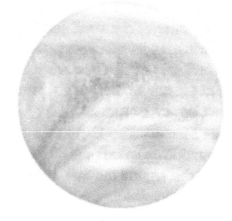

Venus, the planet closest to the Earth

A view of the surface of Venus

Facts about Venus

- Mean distance from the Sun: 0.72 au
- Minimum distance from the Earth: 0.27 au
- Planet's day: 243 days of Earth
- Its sideral period (year): 224.7 days of Earth
- Diameter: 12,104 km.
- Mass: 0.82 times of the Earth's mass
- Surface temperature: 480° Celsius
- Gravitation: 0.88
- Density (water = 1): 5.25
- Main gas in the atmosphere: Carbon dioxide.

EARTH

Our home planet, Earth, is the third planet of the solar family. It is the only planet in the solar system where life exists. Earth too, like other planets, revolves round the Sun. It spins on its axis, one end of which is called the north pole and the other the south pole. One half of the Earth that receives sunlight is hot during summer while the other half has winter during that period. Thus the seasons change alternately.

Yuri Gagarin, a Russian astronaut, was the first to orbit the Earth once in his spacecraft *Vostok* in 1961. He looked at the Earth from space. From space or the Moon, the Earth appears full of greenery and blue water of oceans similar to that as depicted on the globe. Many of its parts hidden under the white clouds are not visible from space.

The Moon is the only satellite of Earth.

Earth, the third (outer) member of the solar system

Facts about Earth

- Mean distance from the Sun: 100 au
- Planet's day: 23 hours, 56 minutes and 4.09 seconds
- Its sideral period (year): 365 days, 5 hours, 48 minutes and 45.51 seconds
- Diameter: 12,756 km (along equator)
- Surface area: 510,065,600 sq.km.
- Mass: 1

- Surface temperature: 22° Celsius
- Gravitation: 1
- Density: 5.517 maximum density
- Main gases in the atmosphere: Nitrogen, Oxygen
- Satellite: 1 (Moon)

THE MOON

The Moon is the only satellite of Earth. Astronomers believe that both Earth and the Moon came into existence separately but at the same time from gases and dust remaining from the Sun's formation. Samples of the Moon's rocks and soil brought to the Earth by the astronauts have shown that the Moon is of the same age as the Earth and was formed about 4,600 million years ago.

Moon, the only natural satellite of the Earth

The mean distance between the Earth and the Moon is 3,84,400 km. Moon's surface area is 37,940,000 sq. km. There is a large number of craters on the surface of the Moon created by meteoroid collisions, and volcanic activity of the past. The Moon has high mountains with gentle slopes. They have neither raised sharp peaks nor steep slopes. Devoid of water and air, the Moon is unable to sustain life. The day dawns suddenly and the night also approaches in the same manner.

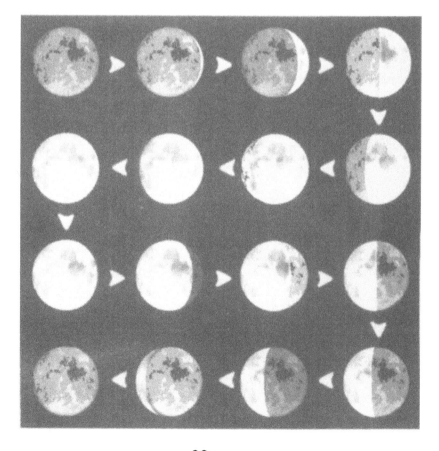

Phases of the Moon

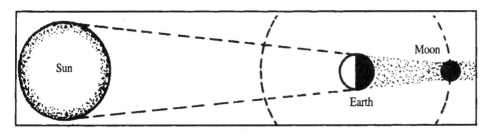

Sun, Earth, Moon

Lunar eclipse

There is no sound either because of total absence of air and atmosphere. The Moon's gravity is about one-sixth of the Earth's gravity. If you can jump up to 1 metre on the Earth, you will jump upto 6.05 metres on the Moon. Similarly, an object that weights 6 kg on Earth, will weigh 1 kg. on the Moon. The day temperature rises to 120° Celsius while the night temperature drops to – 160° Celsius on the Moon. On July 20, 1969, the *Apollo 11* astronaut Neil Armstrong became the first person to set foot on the Moon. Several other explorations from the USA followed, during which many experiments were conducted. The data obtained has vastly increased our knowledge of the Moon.

The Sun's light falls only on one side of the surface of the Moon. The other side is always in the dark. The Moon revolves round the Earth in 27.3 days. It has no light of its own. The Moon looks bright when the sunlight is reflected from its surface. The Moon does not always look the same. Its shape keeps on changing during the month. These different shapes of the Moon are called its phases. The Moon is not visible at all when it comes between the Earth and the Sun. We have full moon when the Sun is on the other side of the Earth.

When the Earth passes directly between the Sun and the Moon, lunar eclipse occurs. It happens only on full moon nights.

Spring tides

Neap tides

The gravitational pull of the Moon causes tides. Although the Sun too is responsible for the tides, but the Moon being nearer to the Earth than the Sun, exerts greater pull on the sea.

23

MARS

Mars is the fourth planet from the Sun. It is about *half* of our Earth's size in diameter. It is also called the red planet because of its reddish orange appearance in the sky. Like that of Earth, the axis of rotation of Mars is tilted in space and its polar regions face the Sun alternately giving it seasons of summer and winter in the northern and southern hemispheres.

Mars, a planet in the solar system

A view of the surface of Mars

Although stories have been written and even films have been made about this planet and its so-called inhabitants, the space investigations have proved that there is no life on Mars.

The pictures received from the space probe *Mariner 9* indicate the existence of deep depressions, dusty valleys and raised landmasses on Mars. There are more volcanoes on Mars than on Earth. Nix Olympia, an extinct volcano on Mars, is three times more in height than Mt. Everest. Its height is 24 km. from the surface of the planet and consists of huge snow caves measuring up to 65 km. in length.

The *Viking* space probes (*Viking-1* and *Viking-2*) sent to Mars in 1976 to explore the possibility of existence of life on the planet have shown that there was no life there. It is no wonder because the temperature on the planet never rises the freezing point. There is no water either.

Mars has two satellites – Phobos and Deimos. Phobos is larger than Deimos.

Facts about Mars
- Mean distance from the Sun: 1.52 au
- Minimum distance from the Earth: 0.38 au
- Planet's day: 24 hours 37 minutes of Earth
- Sideral period (year): 687 days of Earth
- Diameter: 6,795 km.
- Mass: 0.11 times of Earth's mass
- Surface temperature: –23° Celsius
- Gravitation: 8.38 times of Earth's gravitation
- Density (water=1): 3.94
- Main gas in the atmosphere: Carbon dioxide
- Satellites: 2

JUPITER

Jupiter, the fifth planet from the Sun, is the largest planet in the solar system. It is 318 times more massive than the Earth. It is so large that 1,300 Earths could fit into it. Made up of gases, it has features of both, a star and a planet. All planets receive energy from the Sun but Jupiter spreads its radio energy, produced by explosion, in long wavelengths. This planet has the most powerful radio waves next to the Sun in our solar system.

Jupiter's atmosphere contains mainly hydrogen and helium. Methane and ammonia are also present in traces. Its atmosphere is like that of pre-historic Earth. Hydrogen, methane, ammonia and water, which were responsible for evolution of life on the Earth, might have possibly set in motion the process of evolution of life on this planet.

The first space probe, *Pioneer X*, reached Jupiter at the close of 1973. It showed that the planet had magnetosphere. The radio waves coming from the magnetosphere still continue to be a mystery for the scientists, who believe that there is some life somewhere on Jupiter.

In 1979, *Voyager-1* and *Voyager-2* passed this planet. The closest picture of the Jupiter was taken from a distance of 1.8 million km. A 30 km. thick ring around the planet was also observed. Existence of volcanoes and elements like sulphur and oxygen on one of the satellites of the planet named Io, were also detected.

Europa, another satellite of the planet and equal to our Moon in size, is covered under ice at several places having 100 km. thick ice layers.

Jupiter is always covered under clouds. There are five bright and four dark grey bands around it. A mysterious oval-shaped red spot about three times larger than Earth is visible on its surface. It is a persistent storm that appears to be a permanent feature of the planet. It covers an area 40,000 km. long and 4,000 km. wide.

Jupiter, the biggest planet

Jupiter's oval-shaped red spot, about three times larger than the Earth, is a persistent storm

Jupiter has 63 satellites, four of which are several thousand kilometres in diameter.

Facts about Jupiter

- Mean distance from the Sun: 5.20 au
- Minimum distance from the Earth: 3.95 au
- Planet's day: 9 hours 50 minutes of Earth
- Sideral period (year): 11.86 years of Earth
- Diameter: 1,42,800 km
- Mass: 317.9 times of the Earth's mass
- Surface temperature: –150° Celsius
- Gravitation: 2.64
- Density (water = 1): 1.33
- Main gases of the atmosphere: Hydrogen, Helium
- Number of rings: 1
- Satellites: 63

SATURN

Saturn, the second largest planet in the solar system, is the sixth planet in terms of distance from the Sun. It is the loveliest among the planets. Like its neighbour Jupiter, Saturn also resembles a ball of gases but smaller in size. It is 95 times more massive than Earth. The magnificent ring system of this planet has made it mysterious. Spread over a distance of about 275,000 km. these rings revolve round the planet. Investigations made by space probes *Voyager-1* and *Voyager-2* have revealed that these rings are made up of innumerable ice-covered particles. The two *Voyagers* helped in measuring these particles, which have diameter of a few centimetres to 8 metres. The rings number more than 1,000.

Like Jupiter, Saturn also has a red spot on its surface although smaller in size. It contains white, elliptical and banded, light dense clouds. It is yellowish in colour.

Saturn faces very strong winds with a speed of about 1,760 km. per hour. Its surface temperature is −180° Celsius.

There are 60 known satellites of Saturn. These appear to be made up of ice. Titan, the

A picture taken from planet's satellite, Rhea

biggest of the satellites and larger than Mercury in size, is estimated to have some atmosphere.

Facts about Saturn

- Mean distance from the Sun: 9.54 au
- Minimum distance from the Earth: 8.00 au
- Planet's day: 10 hours 14 minutes of Earth
- Sideral period (year): 29.46 years of Earth
- Diameter: 1,20,000 km.
- Mass: 95.2 times of the Earth's mass.
- Surface temperature: −180° Celsius
- Gravitation: 1.15
- Density (water = 1): 0.71 minimum
- Main gases of the atmosphere: Hydrogen and Helium
- Number of rings: More than 1,000
- Satellites: 60

Saturn. a planet of our solar system

URANUS

This planet was discovered by William Herschel in 1781. It is much smaller than Jupiter and Saturn but much larger than Earth. It is 15 times more massive than Earth. It looks vaguely green through the telescope. It is mainly made up of methane gas. It is a cold planet and its surface temperature drops to –210° Celsius.

Astronomers discovered in 1977 nine dull rings around Uranus, spread within a limit of 64,000 km. This is the limit wherein even a massive satellite would break to pieces by its tidal forces. As of now, there are 13 rings in all.

Uranus orbits the Sun every 84 years. Its day equals 16 hours 10 minutes of Earth.

Uranus has 5 main satellites — Miranda, Ariel, Umbriel, Titania and Oberon, with the total number being 27.

Facts about Uranus

- Mean distance from the Sun: 19.18 au
- Minimum distance from the Earth: 17.28 au
- Planet's day: 16 hours 10 minutes of Earth
- Sideral period (year): 84.01 years of Earth
- Diameter: 50,800 km.
- Mass: 14.6 times of the Earth's mass
- Surface temperature: –210° Celsius
- Gravitation: 1.17
- Density (water = 1): 1.7
- Number of rings: 13
- Main gases of the atmosphere: Hydrogen, Helium and Methane
- Satellites: 27

Uranus, a cold planet

NEPTUNE

Neptune is the eighth planet from the Sun. It was discovered by Adams and Leverrier in 1846. It is a cold planet that looks green and has a surface temperature of about $-220°$ Celsius. It is 17 times larger than Earth. Though not yet confirmed but it is believed that like Uranus, Neptune also has rings around it. It has a day equal to 18 hours and 26 minutes and a year equal to 164.8 years of the Earth.

Neptune has two satellites – Triton and Nereid. Triton is bigger than Pluto with a diameter of 3,700 km.

Facts about Neptune

- Mean distance from the Sun: 30.06 au
- Minimum distance from the Earth: 28.80 au
- Planet's day: 18 hours and 26 minutes of Earth
- Sideral period (year): 164.8 years of Earth
- Diameter: 48,500 km.
- Mass: 17.2 times of the Earth's mass.
- Surface temperature: $-220°$ Celsius
- Gravitation: 1.2
- Density (water = 1): 1.77
- Main gases of the atmosphere: Hydrogen, Helium and Methane
- Satellites: 13

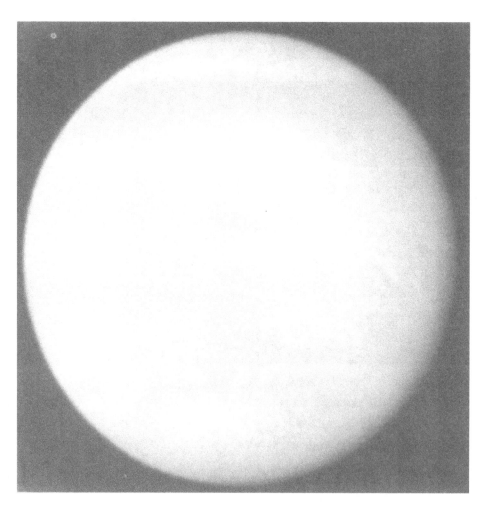

Neptune, greenish cold planet

PLUTO

After the discovery of Neptune, astronomers continued their efforts in search of another planet still farther in space. It was in 1930 that C.W. Tombaugh discovered Pluto which is slightly smaller than the planet Mercury. The Sun shines over it for about 6 hours 30 minutes only. It is very cold. Its surface temperature is −230° Celsius. The Sun looks like a bright star from the surface of this planet. Pluto has no atmosphere. It is like a rocky ball.

This planet has three satellites. Its axis bisects Neptune's orbit and therefore it is presumed to be a satellite broken away from Neptune.

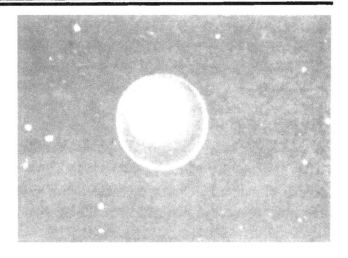

Pluto, a cold planet

Facts about Pluto

- Mean distance from the Sun: 39.44 au
- Minimum distance from the Earth: 28.72 au
- Planet's day: 6 days, 9 hours of Earth
- Sideral period (year): 247.7 years of Earth
- Diameter: 3,000 km.
- Mass: 0.002-0.003 times of the Earth's mass
- Surface temperature: −230° Celsius
- Main gases of the atmosphere: Frozen Methane and Nitrogen.
- Satellites: 3

Note:−

From the time of its discovery in 1930 to 2006, Pluto was counted as the solar system's ninth planet. However, with the discovery of many similar objects, on August 24, 2006, the IAU (International Astronomical Union) defined the term 'planet' for the first time. This definition reclassified Pluto as a member of the new category of dwarf planets along with Eris and Ceres, and given the number 134340.

The Sun looks like a star from Pluto's surface

ASTEROIDS

Planets and their moons are major members of the solar system. There are minor members called asteroids. An asteroid is an irregular lump of rock that orbits the Sun. They are in large numbers in a belt between the orbits of Mars and Jupiter. They lie at a distance of 2.2 to 3.3 au. This belt is estimated to have some 40,000 to 50,000 asteroids. Some of them are so tiny that their diameter cannot be measured with the help of prevalent techniques. Ceres, the biggest of these, has a diameter of 1003 to 1040 km. It was discovered in 1801. 4 Vesta, the brightest asteroid, is the only asteroid that can be seen with the naked eye. Its diameter is 555 km. Hermes, another asteroid which was at a distance of 7,80,000 km. from the Earth, is now extinct.

It is not known how asteroids came into being. Some astronomers think that they are broken pieces of some planet which once existed between Mars and Jupiter. Others consider them as pieces broken off Mars and Jupiter themselves. Some asteroids may be chips and fragments of comets.

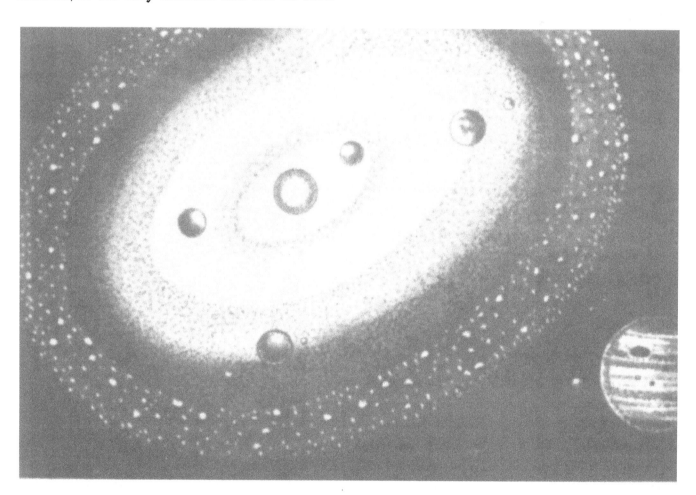

Most asteroids lie in a belt between Mars and Jupiter

METEORS AND METEORITES

Sometimes we see a luminous object moving very fast across the sky leaving a streak of light behind it and which suddenly disappears. It is generally called a falling star. But we know that stars never fall. These tiny bodies which appear to fall are not stars but meteors. They enter the Earth's atmosphere with a tremendous speed and burn up as a result of frictional heating. This heat causes the meteoroid to glow. Meteors also are members of the solar system.

Meteors do not reach the Earth. Most of them either burn up and disappear on their journey or are converted into vapour and dust. Where a meteor does not burn up completely, it falls down and hits the Earth. When the meteoroid hits the Earth's surface, it is called a meteorite.

Craters have been formed on the surface of the Moon, Mars and Mercury as a result of meteorites hitting them. The biggest such crater on Earth in Arizona was created by some

Meteorites

meteorite. Its diameter is 1,265 metres and depth 175 metres.

Around 25 million visible meteors are estimated to enter Earth's atmosphere everyday. They have a speed of 35-95 km. per second. An ordinary meteor takes approximately one second to convert itself into vapour. About 500 meteorites hit the surface of the Earth in a year's period.

The largest known meteorite found in 1920 at Hoba West near Grootfontein at south-west Africa weighed about 60,000 kg. It hit the ground during pre-historic age.

Calcutta Museum has some meteorites as exhibits.

Millions of meteors enter Earth's atmosphere everyday

COMETS

A comet is a heavenly body with a long glowing tail of light behind it. It is also called 'tailed star'. Long ago people feared it because they considered it inauspicious. But now people do not think that way because they have known the truth.

Comets also are members of the Sun's family like other heavenly bodies. They have a definite orbit like that of Earth. However, they have a shape different to that of Earth. About 1,000 comets have passed by the Sun during 100 years. Of them, some are bright enough to be seen without a telescope. The most important among them is Halley's Comet, which passes by the Sun every 76 years. It was discovered by the English astronomer Edmond Halley in 1682 and was named after him. It was last seen in 1986.

A comet has three parts – nucleus, coma and tail. Nucleus is the brightest part of the comet. It may have the diameter of 100 to 10,000 metres. The nucleus of Halley's comet has a diameter of 5,000 metres. Composed of ice, dust and gas, the dirty snowball called nucleus shines through the centre of its head. The part surrounding the nucleus is called coma. It is made up of gases and dust and

A comet seen in Arizona

may have a diameter of more than 2 million km. Coma is surrounded by the clouds of hydrogen gas. Tail is an important part of comet. It is of two varieties – dust tail and plasma tail. The length of a dust tail may vary from 1 to 10 million km., while that of a plasma tail, composed of extremely hot ionised gases, up to 100 million km.

A comet develops the tail as it comes near the Sun. The sunlight pushes away some gas from its head. It is this gas which starts showing and looks like a tail. As the comet approaches the Sun, it rushes with a great speed along with its bright shining tail. The tail of a comet always points away from the Sun.

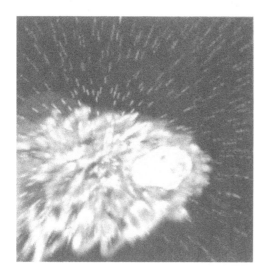

A comet in space

EXPLORING SPACE

Space age began on October 4, 1957 when Russia launched the first artificial satellite, *Sputnik-1* and one month later, *Sputnik-2* was launched with a dog, 'Laika' aboard, the first living creature in space. This suggested that human beings might survive in space.

The first US satellite, *Explorer-1* was put into space on January 31, 1958. Thus, the Russian *Sputniks* and US *Explorer* pioneered the space exploration programme.

The first manned Russian satellite, *Vostok-1* was put into space to go round the Earth on April 12, 1961. *Vostok-1* carried Col. Yuri Gagarin, who became the first cosmonaut to complete an orbit of the Earth and to observe the Earth and the sky from space.

The first woman to go into space in 1963 was USSR's Lt. Col. Valentina Tereshkova. She completed 48 orbits in 2 days, 22 hours and 21 minutes in *Vostok-6*.

In March 1965, Aleksei Leonov aboard the Soviet *Voskhod-2* became the first man to walk in space.

Edward White, the US astronaut of the spacecraft *Gemini 4* was the first US astronaut to make a space walk. He came out and stayed outside his craft in space for 21 minutes.

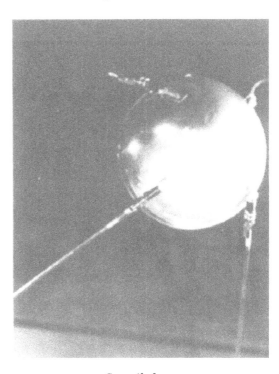

Sputnik-1

Vostok

Gemini

Russian cosmonaut, Col. Yuri Gagarin

Apollo astronaut

In 1965, began the series of two-manned *Gemini* flights. The team of astronauts for the *Gemini* programme practised rendezvous manoeuvres, docking procedure and space walk in preparation for the Apollo missions to the Moon.

Apollo-11

Spaceman Edward White stayed outside his craft and walked in space for 21 minutes

JOURNEY TO THE MOON

The three-manned *Apollo-11*, which had enough space for astronauts to not only move about but also to stand erect within it, set out on a journey to the Moon on July 16, 1969. The special feature of *Apollo's* landing on the Moon was that its two astronauts could touch the surface of the Moon with the help of the four-legged lunar module *'Eagle'*.

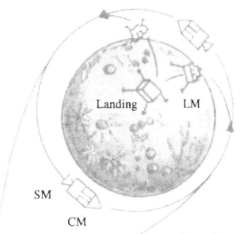

Trans-Lunar trajectory

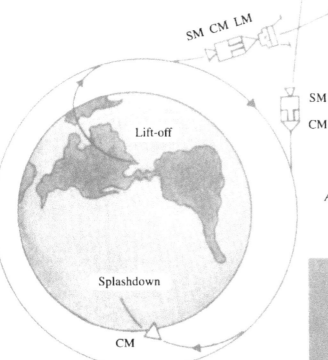

Apollo's journey from Earth to Moon

- SM - Service Module
- CM - Command Module
- LM - Lunar Module

Neil Armstrong

35

At 10.56 p.m. on July 20, 1969, Neil Armstrong set his foot on the surface of the Moon. He was the first human being to put his foot on the Moon.

In his radio communication sent to the Earth Neil Armstrong described it as "one small step for (a) man, one giant leap for mankind". Thereafter, Edwin Aldrin stepped out on the Moon's surface. Michael Collins remained aboard the *Apollo 11* command module in orbit. The two astronauts spent around 2 hours 30 minutes on the Moon. During this period, they collected samples of Moon's rocks and soil weighing over 20 kg.

Mineral & rocks on the Moon

Surface of the moon

36

The total estimated expenditure on this US space programme at sending men to the Moon, with *Apollo 17* was around 2,554.1 million dollars. With a view to reduce expenditure 'space shuttle' was invented by the US scientists. Space shuttle is a reusable manned space vehicle with a speed of about 28,000 km. per hour. The first space shuttle *Columbia* was launched on April 12, 1981. The USSR launched a similar space vehicle named *Buran* in September 1988 which returned to the Earth safely after successfully completing the mission.

After the success of this mission, the following persons made journey to the Moon

Neil Armstrong set his foot on the Moon on July 20, 1969

under the *Apollo* programme. They brought with them rocks and soil weighing 3,800 kg.

Apollo 12 : Conrad, Bean, Gordon
November 14, 1969

Apollo 14 : Shepard, Mitchell, Roosa
January 31, 1971

Apollo 15 : Scott, Irwin, Worden
July 26, 1971

Apollo 16 : Young, Duke, Mattingly II
April 16, 1972

Apollo 17 : Cernan, Schmitt, Evans
December 7, 1972

■■

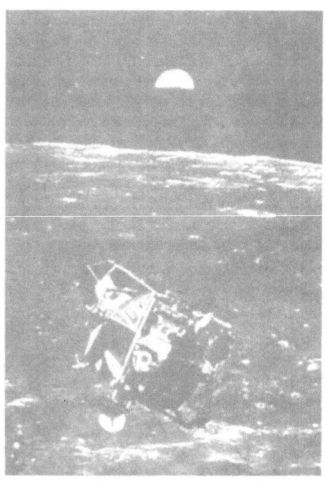

Return journey from the Moon to the Earth

Space shuttle Columbia

Apollo-11 spacemen return to the Earth

38

ORIGIN OF THE EARTH

The earth is one of the nine planets of the solar system and revolves around the sun through space. It has all the favourable conditions required to sustain life. The earth is the only home for human beings, animals, plants and other organisms. Scientists theorise that the earth was born around 4.6 billion years ago out of the clouds of dust and gases. However, before coming into the present shape, it was a fire ball surrounded by the clouds of hot gases. It took millions of years to cool down and the envelope of hot gases turned into clouds. These clouds rained for a long time and the rain water accumulated in the low lying areas of the earth which after a long time turned into oceans of today. In the beginning, the earth was a giant land mass called *Pangaea*.

This land mass gradually split into different land masses which are called *continents* today.

Mountains and volcanoes have resulted due to the upheavals in the interior of the earth. In the process of cooling, the upper surface of the earth became solid which we call as the earth's crust. About 570 million years ago, the life began on the earth in the form of micro-organism. The first 345 million years saw the development of aquatic life. For the next 160 million years reptiles came into existence and subsequent 65 million years saw the development of mammals. The development of man is an event of one million years old but we know about its existence for only 10,000 years ago.

Initially, the earth was made up of molten rocks. As it cooled, rocks, crust, atmosphere and oceans were formed

EARTH AND THE ATMOSPHERE

Inside the earth

According to scientific studies, the earth from the top to centre can be divided into three parts: *Crust*, *Mantle* and the *Core*. Core is further subdivided into *outer core* and *inner core*.

The outer skin of the earth is called crust. The thickness of the crust varies from 5 to 10 km under the oceans and to about 30 to 40km under the continents. The crust is made up of three kinds of rocks — igneous, sedimentary and metamorphic. Beneath the earth's crust are the mantle, the outer core and the inner core. The mantle is a thick layer of solid rocks and goes down to about 2900 km. The rocks in the mantle are made up of silicon, oxygen, aluminium, iron and magnesim. Its temperature increases from 870°C to 2200°C as we go down. The outer core is about 2200 km thick and is made of melted iron and nickel. Its temperature ranges from 2200°C (upper most part) to 5000°C (deepest part). The ball shaped inner core is about 5150 km below the earth's surface. The centre of the inner core is about 1200 km below the inner boundary of outer core. It is made of solid iron and nickel.

Atmosphere

The envelope of air which surrounds the earth from all sides is called atmosphere. The atmosphere is held due to the gravitational pull of the earth. The outer layer of the atmosphere is called the *Exosphere* which extends up to an altitude of 2000 km above the sea level.

All gases of the atmosphere combined together constitute air. Atmosphere contains 78% nitrogen, 21% oxygen and remaining 1% consists of argon, carbondioxide, neon, helium, ozone and hydrogen. Air also contains water vapour and particles of dust.

In the lower layers of the atmosphere, the ratio of different gases remains relatively constant but the percentage of water vapour goes on changing. Clouds float in the lowest part of the atmosphere called *Troposphere*. The air gets thinner, as we go above the surface of the earth. The atmosphere has many parts such as the *Troposphere, Stratosphere, Mesosphere, Inosphere* and *Exosphere*.

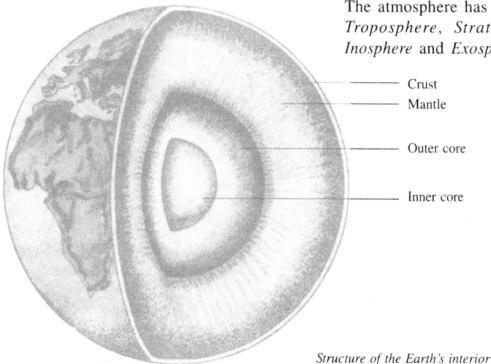

Crust

Mantle

Outer core

Inner core

Structure of the Earth's interior

LITHOSPHERE OR CRUST

The uppermost layer of the earth is called litho-sphere or the crust. Its average thickness is 30 km to 40 km under the continents, 5 km to 10 km under the ocean floor and upto 60 km under the mountains. At certain places below the pacific floor, its thickness is very little. The earth's crust consists of two layers — the upper layer and the lower layer. The upper layer is called SIAL (silicate + aluminium). It is made up of granite-like rocks which mainly consist of silica and alumina. Below the upper layer lies the lower layer called SIMA (silicate + magne-

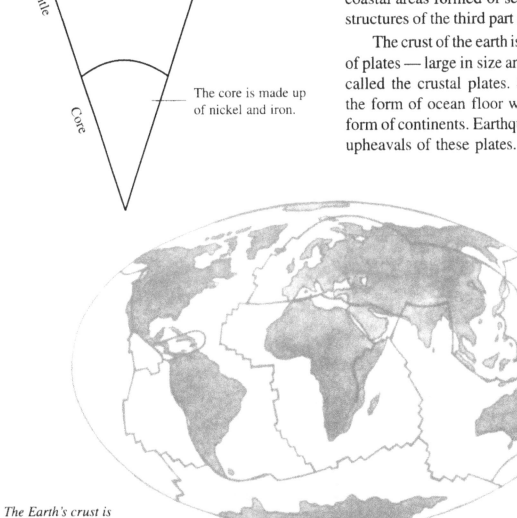

temperature prevails in this region. It is the same material of which the rocks on the surface of the earth are made.

There are three kinds of rock structures in the upper continental crust of the earth. Of them, the most abundant are the Pre-Cambrian crystalline shields. In the second place are the coastal areas formed of sedimentary rocks. The structures of the third part are folded mountains.

The crust of the earth is made up of a number of plates — large in size and of varied thickness, called the crustal plates. Some of them are in the form of ocean floor while others are in the form of continents. Earthquakes occur due to the upheavals of these plates.

The Earth's crust is made up of large plates

43

CONTINENTS

A little less than one-third of the earth's surface is land mass and the rest is covered by water. There are seven large continuous land masses called continents. They are: Asia, Europe, Africa, North America, South America, Australia and Antarctica.

According to the German Geologist, Alfred Wegener, once, the earth was a single land mass known as *Pangaea*. Nearly 200 million years ago, this land mass broke into two parts — Laurasia and Gondwana land.

According to Wegener's Continental Drift Theory, these two large land masses continued to drift away from each other. Present North America and Eurasia (Europe and Asia) emerged out of Laurasia, while the South America, Africa, Australia and Antarctica owe their existence to the Gondwana land. Thus the seven continents of the world appeared.

The theory also explains that not only the continents but also the crust plates, which include both the continents and the oceans, drift. Thus the continents are undergoing changes even now. Infact, earthquakes are also caused due to the drifting of these crust plates.

200 million years ago

130 million years ago

The earth at present

Changes in the continents

44

ROCKS

The rocky crust of the earth is made up of three kinds of rock formations: igneous, sedimentary and metamorphic rocks.

Igneous rocks are made up of hot magma. The magma that is pushed out due to volcanic activities forms igneous rocks on cooling. Igneous rocks are the oldest rocks found on the earth. The main igneous rocks are granite, basalt and volcanic rocks. The volcanic rocks are formed of the hot liquid lava thrust out from the valcano.

Sedimentary rocks are made up of deposits of grains of rocks, shells and skeletons under the sea bed. Coal also is a sedimentary rock formed as a result of the forests buried under swamps in the ancient times. Limestone, gypsum, sandstone and clay are the examples of sedimentary rocks. Igneous rocks account for 95% and sedimentary rocks 5% of the volume of the earth's crust.

The metamorphic rocks are formed by the actions of heat and pressure on igneous and sedimentary rocks. Marble is a metamorphic rock that is formed as a result of action of heat on limestone. Similarly, clay is transformed into slate. Precious minerals like garnets and rubies are found in several metamorphic rocks. Rocks are very useful for us. Marble is obtained from rocks which is used for making the statues of gods and goddesses and are also used in the construction buildings, temples, etc. The stones that are derived from these rocks are used in building, roads. The clay transforms into slate which is also used in the construction of buildings.

Sedimentary rocks

Clay transforms into slate

Limestone transforms into marble

Igneous rocks

EARTHQUAKES

Earthquake produces tremors or vibratory shocks in the earth. Sometimes the tremors are so weak that people cannot feel them. But at times, they are so violent that long cracks are formed in the earth's crust, buildings collapse, monuments topple, and people perish. Many sudden tremors and shocks felt in the earth's crust are called *earthquakes*.

There are many reasons behind earthquakes. According to seismologists, the outer layer of the earth is made up of many thick plates. All these plates are in slow, continual motion with respect to each other. Currents within the hot, molten interior of the earth, produced by thermal convection and the earth's rotation, are thought to underlie plate movement. In some areas the plates are being driven apart as new molten material is forced upward between plates. In other

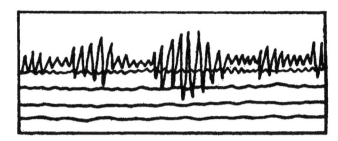

This is how the seismograph records an earthquake

region, the plates slide past each other. In a third kind of situation, plates push direclty into each other, causing one plate to slide beneath the other. The difference in motion between plates, causes rocks to fracture along cracks, creating earthquakes.

A few local earthquakes are volcanic in origin. These are produced by movements of underground molten magma straining and fracturing adjoining surface rocks. In many places tremors have been felt as a result of deep digging by man.

The areas prone to earthquakes are known as *seismic belts*. *Assam in India* is one of the world's most earthquake-prone areas. The instrument used for the study of earthquakes is called the *seismograph*. *Richter scale* is used for recording the intensity of earthquakes. The Indian Institute of Technology (IIT) at Roorkee has a seismo graph. However, it is very difficult to predict an earthquake.

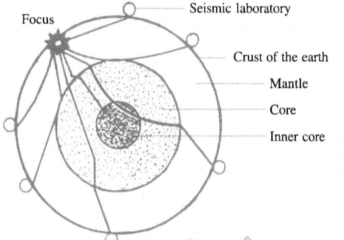

Focus — Seismic laboratory
— Crust of the earth
— Mantle
— Core
— Inner core

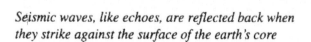

Seismic waves, like echoes, are reflected back when they strike against the surface of the earth's core

MOUNTAIN FORMATION

Most of the mountain ranges have been formed due to major changes that have taken place in the earth's crust over the ages. They are classified in four categories on the basis of the process of their formation. The four categories are: fold mountains, block mountains, residual mountains and volcanic mountains.

Fold mountains are made up of many layers or rocks. They are formed due to violent contractions and pressure inside the earth. Because of this, the rocks turn wavy and overlap each other and their layers get folded to rise above the earth's surface to form mountains. The Himalayas, the Andes, the Rockies and the Alps came into existence as a result of folding process.

Mountains formed by vertical faults are called block mountains. Great blocks of rock are uplifted above the surrounding terrain because of vertical movements along faults. The mountains rise as great tilted blocks. The Vosges in France and Black Forests in Germany are the examples of block mountains.

Residual mountains are formed due to denudation and erosion by which high plateaus are gradually shaped into peaks and ridges. Denudation and erosion is caused by natural agents such as wind, water, snow, etc. The Catskill range in the southern New York state is an example of residual mountains. The Grand Canyon in Arizona, USA, is the result of erosion by Colorado.

Volcanic mountains are formed due to the accumulation and solidification of lava, ash and debris erupted from the earth's interior. They are basically cone shaped with a crater at the top. The *Fujiyama* in Japan is an example of volcanic mountains. (See Volcanoes) Hood and Rainier of America are also example of such mountain ranges. The hight of the mountains increases only a few millimetres every year.

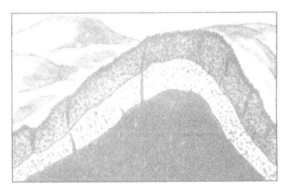
Sedimentary rocks rising up

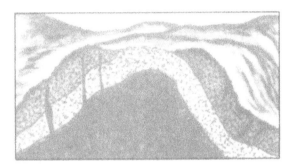
Erosion of rocks

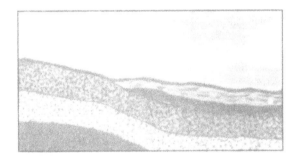
Layers of rocks on the sea floor

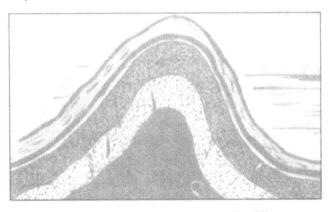
Sedimentary rocks rising up again after 10 million years

47

FOLDS AND FAULTS

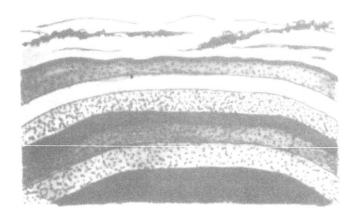

Continuously operating internal and external forces influence, alter and give new shapes to the earth's crust.

The internal forces cause the slow or secular movements lasting for millions of years as well as the sudden movements like earthquakes. The slow movements are responsible for folding and faulting which give birth to mountains, plateaus and rift valleys.

Fold

Folding is a process which produces bends or folds in rock. Folding is caused by great compressional (side ways) forces acting on layers of rock in the earth's crust.

Folds are of various kinds such as symmetrical fold, asymmetrical fold, monoclinal fold, overfold, recumbent fold and nappe.

Faults

When the rocks within the earth's crust break and move apart due to tension and compression, it is called faulting. Faults can move horizontally or vertically from a few centimetres to many kilometres.

Faults are mainly of three kinds — **Normal faults, Reversal faults** and **Tear faults.**

Fold mountain

Tyes of Folds

48

Faults shape the landscape by their movements. When rocks on one side of the fault shift vertically, a ridge is formed on the uppermost surface. A block of land that rises between the two faults is called a *horst*. A large garben between two faults is called a *rift valley*. The *Black Forest Mountain* of Germany is of this type.

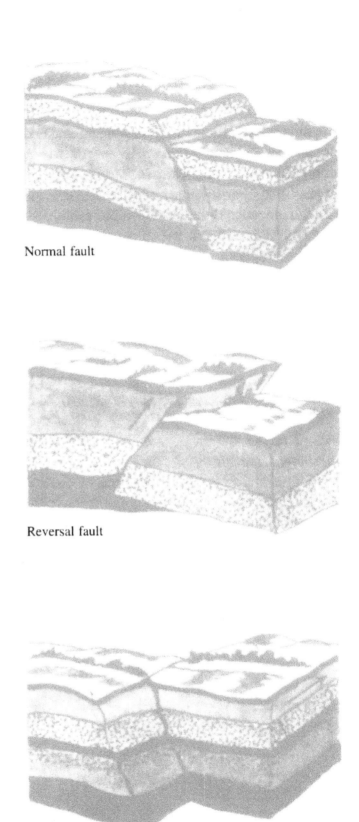

Normal fault

Reversal fault

Tear fault

Rift valley

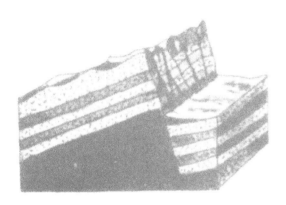

Block mountain

VOLCANOES

The mountains which throw up fire, smoke and cinder are called *volcanoes*. When volcanoes erupt, the magma (melted rock) reservoir in the layers of the earth's interior forces itself out through the crust to the surface of the earth. The liquid substance that is forced out as a result of volcanic eruption is called lava. The lava is a mixture of hot volcanic cinder, pieces of rocks and steam. The number of active volcanoes on the earth is estimated from about 500 to over 800. Most of the volcanoes are funnel or cone shaped. Their mouth is called crater. The crater is connected to a pipe-like opening through which magma from the earth's interior is thrown out. By the action of the gases and steam accompanying the magma, the rocks at the top and slopes are blown up to form a hole. Mt Fujiyama, in Japan, is an example of such a volcanic mountain. Some volcanoes have dome like tops. They are called shield volcanoes. Mauna Loa in Hawaii Island is an example of shield mountain.

On the basis of the amount of their activity, the volcanoes are classified as (i) active, (ii) dormant and (iii) extinct. An active volcano is always erupting. Hawaii Island's Mauna Loa and Sicily's Mt. Etna are examples of active volcanoes. A dormant volcano is temporarily inactive. Examples of such volcanoes are Mt. Fujiyama in Japan and Mt. Vesuvius in Italy. An extinct volcano remains completely inactive for hundreds or thousand of years. Mt. Kilimanjaro in Africa and Mt. Aconcagua in South America are examples of extinct volcanoes. The world's highest extinct volcano is in Argentina Which is about 6960 metres high.

Volcanic eruptions sometimes cause large scale destructions. Pompeii and Herculneum were buried under eruption from Visurius. Volcanic eruption in 1883 blew off two-third of Krakatoa island of Indonesia. It was the greatest explosion during the last 3000 years. The magnitude of this explosion was equal to that of 1500 megaton of TNT and it was heard as far as 500 kilometres. Besides this, the explosion resulted in tidal waves in all the seas and oceans across the globe.

A volcano is a vent or fissure in the earth's crust through which hot solids, gases, smoke and liquids emerge out violently. A volcano also refers to the mountain that forms around the hill.

Gas, steam and dust

Crater

Lava flowing

Magma

Batholith

DESERTS

A desert is a barren region that has vast expanses of sand. Which is yellowish in colour. It receives very little rainfall. In deserts, only a few varieties of plants and animals can exit. These are those species that require little water. They can survive for a long time even without water. In most deserts, days are very hot but the nights are cold.

The deserts are divided into hot deserts and mid-latitude deserts. Both these kinds of deserts are found on the western parts of the continents between latitudes 20° and 30° while the mid-latitude deserts occur in the interiors of the continents between latitudes 30° and 35°. Of the former kind, the best known deserts are the Sahara and the Kalahari in Africa, the south-west American desert, the Great Australian desert, the Chilean desert in South America and Arab-Thar desert in the south-west Asia. The latter type includes the deserts of Turkestan and Gobi. The Sahara desert in Africa is the largest desert of the world.

Both in hot as well as in the mid-latitude deserts, sand dunes are formed and spoiled due to the action of strong winds. They may be up to 510 metres high and 900 metres long. The highest sand dunes are found in the Sahara desert in Algeria. The Thar is the biggest desert of India. The canel is called the ship of the desert and cactus plants grow in large numbers in a desert. ■

Sahara, the world's biggest desert

GEYSERS

A geyser is a jet of boiling water and steam issuing from the earth in a few volcanic regions. The jet of water may persist from a few moments to an hour or more. Beneath the geyser, there is some hot rock and a narrow fissure connects the rock to the mouth of the geyser. Water reaches the rock through this fissure where it is heated to about 100° Celsius. Now the steam pressure inside, forces the hot water to come out to the surface of the earth. The flow of hot water continues till such time as cold water does not return to the fissure. The cold water is again converted into steam and as the pressure rises, the hot water is thrown out.

The jet of hot water may rise in the air to a height of 100 metres or more. In some cases it rises to only 1 or 1.5 metres. Geysers are known to exist only in Iceland, New Zealand and Yellowstone National Park in the USA. Of these three, Yellowstone has by far the most impressive display of geysers. These geysers are a delightful sight for the viewers and are popular tourist spots.

The Old Faithful geyser

HOT SPRINGS

The temperature of water of a hot spring is higher than that of ordinary springs. According to scientific findings, ground water temperature rises with increasing depth. The rate of increase in generally about 1°C for each 25 metres of increase in depth. When underground hot water flows out regularly through some crack or fissure, it is called a hot spring. In some volcanic regions, the steam formed by the heat of the molten rocks is forced out in the form of hot water.

Many minerals get dissolved in hot spring water. The hot spring water containing dissolved sulphur is good for skin diseases. The water of some springs is very hot.

The Maoris of New Zealand cook their food in the local hot spring water. In Haryana, a place called Sona has an hot spring which is quite famous and Manali also has a number of hot springs. Among the hot springs in Dehradun, the capital of the newly formed state of Uttaranchal, the *Tapt Kund* is a hot water sulphur spring which is quite famous throughout India.

A hot spring in Iceland

CRATER LAKES

Crater is formed after a volcanic eruption

As a result of volcanic eruptions, very broad and shallow craters are formed. They are surrounded by a wall-like structure made up of lava. When there is no eruption for a long time, many craters become filled with rain water, forming lakes. These are called crater lakes.

Crater lakes are normally on the top of mountains. The water in these lakes is pure and looks blue. Crater lakes are very beautiful.

The largest and the most picturesque crater lake is situated on the Cascade ranges in the USA. It has been developed into a Crater Lake National Park. It is over, 1800 metres above the sea level. Thousands of tourists visit this place every year.

Crater lakes are the craters of extinct volcanoes filled with water

Crater lakes are formed at the top of mountains.

EROSION

Erosion is a process of scraping, scratching, grinding and transportation of the soil and rocks on the earth's surface by running water, moving ice or glacier, waves and currents of the sea, rain and wind. The existing beautiful landscapes on the earth have mainly been carved and shaped by erosion.

Erosion forms soil, finds minerals and adds air and moisture to land. Lakes, waterfalls, caverns, valleys and natural bridges are created by erosion. Grand Canyon in the USA is an excellent example of such erosion. Various measures are undertaken these days to stop soil erosion. Erosion is caused by many agents. Valleys are result of glacial erosion. River erosion transports soil from one place to another.

Sand dunes in deserts are formed by wind erosion. Erosion by sea waves carves out caves.

Erosions are of various types. The process of grinding the surface rocks by debris or weight is called **mechanical erosion**. The wearing action of moving materials is called **corrosion**. When the blocks of rocks are broken into smaller pieces during transportation, it is called **attrition**.

Queer shape of rocks after their erosion

OCEANS

Oceans cover around 71 per cent of the earth's surface and the remaining 29 per cent, is the land area. They are estimated to contain 1.3×10^{18} tons of water. There are five oceans: the Atlantic, Pacific, Indian, Arctic and Antarctic. There are smaller bodies of water called seas, sounds, bays and gulfs. All of these are connected to each other.

The Pacific Ocean is the largest and the deepest. It covers an area of about 16,62,40,000 sq. km. The second largest is the Atlantic Ocean which has an area equal to half of the Pacific Ocean. The Indian Ocean is the third largest. It extends from Kanyakumari (Cape Comorin) in India to the South Pole, Antarctica. The Arctic Ocean surrounds the North Pole. It is completely frozen and unnavigable. Some people consider Antarctica also to be an ocean.

The ocean water is saline and not fit for drinking. The average salinity is 3.5 per cent. The main source of salinity of oceans is the water of the rivers that carry with them billions of tons of dissolved minerals every year.

Oceans exercise a great influence on climate. They are the main source of rainfall on the continents. Ocean currents regulate the temperature in the coastal regions. The sea breeze and the land breeze keep moving in their set directions. Due to the heat of the sun, the water in the oceans evaporates into smoke and convert into clouds that cause rainfall to make the soil fertile.

The oceans cover around 71 per cent of the earth's surface

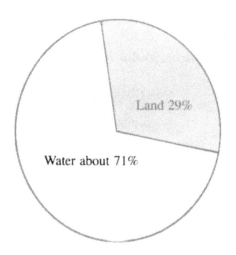

RIVERS

A river is a large stream of water that flows from high land to low land. The water in rivers comes from rain, snow melt, lakes, springs and water falls. The river water eventually flows into oceans.

There are many kinds of rivers such as — the swift flowing rivers, the slow moving rivers, the straight rivers, the meandering rivers, the large and the small rivers. The speed of flow of water in a river depends on the steepness. The rate of flow at the mountain slopes is higher than that in the plains. The course of river near its source is narrow but it widens as it moves downstream.

Some rivers owe their origin to melted ice in the mountains while others to the glaciers. A river is in youthful stage in mountainous region (upper course), in mature stage in flat valleys (mid course) and in old stage in delta region (lower course).

In the upper course the river flows zig-zag through the mountain blocks. Further downwards it develops land forms like gorges and canyons. In mid course, the river slows down but gathers larger volume of water from its tributaries and takes many turns. In the lower course, the river widens its bed and wends its way leisurely. Here, it forms flood plains, ox-bow lakes and deltas. The delta region consists of very fertile soil and silt. Rivers are a very useful means of transport and communication. Ships and boats sail on the rivers. Dams are constructed on big rivers to generate electricity, popularly known as Hydroelectric power. Canals are cut from these big rivers to divert the water for irrigation purposes.

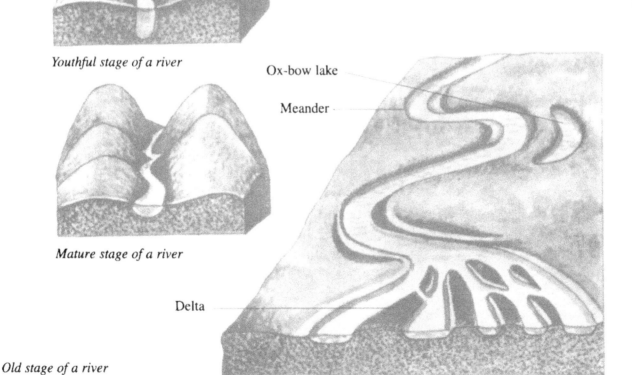

Youthful stage of a river

Mature stage of a river

Ox-bow lake

Meander

Delta

Old stage of a river

57

WATERFALLS

Water falling from the height of mountains is called waterfall. If the volume of falling water is large and it falls from a considerable height, it is known as cataract. A waterfall with a small volume is called a cascade.

Waterfalls are usually created due to erosion by a river in its upper course. When the river passes from more resistant to less resistant rocks, erosion is more rapid in the less resistant formation. Then the river flows down from a

Niagara Falls in Canada

height as a waterfall. The **Niagara Falls** and the **Great Falls of Yellowstone** were formed in this way.

Where the drainage area of a river acquires higher level as a result of some natural phenomenon, the river water falls from a height as a water-fall. Sometimes the flow of a river is blocked due to landslide. Consequently as the volume of water increases, it overflows the blocks of rocks to form a waterfall.

Waterfalls normally occur in mountainous regions. They are also found in regions eroded by ice and glaciers. The highest waterfall in the world is the Angel Falls with 979-metre drop in Venezuela. This was discovered in the year, 1835. The Niagara Falls of America is one of the most famous waterfalls of the world. Half of it is in America and the other half is in Canada. It is a very popular tourist spot as it is the broadest or widest waterfall of the world with a width of about 11 kms. There are certain water falls that generate electricity. The waterfall near Manali in India is very beautiful and worth seeing.

Waterfalls are usually created due to erosion by rivers

UNDERGROUND WATER

The rain water and the water from melted ice that reaches under the surface of the earth through pores and cracks is called underground water. Some underground water may originate from the steam rising from molten rock materials deep within the earth. Erosive action of water on limestone rocks turns their pores into such big sink holes that even a river if that is drained into them, would disappear. Fellbank river in Yorkshire sank into a 111-metre deep hole.

Springs and artesian wells owe their existence to the flow of underground water. Artesian wells are of great economic importance in the mountain-ringed basins in the semi-arid climatic regions of the world. There exist more than 18 thousand artesian wells in the world's largest artesian basin in Australia. Artesian wells have been sunk in the Terai region of India for irrigation and supply of drinking water. The underground water is very useful in our daily lives. It not only provides us the drinking water but is also a valuable source of water for irrigation purposes.

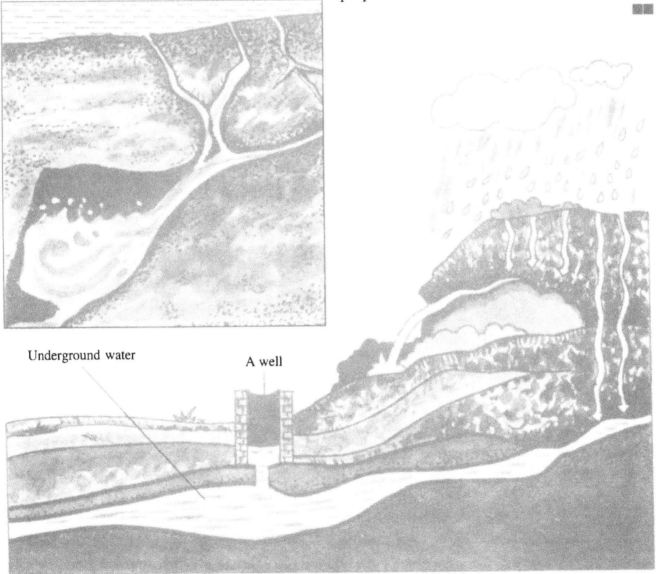

Underground water   A well

CAVES AND CAVERNS

A cave is a natural opening in the earth that is big enough for a person or animal to enter. Most caves are formed by the erosive action of rain water or underground water on rocks such as limestone or dolomite. The water contains dissolved carbondioxide which forms dilute carbonic acid. This acid dissolves the rock, forming passages and large open spaces. Caverns, caves and joint-galleries are the result of this process.

The other types of caves are lava caves, ice caves and sea caves. Lava caves are found near the base of a volcanic mountain. Ice caves are formed within glaciers. Sea caves are formed in the coastal rocks by the action of sea waves.

The early man lived in caves. Even today, some groups of people in Spain and the Philippines live in caves. Kentucky caves in the USA are well known such as the *Mammoth Cave System* which is the deepest cave in the world (563, 270 metres) deep and the *Fisher Ridge Cave System* is the world's longest cave, about 116 kms long. In India, the Ajanta and Ellora caves are quite popular.

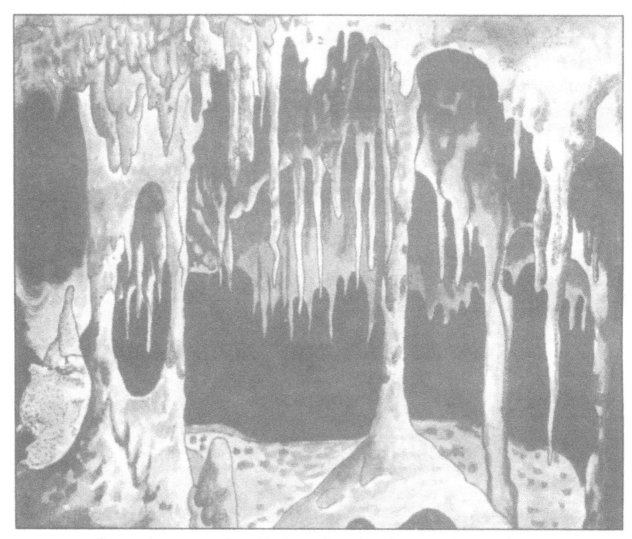

Caves and caverns are formed by the erosive action of rain or underground water

WEATHER AND CLIMATE

The major elements of weather and climate are *atmospheric pressure, temperature, humidity, rainfall, snow fall, velocity and direction of winds and distance from seas and mountains.* The condition of these elements at a certain time at a place is known as the weather of that place at that time. The average weather conditions of a region over a longer period of time is called the climate of that place. Weather denotes the day-to-day conditions of the earth's atmosphere while climate denotes the conditions over a longer period of time.

The world has been divided into 13 major climatic regions. For convenience, these climatic regions have been grouped into three on the basis of their latitudinal position and extent. These are — *the low latitude climatic region, the mid-latitude climatic region and the high latitude climatic region.*

Low latitude climatic region

It includes the humid tropical region, trade wind coastal region, tropical desert and steppe, tropical monsoon and savanna regions. These regions have very high temperatures and heavy rainfall.

Midlatitude climatic region

It includes China type, West European type, Mediterranean, mid-latitude desert and steppe and Manchuria type climatic regions. These regions are not very hot but are very cold in winters.

High latitude climatic region

It includes Taiga type, Tundra type ice-cap and high mountain type regions. In these regions, the temperatures go down severely and in summers too, it remains quite cold.

According to climatic conditions, we decide the architect of our houses, clothes, foods and means of transport.

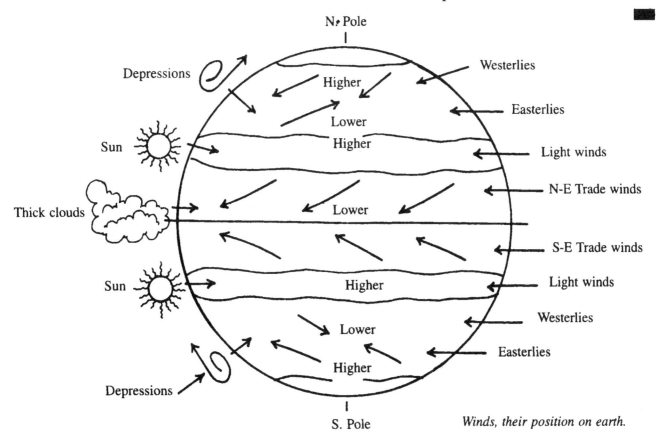

Winds, their position on earth.

61

FOG, MIST AND SNOWFALL

Fog

Fog is a cloud-like mass of water droplets which are formed near the ground or on the surface of a body of water. Fog limits the visibility of objects to less than 1 km. Mist is less intense than fog and has visibility more than 1 km.

There are several different ways that fog is formed. Radiation fog occurs on clear cool nights, if the temperature falls to the dew point. When this happens the water vapour condenses and forms fog. When warm, moist air blows over colder air, advection fog forms. Frontal fog occurs at a weather point, where two air masses of different temperatures meet. *Sea fog* is *denser* than the *land fog*. Those cities which have many industries have denser fog due to the smoke emitted from these industries. Even the more populated metropolitan cities such as Mumbai, Kolkata and Delhi have denser fog than the other smaller cities of India. In 1955, a chemical process was developed which is used to clear the fog to a certain extent.

Snowfall

In clouds, when the temperature falls below the freezing point, i.e., Zero degree Celsius, precipitation takes place in the form of ice crystals. The crystals combine together to form snowflakes. They are usually hexagonal, needle shaped or star shaped. These snow flakes fall to the earth if the temperature in the lower atmosphere is below the freezing point. It is called snowfall. Snowfall is common on higher peaks of mountains and higher latitude regions.

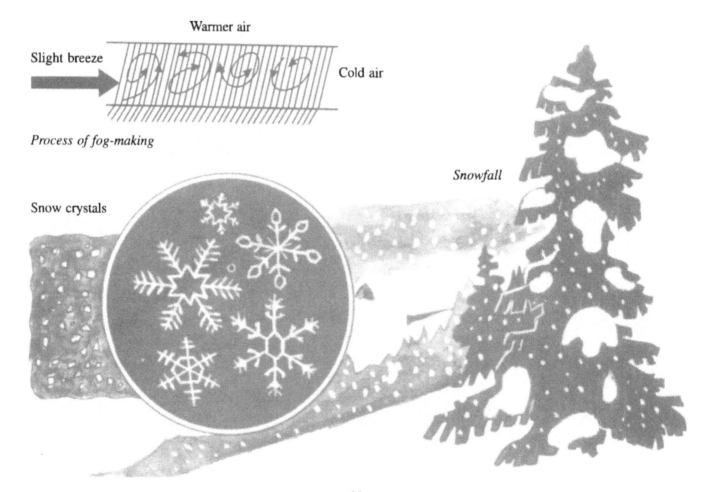

Warmer air

Slight breeze

Cold air

Process of fog-making

Snow crystals

Snowfall

CLOUDS

Evaporation of water from the oceans, lakes and rivers by the heat of the sun is a continuous process. Air containing water vapour is lighter than the air and so it continues to rise up. At considerable height where the temperature is lower, the water vapour condenses to form tiny droplets that look like smoke. It is called cloud.

Clouds are of various appearances, and shapes and exist at different heights. On the basis of these features, they are classified into three main groups. These groups include ten types of clouds.

Low clouds

These are formed at a height up to 2.5 km. They include the stratus — like dark grey sheet, the cumulus — woolly bunchy, the cumulonimbus — heavy dark woolly, the nimbostratus — grey rainy, and the stratocumulus — grey-white layered woolly.

Medium clouds

They are formed at a height of 2.5 km to 4.5 km. This category includes the altostratus — white woolly sheets of layers, and altocumulus — white woolly.

High clouds

These are formed at a height above 4.5 km. They include the cirrus — white patches, the cirrocumulus — white layered patchy woolly and the cirrostratus — transparent fibrous thin white veil.

Clouds give us life-giving rains and snow fall which helps us to cultivate our land for agriculture.

Cirrus clouds

Stratus clouds

Cumulonimbus clouds

Cumulus clouds

63

RAIN

The sun's heat is constantly causing evaporation of water from rivers, lakes, oceans and other bodies of water. The water vapour is always present in the air. As the air rises, it cools and cannot hold much water vapour. At a temperature, called the dew point, the water vapour condenses on tiny particles of soot, dust and salt in the air, forming tiny water droplets. The water droplets combine together, forming clouds or fog. With further cooling, the water droplets combine and gain weight. They fall to the ground due to gravity as rain drops.

Rain is caused only after the humid air rises up and condenses. This upward movement may take place in several ways. When the humid air rises upwards after striking against some natural barrier, like a mountain and causes rainfall. It is known as orographic rainfall. In this type of rainfall, the windward slope of a mountain range gets more rain. When the warm and humid air-masses converge with the cool air-masses, they cause stormy conditions resulting in condensation and rainfall. This kind of rainfall is known as cyclonic rainfall. The third kind of rainfall is caused by convectional ascent of the warm and humid air. As the humid air rises up, it expands and cools followed by condensation and rainfall. It is known as convectional rainfall.

Rainfall is measured by means of an instrument called 'rain gauge'. Cherrapunji in Assam (India) has been recorded to have the highest rainfall in the world. Here the average rainfall in a year is more than **1200 cms**. In **1861**, the average annual rainfall went upto **217.5 cms**, which is a record in itself.

Hydrologic cycle

CYCLONES

A cyclone is a storm with strong spiralling winds rotating around a moving centre of low atmospheric pressure. The centre of the cyclone is known as its eye. The movement of cyclonic winds is counter-clockwise in the northern hemisphere and clockwise in the southern hemisphere.

As the cold air-mass and warm air-mass come into contact with the cyclonic winds, it results in heavy rains and snowfall. Such cyclones are called frontal cyclones.

Tropical cyclones occur in warm water regions of the oceans near the equator. These are known by different names. Over the Atlantic Ocean they are known as hurricanes but over Pacific and Indian Ocean they are called typhoons. Hurricanes and typhoons cause havoc to ships and coastal settlements killing people and destroying properties.

A tornado is a small but violent cyclone that develops in a large thunderstorm lasting for a short time. It has dark funnel-shaped cloud. Along its track trees are uprooted and buildings destroyed. The dust and debris pulled by the tornado from one place are heaped over some other place. During a tornado, the velocity of the wind goes upto 200 kms per hour.

On the 13-14th November, 1970, Bangladesh faced the most devastating cyclone which took away more than 10 lakh lives. Every year, lakhs of people die due cyclones, storms and tornadoes, across the globe.

Tornado — a cloudy water spout

Destruction caused by a cyclone

GLACIERS

Glacier

A glacier is a huge mass of snow and ice which moves slowly down a valley. Glaciers generally occur in polar region and on high mountains. The speed of their movement is slow on mountain slopes. Glaciers differ in their size and velocity of movement. Usually they move about half a metre or one metre a day. But in Greenland, they may move up to 15 metres a day due to their heavier loads (debris). Glaciers cover about 10% of the earth's land surface. Glaciers are generally found on the mountain peaks and mountain passes.

In the coastal regions of Greenland and Antarctica large masses of ice broken off from the glaciers float on the sea. They are called icebergs. Normally about one-tenth part of an iceberg is above the sea level and the remaining under the water. Some icebergs are up to 90 metres above the sea level while 810 metres hidden under the sea water. Icebergs are of enormous size weighing up to 200 million tons.

The largest glacier in the world is *Lambert* about 515 km long and 70 km wide. It lies in the Australian Antarctica region. Other famous glaciers in the world include Zermatt in Switzerland, Laom in Norway, Votion in France and Nishkanevelly in America.

Iceberg

ARCTIC REGIONS

The Arctic Ocean lies on the northern border of the earth. In its centre is situated the North Pole. Surrounded by land masses on all sides, it is completely frozen. On three sides it is flanked by northern boundaries of Asia, Europe and North America, while Greenland and other small islands exist on its fourth side. It is covered with snow all the year round and icebergs float on open seas.

Hardly any vegetation exists there because it is covered with snow. Very few people live there — the Eskimos in the northern America and Greenland, and the Lapps in Europe. They have learnt ways to survive in this cold region. They build dome-shaped snow houses called 'igloos' by stacking slabs of snow. They live on hunting and fishing. Reindeers, seals, walrus and Arctic bears are also found there. They use sledges that move very quickly on the snow.

Eskimos and their Igloos

PETROLEUM

The word 'petroleum' means 'rock oil'. The word 'Petro' means rock and 'Oleum' means oil. Petroleum is a thick black liquid found in huge reserves in subterranean rock strata.

It is believed that marine animals and plant bodies were buried under rocks due to earth upheavels millions of years ago. They were decomposed due to high pressure, heat and bacterial activities and transformed into petroleum.

Petroleum is taken out from inside the earth by drilling deep wells. It is called crude oil. The crude oil is refined in refineries by fractional distillation. During the process of refining we get many products like petrol, diesel, kerosene, naptha and tar. These petroleum products are brought to the oil refineries and industries to clean them from the impurities and make them pure for daily usage. Besides these petroleum is used to prepare plastics, hydrocarbon, medicines, etc.

It is also used to produce natural gas, LPG, etc. Scientists of today have devised many experimental methods to produce petrol but this is not very economical and is very expensive.

The major petroleum producers are the Soviet Union, Saudi Arabia, USA, Mexico, Venezuela, China, Kuwait, Iran and Iraq. In India, petroleum is primarily found in Mumbai, Gujarat and Assam.

Petroleum is formed by decomposition of dead animals and plants buried under rocks

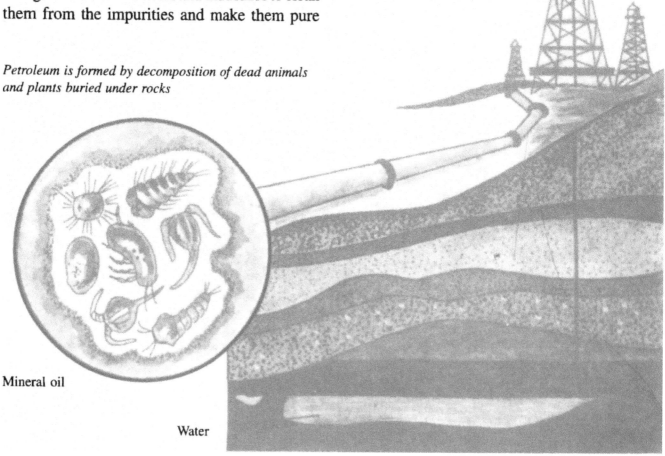

Mineral oil

Water

68

03

LIFE ON EARTH

GEOLOGICAL TIME

The Geological Time Scale is a system of classifying different developmental stages of the earth. According to the Geological Time, our earth is about 4.6 billion years old. The surface of the earth has been undergoing changes constantly since its creation. First, with the formation of atmosphere, the erosion of rocks began by the action-of water and wind. Then, started the stratification (deposition) of the eroded material which was in the form of sand and mire. The layer by layer deposition occurred under the shallow seas at the continental edges. Fossils of some organisms have been found under these layers. These organisms either lived in the continental edges or were carried to them by water currents. They got deposited alongwith the sediments at the bottom of the seas. As time passed the sediments hardened to form sedimentary rocks and the organisms buried under them turned into fossils.

The earth's crust is mainly composed of sedimentary rocks. Various layers of these rocks are formed in sequence of their origin. Younger layers (of recentorigin) are in the upper part while the older ones (formed in the distant past) constitute the lower parts. Therefore, the fossils found in the lower strata are considered to be older than those found in the upper ones.

Geologists have estimated the age of the earth and its different rocky layers. Their estimates are based on the rate of accumulation of salt in the oceans and half life of the radioactive substances such as Uranium and Thorium. The age of the rocks of recent origin is determined on the basis of the rate of erosion and deposition of sediments. By comparing the rock strata found in different places and the fossils found in them the ages of various strata are arranged in an order. By adding up the ages of all the rocks in that order, the Geological Time is calculated.

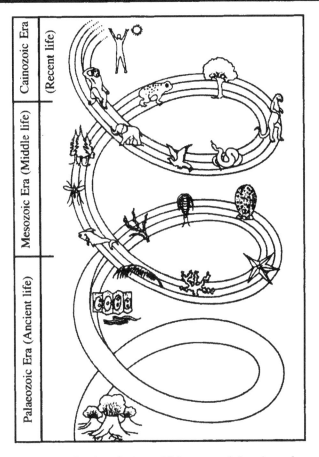

The path of evolution of life on earth has been long

The Geological Time of the earth is divided into two main parts—the Cryptozoic Aeon and the Phenrozoic Aeon. The Cryptozoic Aeon is the age of hidden life and deals with the beginning of the earth's creation. It is also called the pre-Cambrian period. The Phenrozoic Aeon encompasses the period beginning from Cambrian age till date: It is the age of visible life on earth. The pre-Cambrian period is divided into two eras—the Archaeozoic and the Proterozoic eras. Similarly, the Phenrozoic Aeon comprises three eras—Palaeozoic, Mesozoic, and Cenozoic eras. These eras are further subdivided into periods and epochs. The life on earth evolved chiefly during Phenrozoic Aeon. Therefore, in this book, the three eras falling under Phenrozoic Aeon form the basis of study of evolution of life on earth.

71

FOSSILS

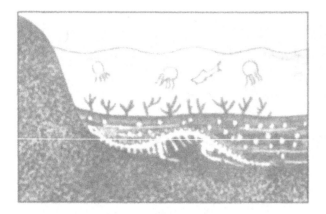

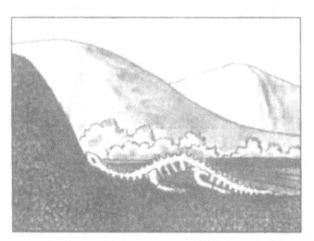

The process of formation of fossils

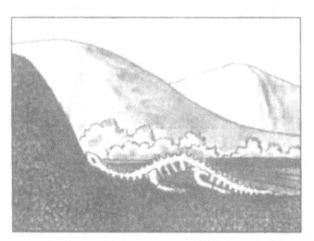

A fossil of trilobite

In ancient times, many plants and animals were buried under the earth due to upheavals occurring on its surface. The bone fragments, shells, skeletons, footprints, trunks of animals and leaves of plants, found embedded in the rocks, are called fossils. The fossils highlight the changes in landmass, weather and seas which have been taking place on the earth since its birth. In other words, the study of fossils tells us about the process of evolution of life on this planet. Georges Cuvier (1769-1832) is regarded as the founder of Palaeontology or the study of fossils.

The process of fossil formation takes a long time. When a plant or animal dies in a lake or sea, it goes to the bottom and soon gets covered with mud or sand. The soft organs of the organisms decay with time. However, bones and shells remain intact. The mud and sand containing bones and shells gradually turns into hard rocks. The minerals dissolved in water enter through the pores into bones and shells and transform them into stone. The study of fossils or Palaeontology helps us to find the age of a rock. It also helps us to learn more about the origin of life on earth.

Fossils are of three kinds. First, those in which the whole body of the organism is found preserved. Second, those which contain bones, shells or moulds of animals; trunk or leaves of plants and trees. The leaves are preserved as a thin layer of carbon inside the rocks. Third, those which are impressions of the movements of organisms in the mud or mire. Sometimes small particles of minerals get deposited on buried logs and transform them into petrified logs. The Forest National Monument in Arizona has thousands of petrified logs. Sometimes, certain plants and bodies of certain animals are covered with snow for thousands of years and they remain intact under the snow.

LIFE IN THE OCEANS

Evolution of life began with the Archaeozoic Era. No fossils in the rocks pertaining to that era have been found. This is because during that time only one celled living organism existed called the protozoa and this was around 15000 lakhs of years ago. Some of the earliest fossils found in the sedimentary rocks belong to the Proterozoic Era. These fossils seem to be of algae or bacteria. But, a large number of different kinds of fossils pertaining to different periods of the Palaeozoic Era have been found. That is why the Palaeozoic Era is considered as the 'age of early life'. During this period, several primitive fishes, aquatic animals, plants, etc. also developed under the oceans and seas.

The Palaeozoic Era began some 570 million years ago and continued till about 340 million years ago. Geologists have divided this era into 6 periods, namely, Cambrian, Ordovician, Silurian, Devonian, Carboniferous and Permian.

During the early Palaeozoic Era, plants and other organisms lived only in seas and oceans. A brief description of different periods of this era is given below:

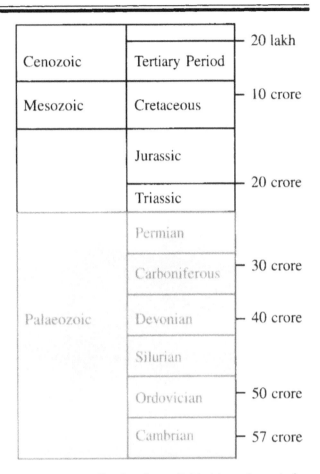

Cenozoic	Tertiary Period	20 lakh
Mesozoic	Cretaceous	10 crore
	Jurassic	
	Triassic	20 crore
	Permian	
	Carboniferous	30 crore
Palaeozoic	Devonian	40 crore
	Silurian	
	Ordovician	50 crore
	Cambrian	57 crore

The Palaeozoic Era has been divided into six periods

The soft-bodied creatures of the cambrian period

73

Creatures of the Ordovician period

Cambrian Period

Cambria is the name of ancient Wales. This period dates from 570 million to 530 million years ago. A large number of fossils belonging to this period have been discovered by geologists. These fossils indicate that soft bodied organisms like jelly fish, starfish, sponges and various kinds of worms inhabited the seas in this period. The close relatives of these organisms exist even today. These soft bodied living beings had started developing some sort of protective shells. First among them were Trilobites— Brachiopods and Arthropods. These shield-headed organisms had many legs. Life did not exist on land during this period. The entire North America was underwater. Oil, Copper, Lead, Asbestos, Marble, etc. were formed during this period.

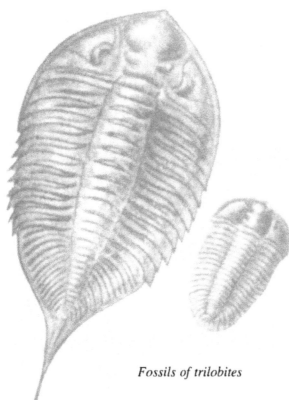
Fossils of trilobites

74

Ordovician Period

This period spans from 500 million to 435 million years ago. Along with the trilobites, strange kind of arthropods, called sea scorpions, lived in large numbers during this period. Some of them were two metre long and had hard nails which helped them devour all living organisms except those with hard shells. They were perhaps the most frightening invertebrates of that time. After the extinction of sea scorpions and trilobites, earthworm like creatures came into being. Simultaneously, echinoderms also made their appearance. Some of their near relatives, like sea urchins and sea lilies, are seen even today. During this period, continents were surrounded by shallow seas. Oil, Gas, Iron, Lead, Zinc, Gold and Silica were formed during this period.

Silurian Period

This period dates from 435 million to 345 million years ago. During this period, corals, lampshells, clams, sea snails and early species of fish were evolved. Jawed fishes were seen for the first time alongwith numerous jawless fishes. This period witnessed rapid growth of vegetation and invertebrates. Continents were surrounded by seas and a large number of volcanoes discharged

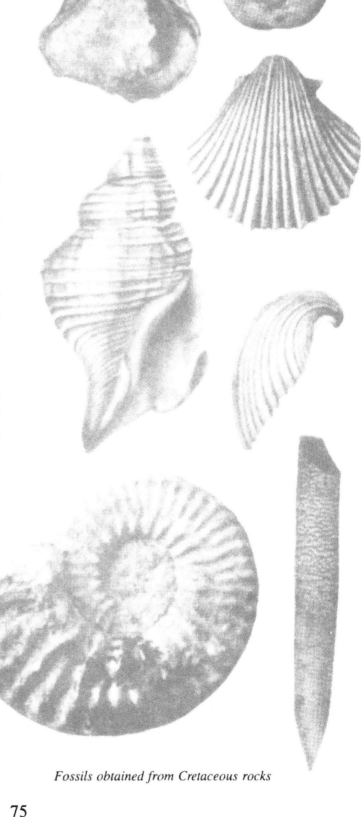

Fossils obtained from Cretaceous rocks

75

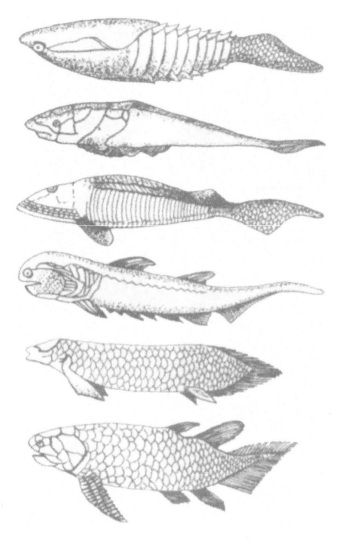

Fishes of the devonian period

gases and lava. lron, oil, gas and silica were formed during this period.

Devonian Period

This period dates from 395 million to 345 million years ago. It saw an abundant, growth of bony fish, some of them had lungs to breathe. Spiders, mites and early wingless insects had also come into being. The end of this period saw the evolution of amphibians. The climate was warm and dry oil, gas, coal and silica were formed.

Carboniferous Period

The period from around 345 million to 280 million years ago is called the Carboniferous Period. During this period trees grew in swamps. Amphibians continued to evolve while insects multiplied at a very fast rate. Reptiles had made their appearance by the end of this period, which also saw the formation of coal, and hence, this period is also called the Carboniferous period.

Permian Period

During this period which spanned from 280 million to 225 million years ago; the seas had turned shallower. The number of reptiles and insects on land was. increasing. For the first time, conical trees, appeared. Ice covered the South Pole.

The first amphibian

THE MESOZOIC ERA

The Mesozoic Era means the era of reptiles. This era is also known as the age of dinosours. During this era dinosaurs not only inhabited the earth in large numbers but some of their species could also fly in the sky. Infact, this was the age when the living organisms that could crawl or creep came into existence. The geologists count the Mesozoic Era from 225 million to 65 million years ago. This era consists of three periods— the Triassic period from 225 to 193 million years ago, the Jurassic period from 193 to 135 million years ago and the Cretaceous period from 135 to 65 Million years ago.

During the Triassic period, the earth was a single landmass which in due course of time separated into two parts. The Northern part, called Laurasia, consisted of North America, Europe and Asia. The Southern Part, called Gondwana land, comprised South America, Africa, India, Antarctica and Australia. The climate during this period was sub-tropical with alternate summer and rainy seasons. A large part of the earth was desert. Plants like ferns, mose and bamboo and trees like conifers and monkey puzzle existed. Reptiles were found in large numbers all over the earth. The first mammals appeared in this period. The formation of coal, zinc and manganese had also started.

Various kinds of creatures such as dinosaurs, dragons, giant lizards, etc. were evolved during the Jurassic period. The climate turned humid and had affected the vegetation. The trees like confiers, ferns and palm were evolving. Some

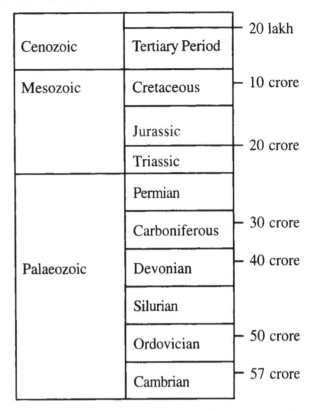

Cenozoic	Tertiary Period	— 20 lakh
Mesozoic	Cretaceous	— 10 crore
	Jurassic	— 20 crore
	Triassic	
Palaeozoic	Permian	
	Carboniferous	— 30 crore
	Devonian	— 40 crore
	Silurian	
	Ordovician	— 50 crore
	Cambrian	— 57 crore

The Palaeozoic Era has been divided into six periods

places were covered with thick forests. Besides large numbers of reptiles, few primitive mammals and early birds had also appeared. During this period formation of coal, aluminium, iron and gold had started. Infact, dinosaurs used to rule on earth during this era and many of them were herbivorous and some were carnivorous.

The continents were formed during the Cretaceous period. Most of the land was flat with no mountain ranges. Oak, magnolia, giant red wood, fig and palm trees were present. Early marsupials and placental mammals came into being. By the end of this period, huge reptiles, i.e. dinosaurs, were on the verge of extinction. Oil, gas, coal, diamond, gold and other minerals were formed. ■■

DINOSAURS

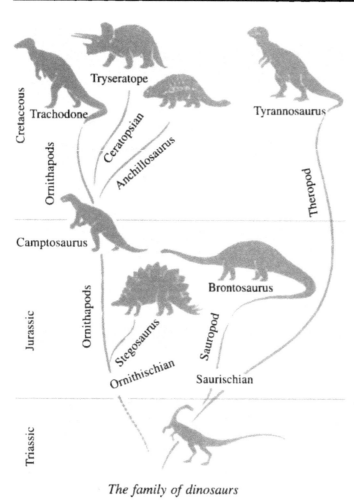

Dinosaurs evolved during the Mesozoic era and their dominance on earth, both land and water, continued for about 150 million years. Some of these reptiles were of the size of a pigeon while others were 20 times as large as an elephant. Some dinosaurs weighed more than 45 metric tons. The word Dinosaur , meaning a giant lizard, was coined by the British biologist Richard Owen. Scientists have divided dinosaurs into two major groups—Saurischian and Ornithischian. The hind part of Saurischian dinosaurs resembled a lizard (lizard-hipped), while Ornithischian dinosaurs had a bird-like hind part (birdhipped). Saurischians are further divided into two groups—the Theropods i.e. carnivores or meat eaters, and the Sauropods i.e. the herbivores or plant eaters. Given below is a brief account of some dinosaur species.

The family of dinosaurs

Allosaurus

Allosaurus was the fiercest carnivorous dinosaur of the Jurassic period. Twelve metres in length and weighing 2 tons, it walked on its hind legs. and its fore legs were shorter in length. With its nails and 15-cm-long knife teeth, it could kill and eat even the largest dinosaur.

Allosaurus and Stegosaurus

Tyrannosaurus Rex

Tyrannosaurus Rex was the largest ever carnivore which lived on earth. Fifteen metres long, 6 metres tall and weighing about 7 tons, it lived till the end of the Cretaceous period. It ran fast and could leap as well. It could easily chew the hardest of bones with its dagger-like teeth. If broken, its teeth were naturally replaced in a short span. No living being on earth at that time could challenge Tyrannosaurus who was more intelligent than other dinosaur.

Diplodocus

It was the largest, harmless herbivore of the Jurassic period. Twenty-eight metres long and weighed 10 metrictons, it walked on its four legs. It had a disproportionately long neck and tail. It

Tyrannosaurus

Eggs of dinosaurs obtained from Cretaceous sandstone in Mongolia

79

remained mostly in water, but kept its mouth above water for breathing. Diplodocus had a large body but a very small mouth. Therefore, it had to eat vegetation all the time to satiate its hunger.

Brontosaurus

Fossils of the hornless ceratopsian dinosaur of the last part of the cretaceaus period found in Mongolia

Stegosaurus

Stegosaurus had a shield-like double plate of bones of one metre height, from one end of its body to the other. Its tail had four big pointed horns. This nine metre long dinosaur was a harmless herbivore and the shield structure of its body protected it from carnivores. Compared to its large body, it had a small, walnut sized brain. That is why its intelligence level was quite low, as compared to other herbivores. Another type of dinosaur belonging to the family of Stegosaurus was Enchilosaurus, though not as big as a Stegosaurus. Its shield-like hard body from head to tail had sharp spear-like bones pointing outwards, protecting it from carnivores.

Plesiosaurs

The three main groups of reptiles found in seas during the Jurassic period were—lichthyosaurs, Plesiosaurs and Pliosaurs. Of them, plesiosaur was the most important. Its tortoise-like body had a neck resembling a snake. It had very sharp teeth and could swim with the help of its feet. It was 12 metres long. Skeletons of Plesiosaurs and lchthyosaurs were first discovered by a 12-year-old English girl, Mary Anning. Besides these, there were many more types and varieties of dinosaurs that inhabited the planet, earth.

Skolasaurus of the Stegosaurus species

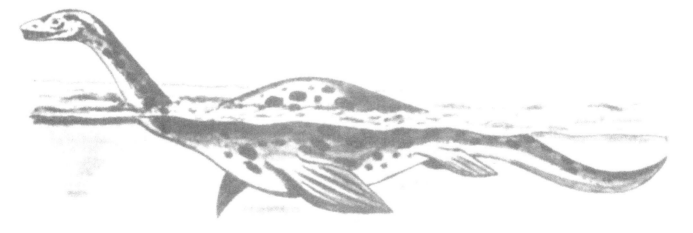

Plesiosaurs

FLYING REPTILES

During the Mesozoic era i.e. the age of dinosaurs, giant creatures flew across the sky, though they were not birds.

During the Permian Period, some 240 million years ago, some of the close relatives of lizards had learnt to glide. They looked like birds and their foreclaws were joined with their hind legs by a thin membrane. When in flight they stretched the membrane to form a parachute like structure. Their long beak had sharp teeth. They also had a long tail.

Among the more developed species of these flying creatures was Pterosaurs which could in fact be termed as a flying vertebrate. They resembled bats and had hollow bones which made their body light. One of the fingers of their forelegs was very long. A membrane stretched from their foreleg to the hind leg. It worked as a wing when the legs were stretched. The remains of Pterosaurs, discovered in March 1975, show that their bodies were not very large. Scientists have estimated their weight to be around 86 kg, but the span of their wings must have been more than 10 metres. When flying they must have looked like a giant demon and these came under the category of flying reptiles.

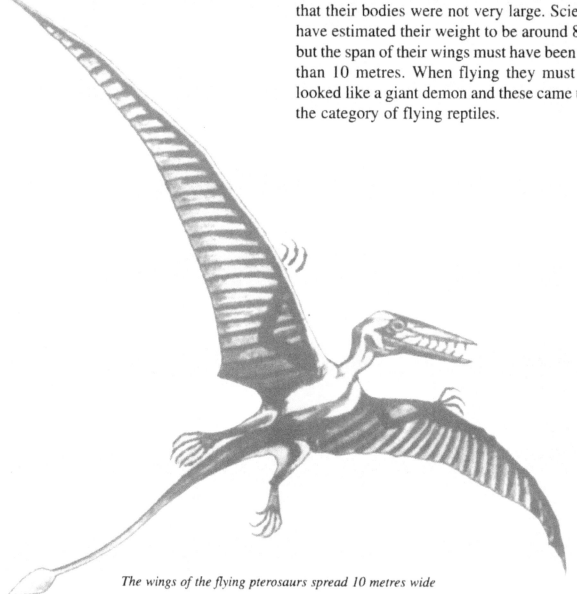

The wings of the flying pterosaurs spread 10 metres wide

ARCHAEOPTERYX—THE FIRST BIRD

The word, Archaeopteryx, means the ancient winged bird. These birds appeared some 140 million years ago at the end of the Purassic period. Their fossils were first found in Germany *in 1861*. Almost as big as a crow (about 20 inches), these birds had feathered wings. Their long hind legs had claws. Foreclaws were attached to the wings. The long beak contained teeth like those of Pterosaurs. Therefore, Archaeopteryx is called the link between the reptiles and the birds.

These ancient birds were perhaps not good at flying. They used to crawl up a tree with the help of foreclaws and feet from where they launched their flight. But one thing is certain that the Archaeopteryx were the first birds and could be called ancestors of the today's birds. They also couldn't fly very high.

The first bird Archaeopteryx

A fossil of Archaeopteryx

CENOZOIC ERA

The Cenozoic Era is referred to as the age of modern creatures. It started after the extinction of the dinosaurs. The Cenozoic Era covers the last 65 million years of the earth's history. It is considered the age of mammals and birds.

The Cenozoic Era has been divided into two periods—the Tertiary Period and the Quaternary Period. The tertiary period lasted for 63 million years while the quaternary period covers the last two million years.

The tertiary period has been subdivided into five segments called epochs—Palaecocene Epoch from 64 million to 52 million years ago; Ecocene Epoch from 54 million to 38 million years ago; Oligocene Epoch, from 38 million to 26 million years ago; Miocene Epoch from 26 million to 7 million years ago and Plicocene Epoch from 7 million to 2 million years ago. Ancestors of man, other mammals, birds, insects and flowering plants evolved during the tertiary period. The seas expanded to cover more parts of the earth. Formation of minerals like oil, coal and aluminium took place. This period witnessed the extinction of giant reptiles and appearance of flowering plants or angiosperms.

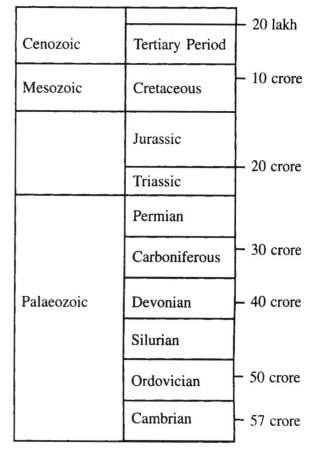

Cenozoic	Tertiary Period	— 20 lakh
Mesozoic	Cretaceous	— 10 crore
	Jurassic	
	Triassic	— 20 crore
	Permian	
	Carboniferous	— 30 crore
Palaeozoic	Devonian	— 40 crore
	Silurian	
	Ordovician	— 50 crore
	Cambrian	— 57 crore

The age of birds and mammals: Cenozoic Era

The quaternary period saw the further evolution of human beings. It is subdivided into two short epochs—the Cold Pleistocene, 2 million to 10000 years ago; and the Warm Holocene or modern age, the one which is going on today. This age witnessed the evolution of most modern forms of life and extinction of huge elephants and hairy rhinoceros.

During the Cenozoic Era tall mountain ranges like the Rockies, Alps and Himalayas were formed.

PLANT EVOLUTION

During the early period of the Palaeozoic Era, there existed simple blue-green algae as well as the modern, complex-structured brown algae. The first land plants were evolved during the mid Palaeozoic Era. This era also saw the evolution of some plants which were quite similar to today's club mosses, ferns and horsetails. At the close of this era, trees formed a major part of vegetation which included the 50 metres tall club moss trees. During the same period, trees like calamite and horsetail also existed which could achieve a height of upto 35 metres and the diameter of their trunk being about 90 cm. In the carboniferous age of this

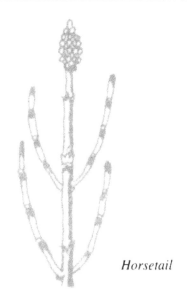

Horsetail

The veqetation of the carboniferous period consisted of club mosses, chelemytes, ferns and seeded ferns

85

era, most of the vegetation was in the form of club mosses, calamites, ferns and seed ferns. When buried under the earth, these trees were transformed into coal, petroleum and natural gas through the action of heat and pressure.

These trees became extinct at the end of this era. Thereafter, evolved several forests of gymnosperms, which were predecessors of modern gymnosperms. This marked the beginning of Mesozoic Era. Various kinds of gymnosperms multiplied rapidly in the early Mesozoic Era. Forests of cycads, maiden hairs and conifers appeared. Gradually other varieties

Tree-fern of the carboniferous period, which was 25 feet tall

Seed-fern of the carboniferous period

of conifers such as pine, yew, red wood and cypress also developed. After some time, they started disappearing and today only 640 species exist.

Flowering plants also started appearing during the closing years of this era. Fossils of this era indicate that many forests of modern trees viz, magnolia, wiloss, oak, palm etc. existed.

The Cenozoic Era was an era of flowering plants during which angiosperms evolved rapidly throughout the world. The entire area from the Arctic to the Antarctica was covered with multi-coloured flowers.

As the Cenozoic Era advanced, the earth started becoming cold and dry. Widespread glaciation occurred around one to one and a half million years ago. It resulted in the extinction of tropical plants near the poles which were now confined only to warm places near the equator. The process of cooling and drying gave rise to pot herbs.

A fossilized trunk of the Redwood tree, in California, which was 80 feet high and 12 feet wide

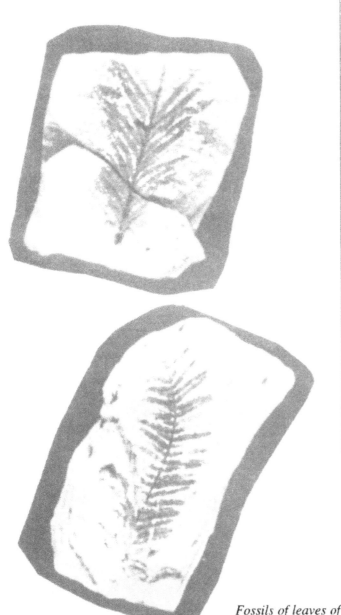
Fossils of leaves of the Redwood tree

EARLY MAMMALS

Fossils have indicated that mammals evolved from reptiles. The early mammals were very small. The first animal that had some features of a mammal was a Diametroden which existed around 250 million years ago. It was 3 metres long with teeth like those of latter mammals. The next in the chain was the 2 metre long Cynognathus which existed 220 million years ago. Its skin was covered with hair which resembled that of later mammals. It was, perhaps, warm blooded. There is no evidence available whether it laid eggs or gave birth to young ones.

The lriconodon that lived 190 million years ago, is considered as the first true mammal. It resembled a cat, its body was covered with hair and it was warm blooded. It is not known whether it laid eggs or delivered young ones, but the mother suckled the new-born.

The mammals evolved from mammal like reptiles which were as small as mice. These reptiles were far ahead of other living beings in their march towards development.

The cat-shaped Triconodon, found 19 crore years back, showed some features of mammals

The early mammals knew how to defend themselves from dinosaurs

88

At the beginning of the Cenozoic Era, about 60 million years ago, there existed three kinds of mammals—Monotremes, Marsupials and Placentals.

Monotremes laid eggs and were cold blooded. Duck-billed platypus and echidna, found in Australia, are the only survivors of this class today.

Marsupials give birth to under developed young ones. The babies crawl into a pouch on their mothers belly. They feed on milk, stay warm and protected in the pouch. Marsupials are warm blooded and are found in Australia and surrounding islands. Kangaroo, Koala, Wombat and marsupial mouse are some examples of this class of mammals. Opossum mouse are found in America even now:

Various kinds of marsupials were found during prehistoric times in Australia, North and South America and Europe. There was the Proceptodon, a giant Kangaroo, during Pleistocene era in Australia. Wolf-sized marsupial lion, Thylacoleo and hippopotamus-sized Diprotodon were also found in this period. All early mammals in South America were marsupials.

Most mammals alive today are placentals. In placental mammals, the foetus and the embryos are nourished in the uterus of the female through a tube called placenta. The young

The Marupial mammal Koala

The first carnivorous mammal Oxyaena

The first omnivorous mammal Barylambda

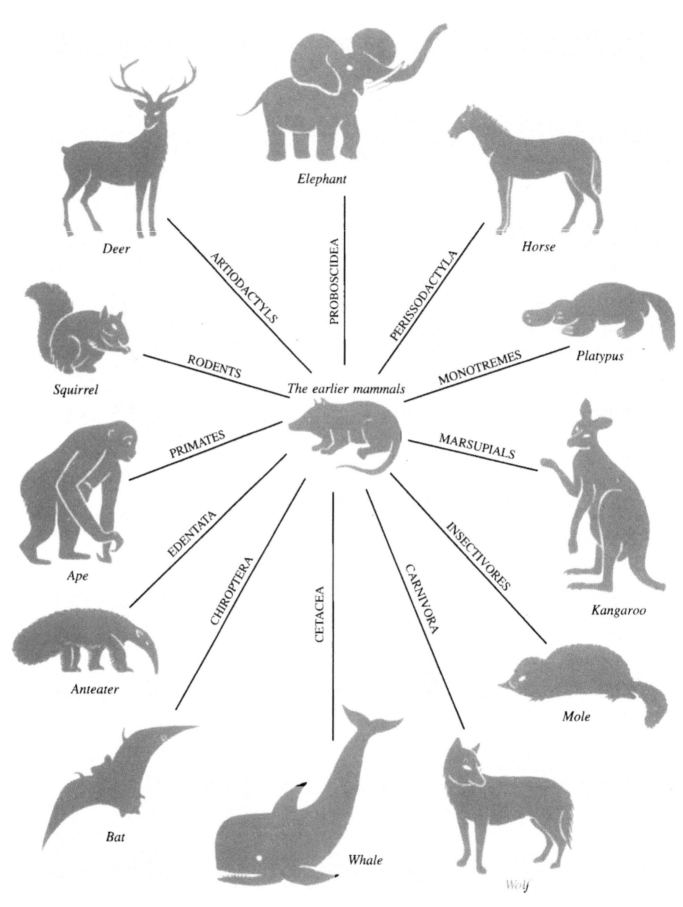

The mammals form the largest group of ceatures, with the greatest of variety

ones when born are quite developed. These mammals are quite intelligent. Of the three kinds of mammals—monotremes, marsupials and placentals—living in the same age, the placental group was more developed than the other two.

The early mammals fed on insects and earthworms. During the process of evolution, some of them became carnivorous. The one metre long Oxyaena, the first carnivorous mammal, lived some 50 million years ago.

Cynodictis, the pre-historic dog, lived 30 million years ago. They were about 30 centimetre long, lived in groups and hunted together.

The sabre-toothed cats were found 26 million years ago. Their teeth were as long and sharp as a dagger which enabled them to tear apart even the thick skinned elephants. There were many species of the 2.5 metre long cats.

Caracass-eating Hyena, measuring 1.5 metre in length, was found around 20 million years ago. It had very strong jaws and teeth with which it could crush thick bones.

Barylambda, the first herbivorous mammals, existed about 55 million years ago. They were 3 metre long, but lived only for a short period before they became extinct. The terocious Brontotherium lived about 28 million years ago. It had a sharp horn on its snout. It was 3 metre long and fed on leaves and juicy fruits.

The prehistoric rhinoceros called Baluchitherium was about 11 metre long and 8 metre tall. It was the largest herbivorous mammal.

Arsinoitherium, a 3 metre long herbivore, was found about 35 million years ago. It had two bony horns on its snout.

Moeritherium, the earliest elephant, lived about 40 million years ago. In size it resembled a large pig. When in danger, it submerged itself in water. It lived on soft and juicy leaves.

Hyracotherium, the early horse, was found 50 million years ago. It was as big as a fox and had long nails instead of hoofs. It had four digits in its forepaws and three in its hind paws. It lived in forests:

During the Ice Age, around one to two million years ago, wolly mammoths were found. These elephant-like animals were 4.5 metre tall. They had curved tusks measuring 3.5 metre. .

Today, there are about 4400 living species of mammals. They range in size from 5-cm-long pygmy shrew to the 31metre-long blue whale, the largest animal on earth.

The early elephant Moeritherium

91

The ancestors of man, monkeys and apes, are our nearest relations

40 cm long

The tree-living mole resembles
the prehistoric Primates

Monkeys of Asia

120 cm long

Lemurs found in Madagascar

The small apes living 2 crore
years back got divided into
two sub-species. The sub-
species that chose to live on
land later evolved into man.

EARLY MAN

Monkeys and apes are regarded as man's close relatives. Monkeys have tails while apes do not. Gorilla, gibbons, chimpanzee and orangutan are members of the ape family. This group of mammals is called Primates.

Most of the ancestors of apes lived on trees, about 65 million years ago. However, one of the groups among them called Ramapithecus, that existed 6 to 14 million years ago, preferred to live on land. Members of this group which lived during the Pleistocene period, about 3 million years ago, are referred to as Hominids. Their remains were found on the Shivalik ranges in India. Hominids were of two kinds. Some of them which became direct ancestors of man were called Homo while the others were called the Australopithecus. Their teeth and jaws were like human beings.

A fossilized skull of Ramapithecus: the jaw and teeth resemble those of man more than of apes

A close relation of man: gorilla

A large number of fossils of Australopithecus have been discovered from river valleys and caves in South Africa. These fossils indicate that the Australopithecus knew how to use its hands to grip articles. It used thick stems of trees and stones to hunt prey. Its brain was half the size of that of modern man, but the Australopithecus could definitely be called an intelligent ape. However, they didn't know the art of making weapons.

Homo Habilis

About 2 million years ago, a small group of men lived by the side of a lake in the eastern part of Africa. They were called Homo Habilis. Compared to the other apes, they were more intelligent and cultured. They possessed a brain more developed than that of the Australopithecus.They hunted in groups and had learnt to make crude stone implements. But they did not know how to make fire and ate uncooked flesh.

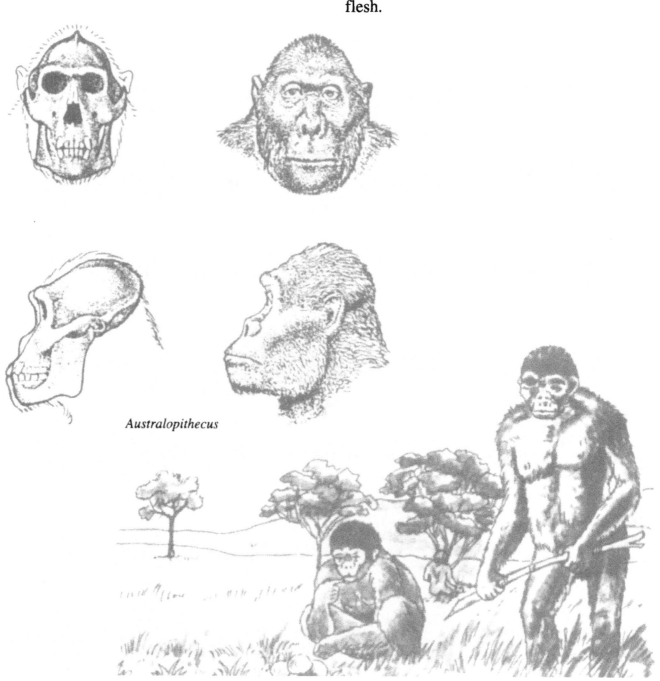

Australopithecus

The Australopithecus could hold objects firmly with his hands

94

Homo Erectus

The Homo Erectus, which could stand erect and walk, evolved about one and a half million years ago. He learnt making fire about three lakh years ago. Initially the Homo Erectus lived in Africa, but gradually they moved to other parts of the world. They lived in huts made of tree branches. The remains of these Homo Erectus were found in the Java islands. They had a brain about 74 percent of the size of modern man. These early hominids were about 1.79 m or 5ft. 10 inches tall.

Neanderthal Man

The Homo Erectus developed into the modern man, called Homo sapiens, some 250,000 years ago.

There were many races of Homo sapiens those days, but now only one is left which is called the Neanderthal Man. They lived in caves in Europe during the Ice Age some 50,000 years ago. They knew how to make fire and could shape working implements out of flint. This race was found throughout Europe and Western Asia.

Homo habilis used stone tools to cut flesh

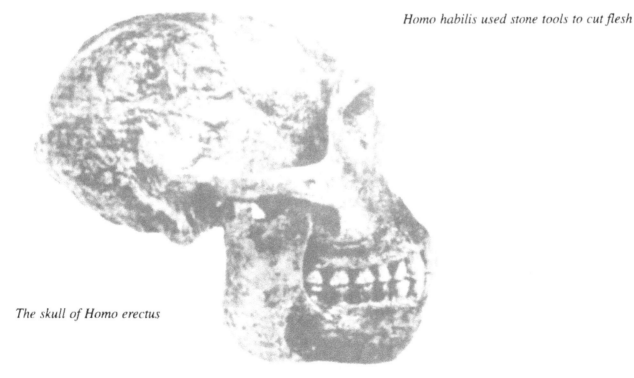

The skull of Homo erectus

95

A large number of skeletons recovered have indicated that they buried their dead along with something which the dead held dear. They were mostly found in Europe and Africa.

Cro-Magnon people

The Neanderthals became extinct some 40,000 years ago. Then, a new group of Homo sapiens quite like the modern man, were evolved. They

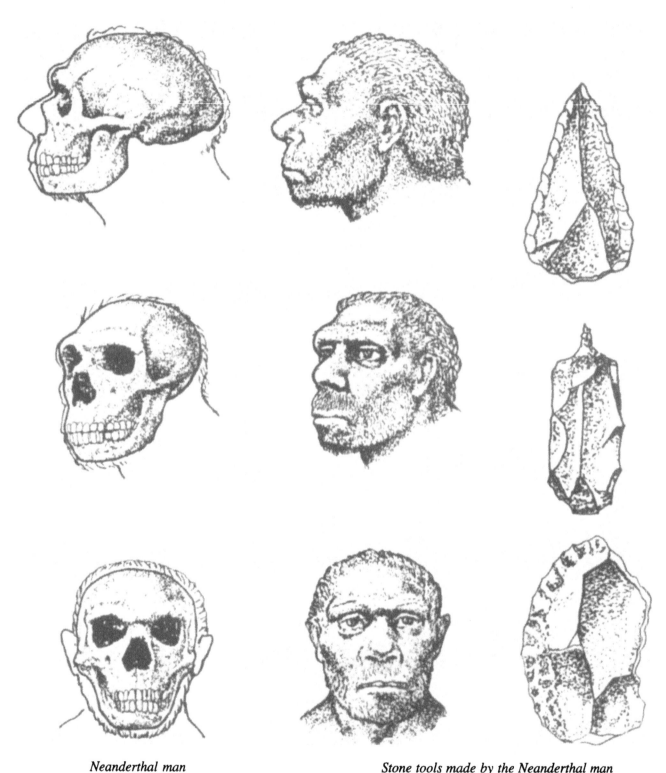

Neanderthal man

Stone tools made by the Neanderthal man

The Cro-Magnon people had developed new tools and weapons according to their requirements

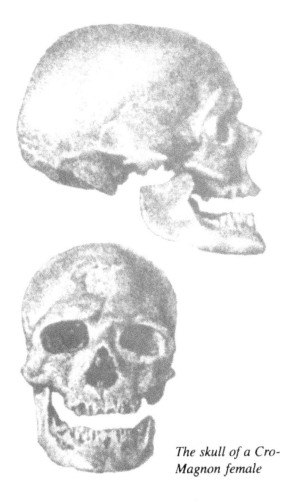

The skull of a Cro-Magnon female

The Cro-Magnon people painted pictures on the walls and roofs of caves

were our direct ancestors. A large number of their skeletons have been found in Europe which tell a lot about their lifestyle.

They hunted their prey with spears fitted with pointed stones. They knew how to cook food. They made use of animal hides. They stitched together animal hides using the needles of stones and bones to make clothes to protect themselves from the cold. They knew the art of drawing. Their drawings were found on the walls and ceilings of the caves they lived in. The colours of these drawings, made some 30,000 years ago, have not yet faded. Their remains have been found in France.

The Cro-Magnon people can be regarded as the first stage of the modern man. Then came the Stone Age when man left caves and started living in the fields. They had started agricultural operations in the Middle East around 1,1000 years ago. Although, not even a million years have passed since the evolution of the modern man, he has become the superior-most among all the living beings on this planet.

The origin of the modern man began about 25,000 years ago. The modern man is considered to be the most powerful and the most intelligent living creature on earth. The American scientist, H.L. Shapiro has said that human beings or *Homo sapiens* of the future will be taller in size with less hair on their heads and will be known as *Homo futuris*.

People had learnt farming about 11,000 years back

People had started living in huts and women used to make chapattis of wheat-flour.

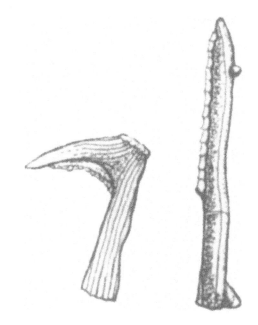

The first tools for farming were made of stone and horns

98

04

THE PLANT KINGDOM

THE PLANT KINGDOM

Scientists know about 360,000 species of the plant kingdom. Most of the plants differ from one another in size, shape and structure. For the sake of study, the plant kingdom has been divided into two sub-kingdoms—**Thallophyta** and **Embryophyta**. In the Thallophyta group, the plants do not develop from an embryo and their morphology is very simple. In the Embryophyta subkingdom, plants develop from an embryo and have complicated morphology. Every subkingdom is divided into phyla and every phyla into subphyla. Subphyla are further divided into subclasses.

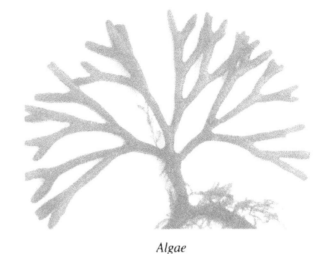

Algae

Classification of Plant Kingdom

I. **Subkingdom – Thallophyta** – the simple plants which do not form an embryo.

Phyla	1.	*"Cyanophyta"* – Blue green
Algae		algae.
	2.	*"Euglenophyta"* – Buglena etc.
	3.	*"Chlorophyta"* – Green algae.
	4.	*"Chrysophyta"* – Yellow green and golden brown algae and diatom, etc.
	5.	*"Phaeophyta"* – Brown algae.
	6.	*"Rodophyta"* – Red Algae.
	7.	*"Pyrophyta"* – Crypto-monads, Dinoflagillates, etc.
Phyla **Bacteria**	8.	*"Schizomycophyta"* – Bacteria.
Phyla **Fungi**	9.	*"Myxomycophyta"* – Slimmold.
	10.	*"Eumycophyta"* – Higher fungi.

Fungi

Bryophyta

Ferns

Angiosperms

Conifors

II. Subkingdom – Embryophyta – embryo-bearing complex plants.

Phyla 11. *Bryophyta* – Atracheata, moss and liverwort, etc. These lack vascular bundles.

12. *Trachaeophyta* – Tracheata. These plants have vascular bundles like fern.

Subphyla 1. *Psilopsida* – Rootless and leafless compound plants. These have become completely extinct.

2. *Lycopsida* – Clubmoss and other related species. Their vascular system is simple and leaves are thick and green.

3. *Sphenopside* – Horsetail and other related species. Their vascular system is simple, leaves are compound, thick and scale-like.

4. *Pteropsida* – Their vascular system is complex and leaves are large and clear.

Class 1. *Filicineae* – Fern plants.

2. *Gymnospermae* – Pine, fir, spruce, etc.

3. *Angiospermae* – Flowering plants as rose, magnolias, lilacs, apple, lily, orchid, etc.

ALGAE

Algae are the simplest kinds of plants. They contain chlorophyll, therefore, they can make their food by photosynthesis. They lack vascular bundles, i.e. they do not possess xylem and phloem. Their thallus lacks true roots, shoots and leaves. Algae include unicellular plants to very big multicellular plants. They do not form embryo after sexual reproduction. Commonly, algae are coloured and they can grow up to 40 or 60 metres in length.

Most algae grow in rivers, ponds, lakes and oceans. Some can live in moist places on land, mountains and near trees. A few species of algae are found on the ice in the polar regions and in hot water springs. Some of the blue-green algae can survive even at 70-80°C. Some algae grow on plants and a few complete their life cycle inside some plants. Many algae are saprophytic and parasitic. The best known algae is probably the seaweed found at beaches.

About 25-30 species of algae are used as food by man. The algae Porphyra of Phiophcae class is a common meal in Japan. In China, Nostoc commune is used as food. In India, algae is used to prepare icecream, chocolate milk, gelatin and beer. Some algae are poisonous in nature and can paralyze a person.

We get carbohydrate, vitamins A, B, C, D, E and other materials from algae. Fish depend on algae for their food. Many algae obtained from oceans are rich sources of iodine, potassium and other minerals. Algae is a good fertilizer also. The red algae grow in large colonies and turn the water red. Some red algae also form islands.

Different species of algae

FUNGI

Fungi are considered to be the oldest plants. They do not contain chlorophyll. Like other thallophytes, they do not contain roots, stems, flowers and leaves. These plants cannot manufacture their own food and so for food, they depend on decaying organic matter, plants and animals. Fungi are, therefore, called **parasitic** or **saprophytic**. Most fungi reproduce asexually, but a few species have a sexually reproductive stage which alternates with an asexual stage. Their cell wall is made up of cellulose of chitin. They produce enzymes to digest food. Some familiar fungi are: yeast, molds, mushrooms, mildews, rusts and smelts. Fungi can grow in all kinds of climates and environments. Some fungi are also found in air and water. Sir P.A. Nichelly is considered to be the *father of fungi* way back in 1729. He had discovered the fungi.

Fungi can be useful or harmful to human beings and other organisms.

Some fungi, such as Agaricus, Morchella, etc. are consumed as food by man. Certain molds, such as Camembert and Roquefort are added to cheese to provide a flavour and to help ripen the cheese. Yeast is used to manufacture bread and alcohol. It is a good source of protein and vitamin B. Some molds are used to produce antibiotics. The famous antibiotic penicillin was obtained from a mold in 1929. Fungi are also helpful in maintaining soil fertility.

Rust, Smut, Mold, etc. cause many dreadful diseases in crops. They also cause diseases in man and domestic animals. They spoil our food. Some species of fungi are poisonous and if they are consumed by man, they may prove fatal. Among these, the highly poisonous fungus is toad stool. From ergot fungus, a hallucinating drug L.S.D. is prepared. Ringworm and athlete's foot are diseases caused by fungus. Many varieties of mushrooms are poisonous and can cause sickness or even death if eaten.

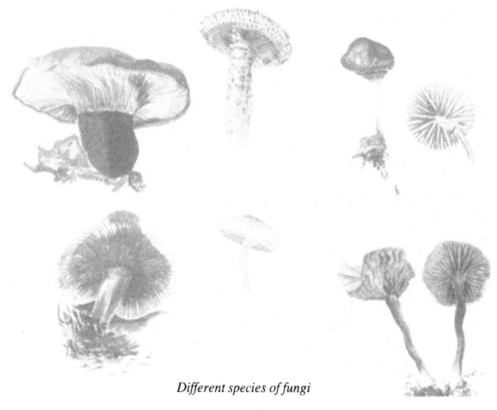

Different species of fungi

BACTERIA

Bacteria are the smallest and simplest living organisms. Most bacteria are unicellular but some may have many cells, up to 20 in number. Their length varies from 2 to 5 microns (one micron is equal to 0.001mm.) Bacteria can be of two types: **the moving type** and **stationary type**. There are four common bacterial body forms – spherical or coccus, bacillus or rod-shaped, spiral and vibro. Experts say that perhaps bacteria was the first living organism to exist on the earth. Bacteria were first seen by Antony Van Leeuwenhoek in 1675.

Bacteria are found everywhere. They may live in soil, in water, in air or in other organisms. Although most bacteria live at temperatures between 10 and 40°C, some may live in ice as well as in hot springs. Bacteria derive their food from other living and nonliving things. Therefore, they are known as **saprophytes**. They grow by multiplication. A single bacteria divides to form two, from two it makes four and so on. Within a few hours, their number exceeds millions, but only a few of them survive for want of food.

Some bacteria are useful to us in many ways. Some are helpful in manufacturing important chemicals of industrial use. Cheese, vinegar, curd, butter, etc. are produced by bacteria. Bacteria are also helpful in the synthesis of vitamins, in decomposition of dead bodies and in maintaining soil fertility.

Some bacteria are also harmful to man. They cause many diseases in man, domestic animals and in crops. They spoil our food and extract nitrogen from the soil. Disease-causing bacteria produce a kind of poison called toxin in our body. This toxin is very harmful for the body tissues of the patient and kills them. T.B., tetanus, leprosy, syphilis, gonorrhea, etc. are a few bacterial diseases. Among these diseases, cholera, fever, diphtheria, etc. are very contagious in nature.

Some bacteria in the bodies of the living organisms develop immunity due to the attack of other bacteria. The person who develops immunity becomes resistant to that particular disease. Active immunity is created by vaccination. Vaccines are mainly produced from original or weak bacteria or by the toxins produced by them. Antibiotics and sulpha drugs are used to control bacterial diseases.

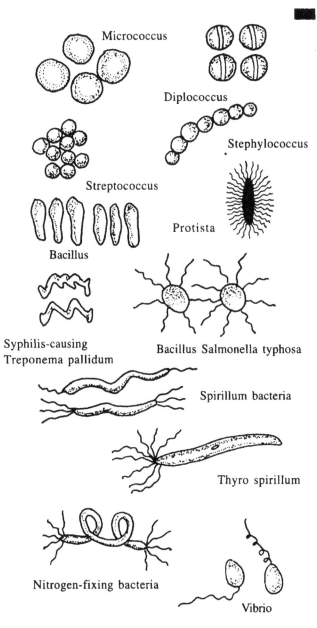

Various species of bacteria

BRYOPHYTA

Bryophyta is the smallest group of embryo-bearing plants. They belong to the subkingdom Embryophyta. These plants are generally small in size and lack vascular tissue. They posses Antheridium (male reproductive organs) as well as Archegonium (female reproductive organs). They do not bear true roots, shoots and leaves. This group includes mosses, hornworts and liverworts. Bryophyta contains some 300,000 species of plants.

Bryophytes are found in all moist parts of the world. They also grow on moist rocks and tree barks. Some of them may grow in hot and dry climate also.

Common bryophytes are not of much economical value. They help in the formation of important land masses. These were the first plants which helped in making land from rocks. Liverworts, mosses and lichens grow on these unnoticed lands. Bryophytes are also helpful in preventing soil erosion. They retain the moisture of the soil, set it at a place, and so prevent its being washed away. In this way, they are also helpful in controlling floods. Big mosses are used for packing. Moss is a vegetative fuel which accumulates in marshy areas and gets decomposed. It slowly becomes the source of energy by carbonization. In many parts of the world, such as Ireland and Scotland, peat is the main source of energy. Sphagnum is cleaned and after washing, used for the dressing of wounds.

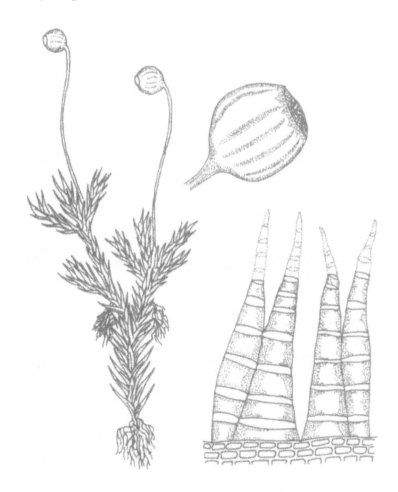

Bryophytes are found in moist areas

106

FERNS

Ferns and their distant relatives, horsetails and clubmosses come under the subphylum Philocophyta of phylum Tracheophyta. These are flowerless plants and do not bear seeds. This group includes about 250 living genus and 9,000 species. Scientists think that angiosperms have evolved from primitive ferns.

People in the seventeenth and eighteenth centuries were surprised when they did not find flowers and seeds on these plants. They thought that these plants reproduce by some secret means at night. They believed that if a person gets a fern seed, he would acquire extraordinary qualities. Shakespare had written in HenryIV that the fern seed had been found and one could walk by becoming invisible. Even today, many people complain to the fern plant sellers that on the lower sides of their leaves, brown patches or worms are developing. In fact, these are not worms or patches but are Sporangnium, which are called sori.

Ferns mainly grow in moist and dark places of forests. They may also be found in open areas. These are found all over the world, but their number is very large in moist tropical forests. These are herbaceous, but sometimes may be found as climbers or trees. The fern tree has branchless straight stems. Their tips bear palm-shaped leaves. Their stems may have heights of up to 18 metres.

Fern plants are very beautiful and their bulbs are edible. Their fibres are used in making pillows etc. Ferns of ancient times helped in the formation of coal. Fern leaves are used to pack vegetables and fruits. Fern is also used to prepare a medicine which kills intestinal worms.

9,000 species of ferns exist

GYMNOSPERMS AND ANGIOSPERMS

Gymnosperms and **Angiosperms** are seed-bearing plants. The gymnosperms bear naked seeds attached to the plants, while angiosperms bear seeds in covers or fruits. Most gymnosperms are evergreen plants, while angiosperms shed their leaves once every year.

The leaves of all seed-bearing plants contain chlorophyll and they synthesize their own food. In gymnosperms, reproduction takes place in cones, while in angiosperms, this process occurs in the flower. Gymnosperms have simple xylem but angiosperms have a complex type of xylem.

The wood of gymnosperms is used for furniture and as fuel. For example, pine, douglas fir, redwood, spruce, deodar, etc. About 300 species are cone-bearing conifers. These trees are mostly found in the northern and southern temperate forests. In these trees seeds are carried through air. There are both male and female trees in this category.

Angiosperms include melon, water melon, cucumber, tomato, beans, grapes, guava, etc. There are two types of angiosperms – **monocotyledons** and **dicotyledons**. Monocotyledons have seeds with one leaf inside, while dicotyledons have two. We get cereals, pulses, fruits, vegetables from these plants. We also get coffee, cotton, spices, oil and medicine from them.

Various kinds of cone – bearing conifers

ECOLOGICAL CLASSIFICATION OF PLANTS

Like a human being or an animal, a plant must adapt to its environment in order to survive. For their survival, plants compete with other plants for light, water, air and soil. The ability of plants to adjust with the environment is called **adaptation**. According to ecological classification, plants are categorized as follows.

Hydrophytes

Hydrophytes are those plants which live in water and adjust with their surroundings. They either remain fully dipped in the water like Hydrilla, Velisineria, etc. or most of their body parts remain under the water like trapa, lotus, etc. Water lilies, sedges, crowfoots are other important water plants. These hydrophytes travel long distances in the water, as the wind blows.

Mesophytes

The land plants which grow under normal conditions are called mesophytes. These plants grow on such places where the climate is neither too dry nor too moist. The atmospheric temperature and relative humidity are also normal. Wheat, pea, tomato, mango, guava, etc. are examples of mesophytes. Their roots are tightly embedded in the soil. The leaves are large and broad and the stem is erect.

Xerophytes

These plants grow in deserts or in dry areas. They can survive even under long dry conditions. To fulfil the water requirement, their roots become very long and the stems and leaves become thick and fleshy. The leaves of these

Different varieties of cacti

plants get modified into spines to control the loss of water by transpiration. Various kinds of cacti, aloes, stone crops etc. belong to this category.

Holophytes

These plants grow in such places where the amount of mineral salts in the water is very high. The stem and leaves of these plants become thick and fleshy. Such plants sometimes bear respiratory roots which come out of the soil and perform breathing function. Rhizophora and Avicenia are examples of such plants.

Epiphytes

These plants grow on other plants, on poles or on wires or they develop over the roofs of buildings. Epiphytes are mostly chlorophyllus and they prepare their own food. These plants get carbon dioxide and water from moist air. Their roots get water and nutrients from the dust deposited around them. Moss, fern, and orchid are examples of these plants. Lentils, the creepers of a pumpkin plant, etc. come in this category.

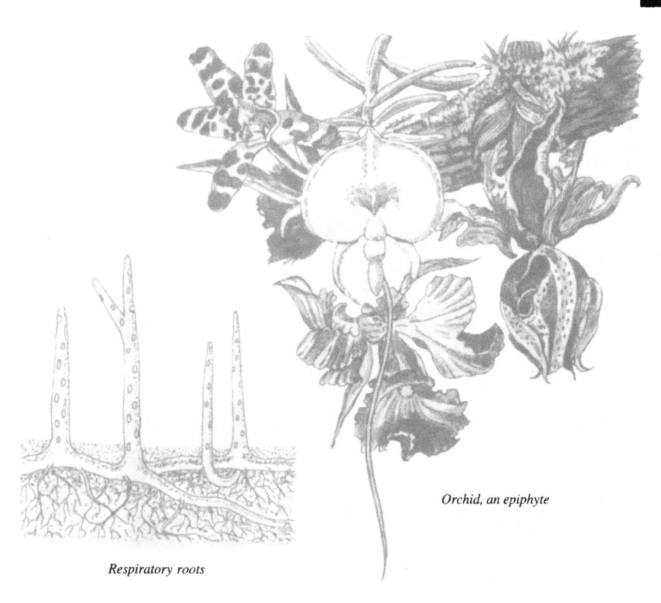

Respiratory roots

Orchid, an epiphyte

USEFUL ROOTS

Roots are not only important for the plants, but they are indispensable to us also. The roots of the plants like akonite, hing, blue flower plant of Kirat family, golden seal, Genron likorice, mashmalo, velerian, etc. are medicinal roots. Roots of some plants are used as food by us. Sugarbeet, carrot, redish, turnip and tapioca are root foods. These roots store food and, therefore, they become thick and fleshy. They have many modifications, such as fusiform *(Radish)*, conical, *(carrots)* napiform *(Beets)*, tuberous *(Sweet Potato)* and nodulated, etc. Sugarbeet, carrot, turnip, etc. are biennial varieties. During the first year of their life cycle, these plants manufacture a lot of food, a major portion of which gets deposited in their top roots. During the second year, this food is utilized in the formation of flowers and the seeds. We make use of these roots for our food during the first year of their life.

Spices and some flavours are also obtained from the roots of various plants. Several dyes are extracted from the roots of Madar and alkana plants. The Turky red dye is obtained from madar. Dyes derived from roots have now been largely replaced by synthetic dyes.

The roots of most leguminous plants, such as beans, sweet clover, and alfalfa contain nitrogen-fixing bacteria which help make nitrogen available to the plant.

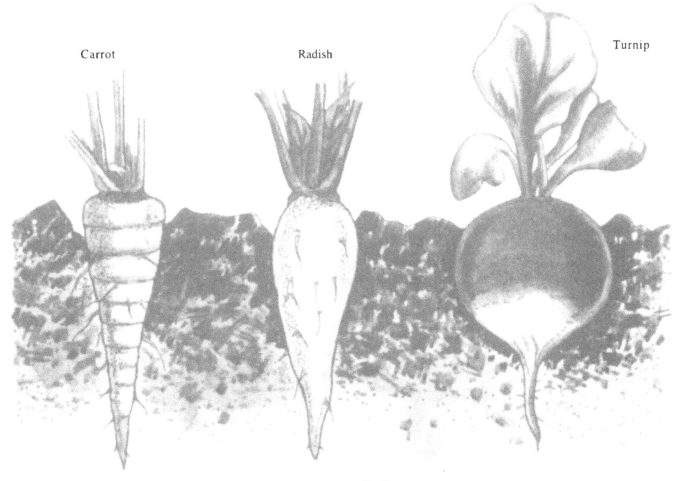

Carrot   Radish   Turnip

Various root foods

111

UNDERGROUND STEMS

The stem is the part of a plant that usually grows above the ground. It has two main functions – firstly, it supports various parts of the plant like branches, leaves, flowers and fruits; and secondly, it transfers water and nutrients from the roots to different parts of the plant's body. Apart from these functions, the stem sometimes performs some special functions also. Storing the food is one of them. These are called modified stems.

Among the modified stems, the stems, which are found under the ground are called **underground stems**. These stems remain in a dormant state but under favourable conditions, plants sprout from their buds. These stems store the food and become thick and fleshy. By the process of vegetative reproduction, they give rise to new plants. Underground stems are mainly of four types – Rhizomes, Tubers, Bulbs and Corms.

Rhizomes are straight, thick and fleshy underground stems. These grow horizontally under the ground. Ginger, turmeric, fern, lotus, canna, surgarcane, etc. are such stems. They possess nodes and internodes. Roots arise from lower part of the stem. When these roots are cut or plucked off, each part gives rise to a new plant.

Tubers are swollen parts of those branches which grow under the ground. These branches become swollen at their tips due to stored food. From the depressions, the aerial roots arise. Tubers do not produce roots. These stems are mostly round in shape, like potato and jerusalemartichoke.

Bulb is a big, round-shaped bud, on the underside of which is present a small stem. Bulbs arise from the small stem present at the base. Onion, garlic, lily and hyacinth are common bulbs.

Corm is a small, round, vertical underground stem. It differs from bulbs because in bulbs the leaves arising from the stem are fleshy, while corm is mainly made up of pillar-type tissues. Arbi, yam and saffron are such types of stems.

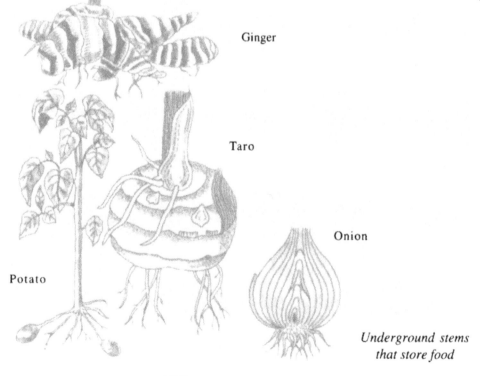

Ginger

Taro

Potato

Onion

Underground stems that store food

112

PHOTOSYNTHESIS

Plants need two types of nutrition – **organic** and **inorganic**. They get inorganic nutrition from the soil through their roots, while organic food is synthesized in their bodies. The process by which green plants synthesize organic food is called **photosynthesis**.

Photosynthesis occurs only in green plants. The green colour of leaves is due to a pigment called **chlorophyll**. Photosynthesis cannot take place in the absence of chlorophyll and water.

Photosynthesis is a very complicated process. It takes place in the presence of sunlight. The reaction occurs as follows:

Water + Carbon dioxide

$$\xrightarrow[\text{Chlorophyll}]{\text{Sunlight}} \text{Sugar + Oxygen}$$

$$6H_2O + 6CO_2 \xrightarrow[\text{Chlorophyll}]{\text{Sunlight}} C_6H_{12}O_6 + 6O_2 \uparrow$$

The process of light suspension takes place at about 25°C to 45°C.

Plants absorb water from the soil through their roots and pass it to the leaves. Carbon dioxide enters the leaves through tiny pores on their surface and in the presence of chlorophyll and sunlight, the given reaction takes place. The glucose synthesized during photosynthesis is utilized by plant cells as chemical energy. Glucose forms amino acids and proteins with nitrogen. Glucose also gets converted into cellulose and it forms various plant tissues. A part of glucose gets deposited in leaves, stems and roots in the form of starch. The starch is accumulated during the night time. This is how the life cycle of a plant goes on.

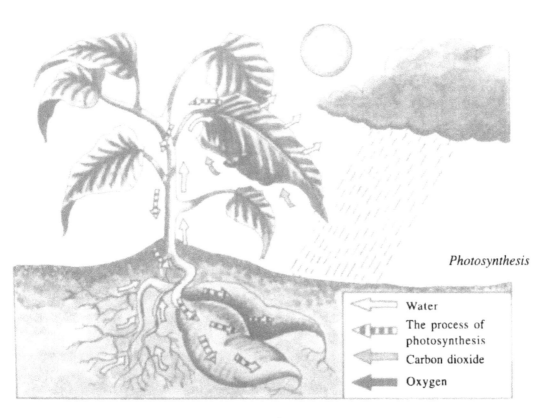

Photosynthesis

⇦	Water
⬅	The process of photosynthesis
⇦	Carbon dioxide
⬅	Oxygen

DEFOLIATION

The plants which shed their leaves every year are called **deciduous trees**. On the other hand, evergreen plants do not shed their leaves every year, but remain always green. Defoliation does not take place for all the leaves at the same time. The leaves which have fallen are replaced by new ones and the tree always appears green. The process of defoliation in both kinds of trees is the same.

The first stage in the leaf fall is the formation of an abscission layer at the place where the leaf joins the twig. Not long after the abscission layer is formed, it begins to loosen and dry out.

After a while, it ruptures. The leaf is held to the twig only by the vascular bundles. As the leaf is attacked by frost or sways in the wind, the vascular bundles begin to weaken. Finally, they snap and the leaf falls. This is called **defoliation**. The tree becomes almost barren during defoliation. A layer of cork cells develops below the abscission layer and provides a protective covering over the spot from where the leaf has been detached.

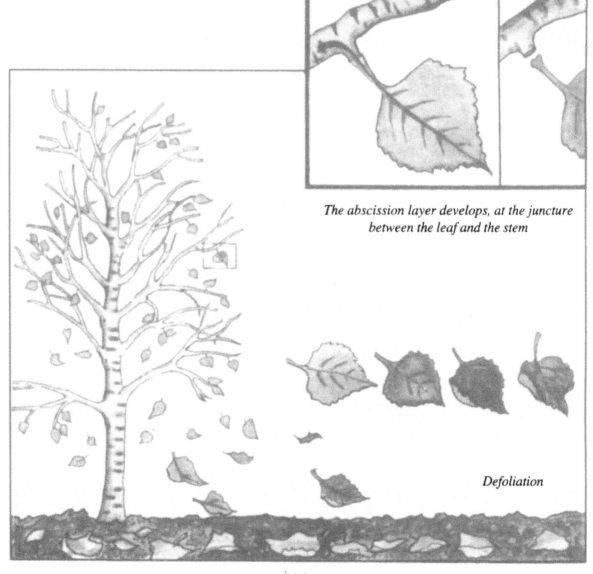

The abscission layer develops, at the juncture between the leaf and the stem

Defoliation

STRUCTURE AND FUNCTIONS OF A FLOWER

The flower is the structure in all angios-perms that is responsible for sexual reproduction. Its function is to produce seeds which will grow into new plants. There are at least 200,000 kinds of flowers with different sizes, shapes and colours. A typical flower has a pedicle, which is attached to the stem. The upper part of the pedicle, which is slightly fleshy, is called thalamus with which different parts of the flower are attached. The four main parts of a flower are as follows.

1. Calyx

The calyx is made of several green leaf-like sepals which protect the developing flower bud. In some cases, the sepals may be brightly coloured. The calyx also protects the flower when it is a bud.

2. Corolla

The corolla is made of several, usually brightly coloured, petals. The petals may take any of several shapes or may not be present. The calyx and corrolla together form the perianth. The

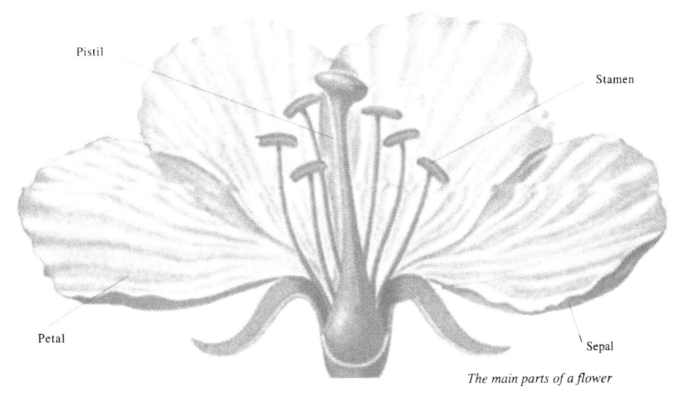

The main parts of a flower

115

corolla attracts insects and other agents for pollination.

3. Androecium

Inside the perianth are the reproductive structures of the flower. The **stamens** are the male reproductive structures. The number of stamens varies from zero to several hundred. A stamen consists of a thin filament supporting a thick knob-like anther. The anther contains pollen sacs which produce **pollen**, the male gametes.

4. Gynoecium

This is the female reproductive structure. It has three parts – the ovary, style and stigma. The stigma is the sticky part, style is a thin tube and the ovary is an enlarged area at the base that produces ovules. After fertilization, fruits and seeds are produced here.

Pollination

The transfer of pollen from anther to the stigma is called **pollination**. The process of pollination is of two types namely **self-pollination** and **cross-pollination**.

In self-pollination, the pollen from its own stamen may reach the pistil or pollen from one flower may land on the pistil of another flower on the same plant.

When pollen is carried from the stamen of one plant to the pistil of a flower on another plant, the transfer is called **cross-pollination**. The most common agents responsible for cross-pollination are insects, wind, water, birds and animals.

Fertilization

Pollination is the first step in making seeds. The next step is fertilization – the union of egg and

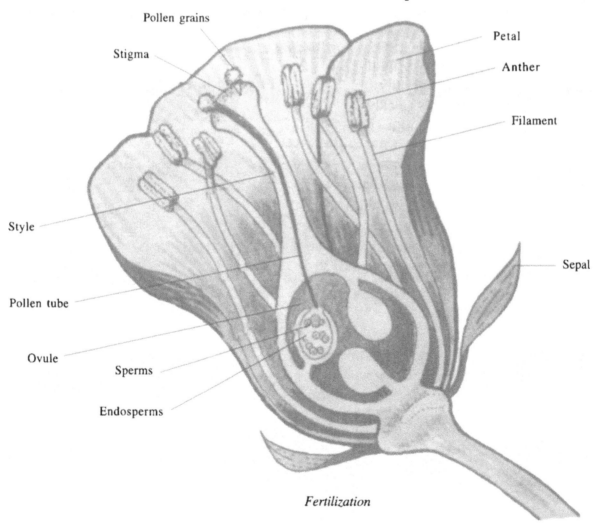

Fertilization

116

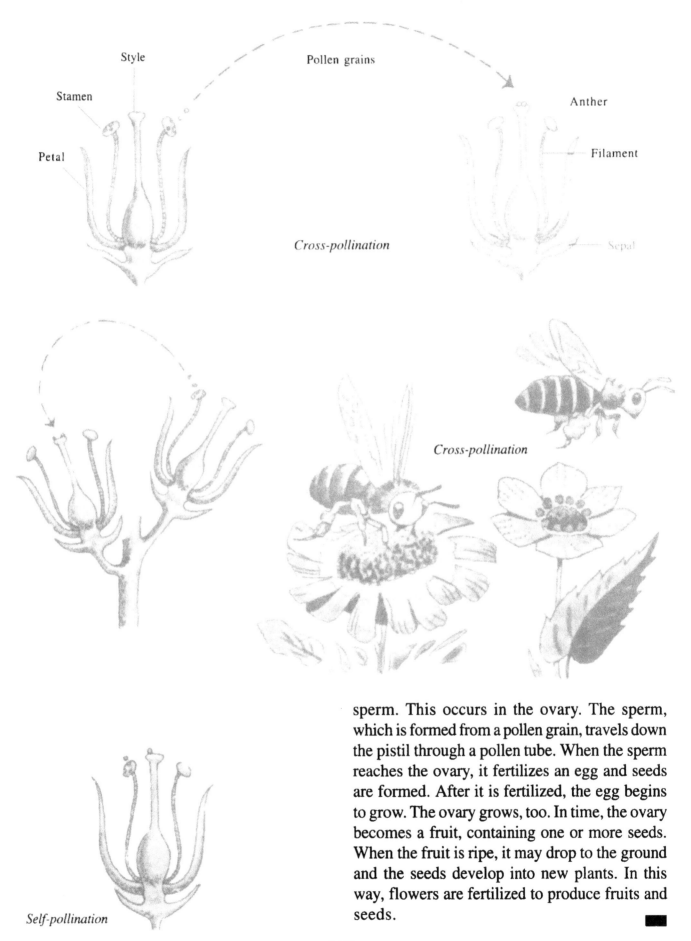

Style

Stamen

Petal

Pollen grains

Anther

Filament

Sepal

Cross-pollination

Cross-pollination

Self-pollination

sperm. This occurs in the ovary. The sperm, which is formed from a pollen grain, travels down the pistil through a pollen tube. When the sperm reaches the ovary, it fertilizes an egg and seeds are formed. After it is fertilized, the egg begins to grow. The ovary grows, too. In time, the ovary becomes a fruit, containing one or more seeds. When the fruit is ripe, it may drop to the ground and the seeds develop into new plants. In this way, flowers are fertilized to produce fruits and seeds.

117

INSECTIVOROUS PLANTS

There is a superstition about insectivorous plants that some can trap large-sized animals, including even man, and digest them. But scientists have yet to discover such a type of plant. The known insectivorous plants eat only small insects, crustaceans and other aquatic animals.

Insectivorous plants grow mostly in marshy places, which are deficient in nitrogenous compounds. These plants do not get nitrogen from the soil so they cannot synthesize protein in their bodies. Therefore, by eating small insects and animals, they meet their protein requirements. These plants also prepare food by photosynthesis. There are about 400 species of insectivorous plants, of which a few are described here.

Pitcher plant

118

Pitcher Plant

The leaves of this plant are tubular or pitcher-shaped and contain a sweet enzyme. When the insects get attracted by the colour and smell of the leaves, they crawl over the hard and drooping surface of the leaves and get trapped inside. After that, the cells of the leaves digest the soft parts of the insect's body. In this group, the leaves of the plant Sarracenia are the longest, more than a metre.

Drosera or Sundew

Sundew is another group of insectivorous plants. These plants are 8 to 20 cm long and green in colour. These plants have circular and flat laminae and hair-like structures on the upper surface of the leaves. These secrete a gummy liquid which shines like dewdrops. Insects, attracted by its smell, sit on the laminae and get stuck to the leaf. The surrounding tentacles also close in and trap the insect. The enzymes present in this gummy liquid digest the nitrogenous substances of the insect.

Venus' Flytrap

These plants mainly have laminae and lobes. From the lobes, there arise 12 to 20 hard teeth of about half-an-inch length. On the surface of the laminae, several thin hairs grow which are sensitive to touch. If an insect comes in contact with these sensitive hairs, the two flaps of the leaf start folding like a book. In the process, the insect is trapped and the glands present on the upper surface of the leaf digest the insect. After digesting the soft body parts of the insect, the leaf opens again. The leaf takes just a second to trap the prey.

Sundew

Venus' Flytrap

TALLEST TREES

The tallest tree – General Sherman

Bamboo

Pine, spruce, fir, cedars, juniper, cypresses etc. are the famous coniferous trees. These trees bear unisexual cones. Male and female cones can be formed on the same plant or on different plants. The leaves of all coniferous plants are simple and they remain on the plants throughout the year. Therefore, these trees are called **evergreen trees**.

The tallest trees of the world belong to this group. The redwood trees are considered to be the tallest of the world. These trees are now found only on the coastal regions of California. Their average height exceeds 90 metres and diameter is up to 5 metres. Their age is around 4,000 years. The tallest tree of this group was found in California in 1963. At that time, its tip was dried and its height was 112.10 metres. In 1970, its height was 111.60 metres. The tree is gradually drying up. The world's tallest tree that is alive and is in good condition is the General Sherman of California, which is 85 metres tall. The trunks (stem) of these trees are used to build and cover roads.

Among the flowering plants, Eucalyptus regnans with broad leaves is the tallest tree of the world and present in Tasmania. It is 99 metres high.

Heights of some other plants:

- **Callie grass** (5.5 metres)
- **Saguaro cactus** (16 metres)
- **Tree fern** (18 metres)
- **Bamboo** (37 metres)
- **Giant kelp** (60 metres)

Callie grass　　*Tree fern*　　*Giant kelp*

PARASITIC PLANTS

Most of the plants prepare their food by photosynthesis, but there are some which depend on some living organisms, dead plants or dead organic matter. Such plants are called **parasitic plants**. The plant from which parasitic plants obtain its food is called the **host**.

Cuscuta depends on other plants. The yellow, thin and delicate stem of cuscuta winds itself around the host. Its haustoria penetrates into the xylem and phloem of the host stem and absorbs food, minerals salts and water from there. The host of cuscuta gradually dies.

Rafflesia is a noteworthy parasitic plant found in south-east Asia. This is very thin like a thread and its flower is the largest in the world. The diameter of the flower is about one metre and it weighs up to 7 to 8 kilograms. They grow on Cissus twigs. These flowers produce a foul smell like that of a rotten dead body. These flowers are poisonous in nature. This flower

The world's largest flower, Rafflesia

Cuscuta

takes 5 to 7 days to bloom. Several other families of parasitic flowering plants are also well known such as the broom-rapes *(Orobanchaceae)* and the dodders *(Cuscutaceae)*. They generally attack the roots of the host plant.

121

TYPES OF FORESTS

Forests are those areas of the land where plants and trees are found in abundance. Though the grass is less in the forests, animals are present in large numbers. Forests are of the following three main types.

Tropical Rain Forests

Such forests are found in hot and humid tropical areas. Most of these forests are present around the equator. These areas remain hot throughout the year and receive high rainfall. Most of the trees of these forests are evergreen, with broad leaves. The world's maximum biomes, or plants and animals, are found in these forests. Only in the Amazon forests, about 2,500 varieties of trees are found. These forests exist extensively in southern Mexico, Central America and Amazon and Orinco rivers of South America. More than 50% of the world's rain forests are found in Brazil, Zaire and Indonesia. These are found in Africa and a few parts of Asia and in the islands of Southern Pacific Ocean also. Tropical deer, monkey, snake and giant lizards are found in the rain forests. One-third part of South America is

Tropical rain forests

122

covered with rain forests. Besides these, a number of birds such as parrots, kites, vultures, humming-birds etc. are also found in these types of forests.

Temperate Forests

These forests are found in comparatively colder and drier areas than the rain forests. The trees of these forests shed their leaves in the cold and dry season once in a year. Oak, beach, chestnut, elm, mopil and tulip trees are common in these forests. Animals like Grey fox, Virginia deer, squirrel, raccoon, bear, etc. are also found in these forests in large numbers. Here, we can find a lot of birds, who eat flesh of animals.

In most parts of north-western America, such forests accur. These deciduous forests are also found in central and northern Europe, some parts of Africa, South America, etc.

Coniferous Forests

Coniferous forests are found in cold areas. Most of the trees of these forests are evergreen gymnosperm trees like pine, spruce, fir, hemlock, etc.

In the USA, coniferous forests are found in the mountainous ranges of mid and northern California and the north-western Pacific region. In these western coniferous forests, various varieties of trees as pine, douglas fir, redwood, western hemlock, etc. are found. Animals like the lynx, deer, sheep, bear, goat, etc. are also found. In these forests, we can find owls, squirrels, rabbits, etc.

These forests also exist in Minnesota, Michigan, Wisconsin and New England, in some parts of Europe and Asia and in a few islands of southern Pacific.

Coniferous forests

Temperate forests

POISONOUS PLANTS

The flowers and fruits of some plants look very beautiful and attractive but they may be poisonous. A few famous poisonous plants are the following.

Strychnos nux-vomica is a poisonous plant found in Asian countries. Its fruit is of the size of an orange and it contains five seeds in it. Strychnine poison is prepared from these seeds. This poison stops respiration and causes death.

The white flowers which flourish on the **hemlock** plant produce an unpleasent smell. A poison called *caniine* is obtained from its roots and seeds. This poison paralyses all the organs of the human body and causes death. The great Greek philosopher, Socrates had been given a cup of hemlock to drink as punishment.

Colchicine poison is obtained from the seeds of **meadow saffron** plant. This poison is used for the treatment of arthritis.

Apart from these plants, foxglove, monkshood, poppy, cherry laurel, thorn apple, caper spurge, laburnum, henbane, holly, honey suckle, buckthorn, toadstool, bittersweet, etc. are also famous poisonous plants. Many varieties of **mushrooms** are also poisonous. Many plants of the subhax group are also poisonous, which produce blisters on the skin coming in contact with it. Ivy is also a poisonous plant and produces allergy in the body.

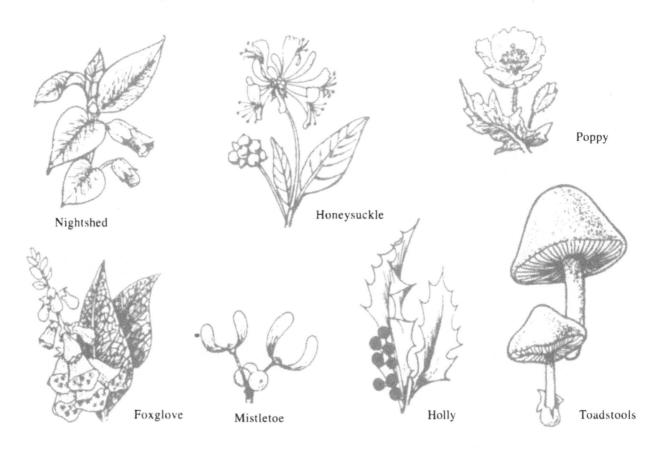

Nightshed

Honeysuckle

Poppy

Foxglove

Mistletoe

Holly

Toadstools

Poisonous plants

USES OF PLANTS

In the primitive ages, man was dependent on plants only for the food, but as he became more civilized and developed, he started using plants for constructing houses, making clothes, generating energy and making weapons. Today, we get the following substances from the plants.

Food Materials

Persons living near seashores use algae for food. Sargassum and Batrachospermum vagia are the main algae used as food. In addition to this, chlorella is used as proteinic food. It gives us water and oxygen also. This algae is used by astronauts during space flights.

Some kinds of fungi are also used as food in India, France, USA, etc.

From the flowering plants, we get cereals, pulses, vegetables and fruits. Apart from these, we get sugar from sugarcane, sugarbeet, palm, mapple, etc. Glucose is obtained from grapejuice. Different oils are obtained from edible seeds like mustard, groundnut, coconut, cottonseed, castor seeds, etc.

Volatile oils used in perfumes are obtained from the flowers of rose, sandal, jasmine, lavender, rosemary, etc.

Condiments

Different condiments are obtained from the different parts of various plants. India, Sri Lanka, Java, Africa, Madagascar are famous all over the world for spices. These enhance the taste of the food. Ginger, turmeric, cinnamon, cassia, cloves, saffron, cumin seed, pepper, cardamom, mustard seeds and caraway are some important condiments obtained from the plants.

Beverages

Many beverages are also obtained from plants. For example, we get tea leaves from the tea plant Camellia sinensis. Tea leaves contain an alkaloid called caffeine. This substance stimulates the

The tea plant – Camellia sinensis

Coffee seeds – Coffea arabica

125

Cotton

Pepper

heart and makes it more active. Tea also contains tannin. The red colour of the tea is due to resin. India, Bangladesh, Sri Lanka, China, Japan, Indonesia, etc. are major tea producers. About half the people of the world drink tea every day.

Coffee powder is made from the seeds of a plant called Coffea arabica. Its green seeds contain 11 percent protein and 8 percent glucose. Coffee contains vitamin Niacin in excess, therefore people who are used to taking coffee regularly do not suffer from the disease pellagra. Brazil is the leading producer of coffeee.

Cocoa is obtained from the seeds of the Theobroma cacao plant. A single fruit contains 40 to 60 seeds. Cocoa powder is prepared by drying and roasting these seeds. These seeds contain 50 to 57 percent of cocoa butter. Alkaloids like caffeine and theobromine are also found in cocoa.

Among the many beverages, only cocoa is the one which contains nutrients in it. It is also used in making chocolates. India, Sri Lanka, Kenya, Ghana, Brazil, Mexico, etc. are its main producers.

Many alcoholic beverages like beer, brandy, rum, whisky, etc. are prepared by the distillation of alcoholic liquor or by fermentation of glucose.

Fibres

String or threads for cloth are prepared from various kinds of fibres. Cotton fibre is obtained from the cotton plant and jute fibre from jute stems. The flax plant gives flax fibre. The hemp plant produces the sunn fibre. Coir is obtained from fruits of the coconut plant. Sometimes, the fibres of trees are used to manufacture paper in industries.

Medicines

Drugs obtained from a few plants are very important in the field of medicine. A fungus called Penicillium notatum is used to prepare penicillin. Similarly, claviceps is another fungus used to make ergot. The bark of the cinchona

126

Cocoa – Theobroma cacao

plant is used to prepare quinine. The stem of Ephedra geraradiana is used to prepare Ephedrine for the treatment of asthma. From the leaves and roots of Atropa belladona, a medicine called Atropine is prepared. Cocaine is extracted from leaves of Erythroxylum cocoa. Morphine is prepared from Papaver somniferum. From the fruits of Strychnos nux-vomica, a medicine called Strychnine is prepared. In India, there are about 4,000 species of medicinal plants. Turmeric *(Haldi)*, ginger *(Adrak)*, garlic *(Lahsun)*, onion *(Pyaj)*, etc. are also used in preparing medicines.

Dyes

Plants also give us various kinds of dyes and colours. From the plant tissues, indigo is prepared. Haemotoxylon is a black colour obtained from an American tree. Yellow and brown coloured fastic dyes are also obtained from American plants. The colours of saffron and chlorophyll are used in food and medicinal preparations.

Orangeish yellow colour is prepared from the palas flowers.

Rubber

The plant Hevea braziliensis produces a milky liquid called **latex**. Latex contains an organic substance which becomes hard upon coming in contact with air and becomes a flexible solid. This is called rubber.

Rubber trees

127

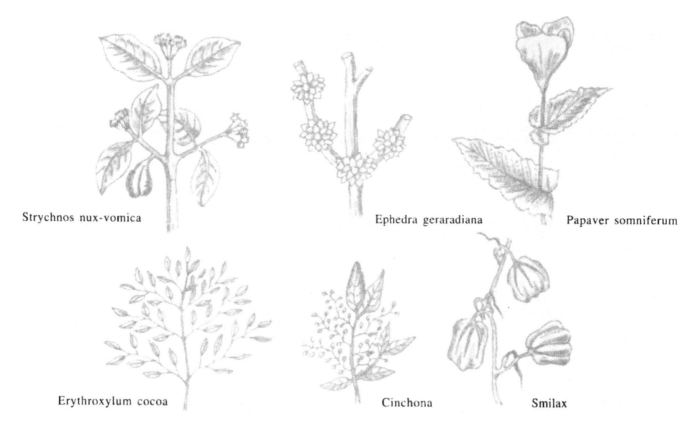

Strychnos nux-vomica   Ephedra geraradiana   Papaver somniferum

Erythroxylum cocoa   Cinchona   Smilax

Some medicinal plants

Resins and Gums

We also get resins and gums from the plants. Asafoetida is a kind of resin used in manufacturing paints and varnishes. They are also used in medicines. Turpentine oil and spirits are also obtained from these plants.

Gum is formed due to destruction of the cellulose of cell wall. This is soluble in water. Gum is used as adhesive and also to prepare drugs.

Other Uses

Wood is used as a fuel and also as building material. Paper is also prepared from wood. Wood is used to make methyl alcohol. Various types of boards are also made from sawdust.

Several kinds of acids, acetone, pitch, tar, oil, cosmetics, balsams, etc. are all gifts of plants only. Coal is also obtained from plants. This raw coal is used after keeping it for a few months. Coal is used to prepare a number of products such as paints, useful chemicals, etc.

05

ANIMAL KINGDOM

ANIMAL KINGDOM

Today, more than 1.2 million members of the animal kingdom have been identified. Scientists have classified them, besides giving them specific names. However, a large number of animals still remain to be classified and named.

Animals have been classified into different groups on the basis of their similarities and dissimilarities. For the purpose of classification, each animal's name is written in two words. The first word indicates its *genus* and the second its *species*. Several genera (plural of genus) having similarities form an *order*. Several orders make a class, and similar classes constitute a *phylum*.

Animals are broadly classified into two groups – the **Invertebrates** and **Vertebrates**. The animals not having a backbone are called invertebrates. Examples: amoeba, corals, worms, snails, insects and starfish. The vertebrate group includes the animals which have a backbone. Fishes, amphibians, reptiles, birds and mammals belong to this group. Given below are some main groups of invertebrates, with examples:

Group	Animal
• Protozoans	: Amoeba
• Poriferans	: Sponge
• Coelenterates	: Jellyfish, Coral, Sea Anemone
• Platyhelminthes	: Flatworm, Liverfluke, Tapeworm
• Nematodes	: Roundworm
• Annelids	: Earthworm, Leech
• Molluscs	: Snail, Slug, Clam, Limpet, Octopus
• Chilopods	: Centipede
• Diplopods	: Millipede
• Crustaceans	: Prawn, Crab, Wood Louse

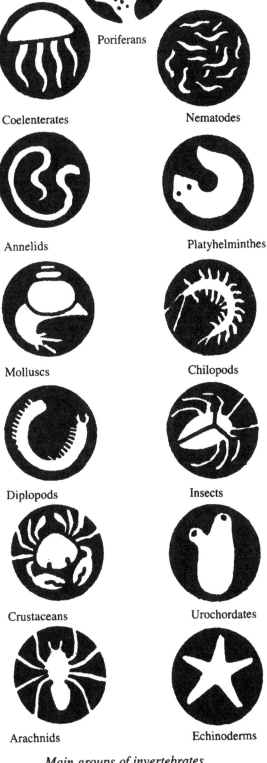

Poriferans
Coelenterates
Nematodes
Annelids
Platyhelminthes
Molluscs
Chilopods
Diplopods
Insects
Crustaceans
Urochordates
Arachnids
Echinoderms

Main groups of invertebrates

Contd...

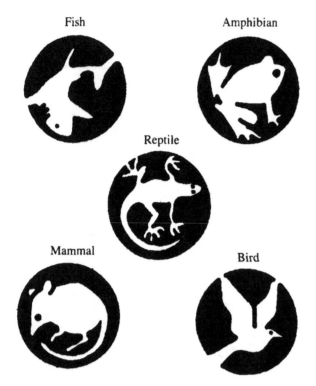

Five main groups of vertebrates

- Insects : Butterfly, Wasp, Ant, Louse, Beetle
- Arachnids : Spider, Scorpion, Mite
- Echinoderms : Starfish, Sea Cucumber, Sea Urchin
- Urochordates : Sea Squirt

Given below are five main groups of vertebrates with their animals :

Group	Animal
• Fishes	: Shark, Trout, Eel, Seahorse
• Amphibians	: Frog, Toad, Salamander
• Reptiles	: Lizard, Snake, Tuatara, Crocodile
• Birds	Crow, Ostrich, Parrot
• Mammals	: *Homo sapiens* (human being), Dog, Bat, Whale, Kangaroo, Platypus

Of the 5 to 10 million living species, 75 per cent are animals, 18 per cent plants and 7 per cent simple organisms. Out of these, we know about the species of over 12,00,000 animals, 3,00,000 plants and 1,00,000 other creatures. Information about the remaining species is yet to be collected. This book provides information about some of the invertebrates and vertebrates.

Species of known animals

- Insects : 9,50,000
- Other invertebrates : 2,27,000
- Vertebrates : 45,000

Numbers of known vertebrates

- Fishes : 23,000
- Amphibians : 3,000
- Reptiles : 6,000
- Birds : 8,600
- Mammals : 4,200

Special Characteristics of Vertebrates

Fishes

Scientific name – Pisces, Live in water, Cold-blooded, Respiration through gills

Amphibians

Live in water as well as on land, Cold-blooded, Respiration through nostrils

Reptiles

Crawling creatures, Live in water as well as on land, Cold-blooded

Birds

Flying creatures, Mainly live on trees, Warm-blooded

Mammals

Live on land, Warm-blooded, Feeding milk to their young ones, Normally every mammal gives birth to young ones.

PROTOZOA AND METAZOA

The invertebrates are divided into two groups : **Protozoa** and **Metazoa**.

Protozoa are single-cell animals. They are very tiny and can be seen only under a microscope, or some of them can be seen through naked eye. All their biological activities take place within their unicellular bodies. Some protozoa are *parasites*, and they cause various diseases like malaria, sleeping sickness and dysentery. Generally, the organisms of this group are found in pools, canals, waterfalls, wet soil, marsh and sea water. Amoeba, uglina and paramaecium, etc. belong to this group. About 20,000 species of protozoa exist today. They are 0.0002 mm to 16 mm in size.

Amoeba: These organisms are found in pools, canals, rivers, mud and rotten leaves. An amoeba's shapeless body is a mass of protoplasm with a single nucleus. The whole body, about 250-600 micron long, is covered by a thin membrane, called *plasmalemma*, through which exchange of gases takes place. One or more projections are seen jutting out from the body. These are called *pseudopodia*, which keep on changing their shape. With the help of pseudopodia, amoeba moves and grabs its food. It has an asexual as well as a sexual way of reproduction.

Metazoa are multi-cellular organisms. Their body structure is more complex than that of protozoa. Metazoa differ from each other in many ways. Therefore, they are divided into several phyla, such as *Porifera, Coelenterata Platyhelminthes, Nemathelminthes, Annelida, Arthropoda, Mollusca* and *Echinodermata*.

Amoeba

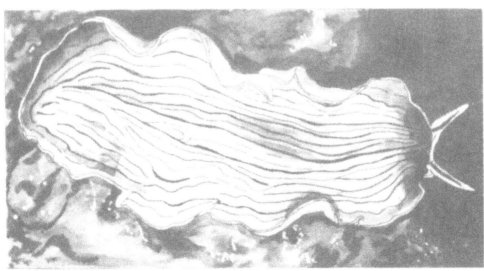

Sea flatworm

MOLLUSCA

Mollusca is a very large family of animals, comprising about 75,000 species. Aristotle gave the word 'Mollusca'. These are soft-bodied animals. The body is divided into three parts – the head, foot and trunk. The whole body of these animals, except the head and the foot, has a cover which is called *mantle*. All around the mantle is a hard *shell* made of calcium carbonate. This shell protects the soft body of the animals. These animals have muscular foot for movement. Their blood is generally colourless. However, some of them may have red, blue or green blood.

Molluscs breathe through *gills* or *air sacs*. Their digestive system is fully developed and excretion takes place through kidneys. Generally, these animals are unisexual.

Molluscs are found in rivers, canals, lakes and seas. They are classified into six main groups – **Cephalopoda, Emphineura, Gastropoda, Scaphopoda, Polyplacophorra and Monotala Sophora**. The cephalopoda group comprises squids, octopus, cuttlefish, etc. In Emphineura group are included oyster, clamps, etc. Snail, slug, etc. are from the gastropoda group. Scaphopoda group includes animals with indented shell. Organisms like chitin constitute the polyplacophora group. Monotala Sophora includes animals such as Neo Dilina.

Snail

About 3,500 species of snails and slugs have been found. They belong to the Gastropoda group of the Molluscaphylum.

The snail has a circular body which is covered by a shell. The body is divided into three parts – head, foot and trunk. Snail has eyes and a pair of tentacles on its head. Respiration takes place through gills and air sacs. It has a flat foot for movement.

A large-sized African Snail with 20 cm shell

Snails are found in damp places, fresh water, trenches and ponds. They are *herbivores* and feed on water plants.

Octopus

Octopuses are found in the deep seas. They are also called *seagiants*. They belong to the Cephalopoda group. More than 150 species of octopus exist today. Unlike other animals of the Mollusca family, octopuses do not have shells. Their sack-like clumsy body has eight arms, which are bent like spring at the ends. The octopuses of large species have about five metre-long arms.

The octopus sticks to its prey with the help of suckers on its arms. Thereafter, it makes the prey lifeless by the poison of its glands. It tears and eats the prey with its parrot-like beak. Besides a mouth, it has two eyes on its head.

Octopuses and squids are more active and clever than other invertebrates. Octopuses use their arms to roam about under the sea. But when they require to move fast, they suck water through a hole into their mantle cavity, and then throw it out like a jet current. For self-defence, the octopus releases a cloud-like blue colour which hides it from the enemy.

Squid

Squid belongs to the Cephalopoda group. It is the largest invertebrate insofaras the size is concerned. A giant squid weighs nearly two tonnes and is about fifteen metres long Its rocket-like body has red, yellow, green and blue spots. Its foot is divided into ten arms, of which two are larger than others. They are about fifteen metres long. There is a bunch of *suckers* at the end of these arms, with which it grabs its prey. There are two big eyes, with a diameter of more than 38 cm (15 inches), on the squid's head. Among all the living organisms, squids have the largest eyes.

A squid is more active than an octopus. It swims on the principle of *jet propulsion*. To protect itself from an enemy, it releases a black ink-like substance through its siphon-like organ situated under its arms. This substance makes the water black-coloured and its enemy cannot see anything through it, enabling it to escape. This is a very strange creature.

Squid is the largest among invertebrates

The sea demon: Octopus

INSECTS

Among animals, insects are found in the largest number. About 3.9 million years ago, insects appeared on the earth. Around a million species of insects have so far been classified and named. Still 4 million insect species are yet to be classified.

Insects belong to the Arthropoda group. They have a segmented body and their legs are jointed. The body is generally divided into three parts – the *head*, the *thorax* and the *abdomen*. They do not have the inner skeleton. However, the tender parts of their body are covered by the *exoskeleton*. Water, light acids, alkali and alcohol do not affect the exoskeleton. These animals have compund eyes, that is, the number of lenses in each eye is not fixed. To the thorax of an insect are attached three pairs of legs. These animals are divided into two main groups – the **winged** and the **wingless**. Mosquitoes, flies, wasps, cockroaches, butterflies, grasshoppers, ants, beetles, moths, etc. are all insects. Insects are found all over the world. They are more a friend than enemy to men.

Butterflies and moths

Butterflies and moths belong to the Insecta group. Over 1,50,000 species of these insects have already been studied. Butterflies are of bright and attractive colours. Their body has three parts – the *head, chest* and *abdomen*. The abdomen is divided into ten segments. Under the tenth segment, there exists in females the reproductive organ, while in males a pair of

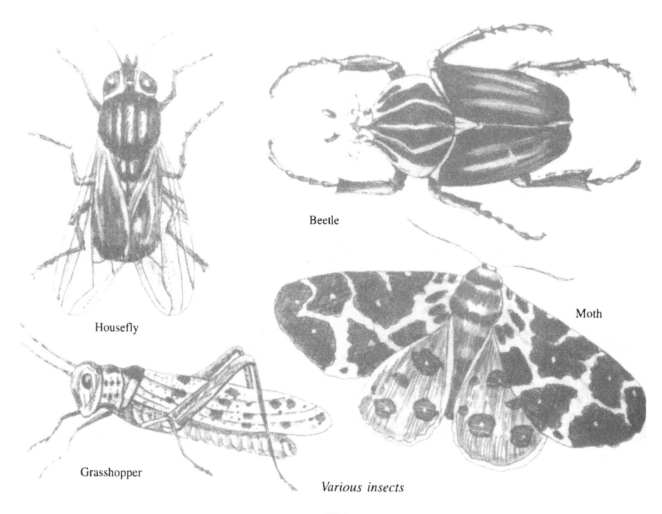

Housefly

Beetle

Grasshopper

Moth

Various insects

136

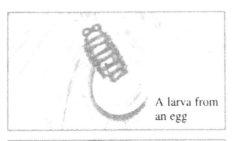

A larva from an egg

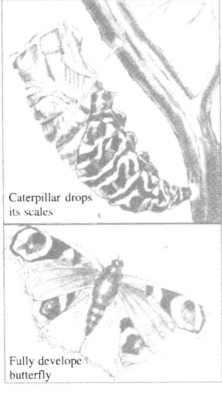

Caterpillar drops its scales

Fully developed butterfly

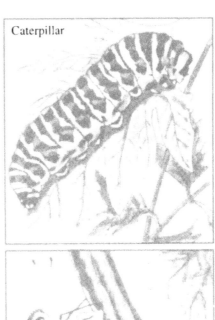

Caterpillar

Pupae develop into a butterfly

Life cycle of a butterfly

claspers. There is a pair of *antennae* on the head. A pair of *maxillae*, which constitute a *proboscis*, is placed in the mouth region. Butterflies suck nectar from the flowers with the help of this proboscis. On the upper side of the thorax are two pairs of wings of coloured scales. The lower side of the thorax is attached with three pairs of legs.

The female butterfly lays round or oval-shaped eggs after copulation. In about ten days' time *larvae* called *caterpillars* are hatched out. The caterpillars have nothing in common with a butterfly. They feed on leaves. After a few days, they stop eating leaves; their movement is also restricted. Now they begin to change into *pupae* or *chrysalis* in the process of evolution. After some days, these pupae evolve into beautiful butterflies.

Moth and butterfly differ from each other in many ways. Butterflies usually fly during the day, while moths fly or move about only at night. Both have different antennae. Moths have smaller ones as compared to butterflies. Though silk moth evolves like a butterfly, its larva has two long glands which produce silk. The sticky substance secreted by these glands transforms into silk after having been dried in the air. The larva moves its head in such a way that the sticky thread ejected from the gland entwines around its body. Thus, a shell of silk coating is spun around its body. It is called *cocoon*. China is famous around the world for its silk. After some days, the larva develops into a pupa, which comes out after cutting the shell of silk coating. Now it can fly. It is called a moth.

CRUSTACEANS

Around 30,000 species belonging to this group of animals are known today. Like insects, the crustaceans are also Arthropods. Generally, these animals have a crust-like exoskeleton and jointed legs. Their head and thorax constitute a joint structure called *cephalothorax*. Their compound eyes are located over a small stalked structure. There are two pairs of antennae which work as feelers. With the help of these, they feel things and objects around them. Crustaceans breathe through gills. These animals are found in ponds, rivers, lakes and seas. Minute transparent animals, crabs and lobsters, crayfish, shrimp, etc. belong to this group.

Hermit Crab

Hermit crab is a queer, decapod crustacean. It does not have a hard exoskeleton; therefore, it uses an empty shell of any other sea animal for its own protection. It withdraws its whole body inside the shell, except its two pairs of legs and pincers. But, when in danger, it even withdraws its legs and pincers inside the shell. A hermit crab spends its entire life living inside an empty shell. If its body outgrows the shell, making it feel suffocated, it discards the shell and occupies some other bigger shell. Wherever it goes, it always carries the shell with it for the protection of its body. Elonela (0.25 cm) is the smallest crustacean.

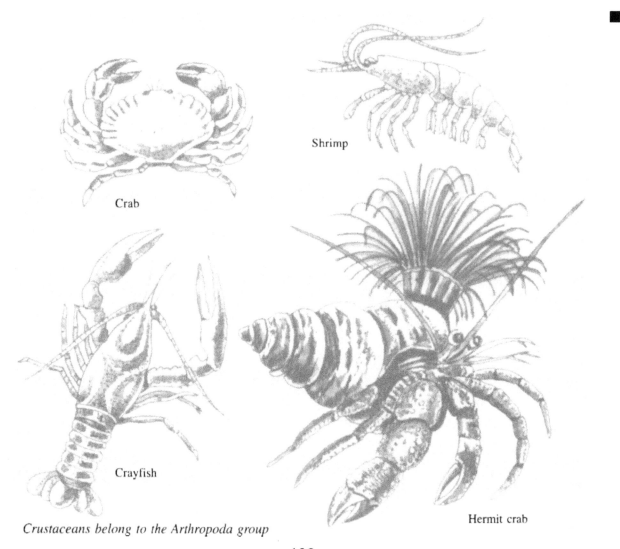

Crab

Shrimp

Crayfish

Hermit crab

Crustaceans belong to the Arthropoda group

138

ECHINODERMATA

Starfish, sea urchins, sand dollars, sea cucumbers, sea lily, etc. belong to the Echinodermata group. About 6,000 species of this group of animals are known today. The skin of these animals is spiky. The spikes are made up of calcium carbonate. Their body structure is such that their head, tail and legs are not visible separately. Their body is circular in shape, from which their arms and other organs project out like spokes of a wheel. The blood circulation system in these animals is different from that of other animals. They have a water vascular system within their body. The nervous system and sensory organs are underdeveloped. They have small sacs called *tube feet*. These help them in movement, respiration, excretion and also in catching their food. These animals are unisexual.

Sea Urchins

The body of sea urchins is a shell of hard plates. Long thorns stretch out from their entire body. These animals have their mouth on the lower part of the body, while the anus is located at the opposite end. The jaws of the sea urchins are strong enough to chew the sea moss easily. They are gifted with *regeneration capability*. That is, if a part of their body is lost, it regrows on its own. People fear going closer to the hatpin urchins because of their 30 cm long, thin poisonous thorns. The pricks of these thorns on human body cause unbearable pain.

Starfish

There are 1,600 known species of starfish. Looking like a five-arm star, these animals have a hard and rough body. They do not have a head. Their mouth is on the lower surface of their body, while the anus lies on the upper surface of the body. A round plate called *madreporite* exists on the upper surface of the body, through which water enters into the body. There are ossicles and spikes on the skin of a starfish. Each arm on the lower surface has a groove with tube feet. These help the starfish in its movement.

Starfish too is gifted with regenerative power. If it loses one of its arms, another grows again on its own at the same place. It has the greatest ability of reproduction.

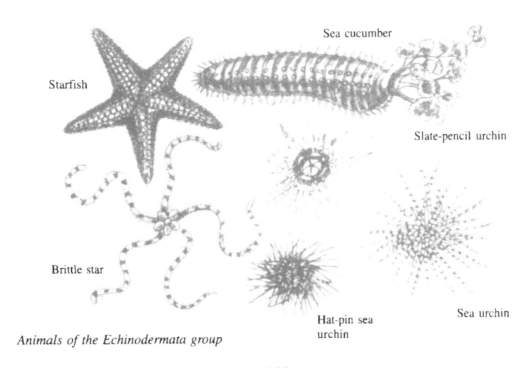

Animals of the Echinodermata group

Starfish

Sea cucumber

Slate-pencil urchin

Brittle star

Hat-pin sea urchin

Sea urchin

PISCES

All species of fishes are included under Pisces. As of now, there are around 20,000 species of fishes. They are vertebrate animals and evolved some 500 million years ago. Fishes have boat-shaped bodies. They swim with the help of their *fins*. Respiration in them takes place through *gills*. They are cold-blooded, that is, the temperature of their blood varies with the rise and fall in the temperature of environment. Most fishes lay eggs while others give birth to their young ones.

Fishes are divided into four categories — **Cyclostomate, Placodermi, Chondrichthyes** and **Osteichthyes**.

Cyclostomate fish have round mouth which works as a sucker. They have neither jaws in the mouth nor scales on their body. There is only one nostril on the head. Petromyzon, a cyclostomi fish, is found in the seas of North America and Europe.

Placodermi means plate-skin. Fishes of this category have a pair of jaws and wings. Some fish of this category resemble the shark.

The fishes in the Condrichthyes category have inner skeleton, made of cartilage. They have jaws in the mouth and a special kind of scales on their body. The external nostril and mouth are located on the lower part of the head. Scoliodon, saw fish and torpedo belong to this category.

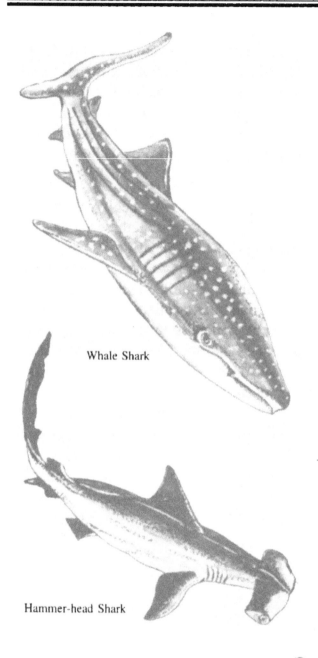

Whale Shark

Hammer-head Shark

Great White Shark

The Osteichthyes category of fishes have an inner skeleton made of bones. The opening of their mouth is located on the front part of the head, while the nostrils lie on their head's back. This category is further sub-divided into: (a) the Choanichthyes; and (b) the Neopterygii. The Choanichthyes have *lungs* to breathe and, therefore, are also called *lungfish*. The land vertebrates owe their evolution to the Choanichthyes. The Neoceratodus, a fish of this group, is found in Australia. The second sub-category, Neopterygii, includes developed fish. Their entire skeleton is made of bones. The gills on either side of their body are covered by a plate. Labes, seahorse and heteropinuestic belong to this sub-category. Some fishes like eel generate electric current. Some fishes like lantern fish produce light.

Ray Fish

Angler Fish

Australian Lungfish

141

AMPHIBIANS

Common Frog

Toad

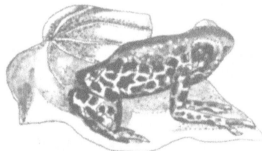

Arrow-poison Frog

Gliding Frog

Amphibians

Amphibians are those animals which can live both in water and on land. These animals evolved some 3.5 million years ago. It is believed that they evolved from the lobe-finned fish. They have lungs, therefore, they can also live on land. The amphibians have been divided into three groups – the **frogs** and **toads**, the **salamanders** and the **caecilians**.

Unlike the present-day small amphibians, the early amphibians were over a meter long. Today, only the giant salamanders of China are as long as 3.5 meters. In Africa too, a giant type of frog measuring 34 cm in length and weighing over 3 kg is found.

The skin of amphibians is smooth, soft and moist. The respiration in their body takes place through gills, skin and lungs. They have a *three-chambered heart*. They have two kidneys. Both the genital aperture and the anus open in the same cavity through separate routes. The hatching of their eggs and development of foetus takes place under water. The amphibians are cold-blooded animals, that is, their body temperature falls and rises according to the temperature of the environment. It is because of this reason that these animals go underground to protect themselves from extreme cold and heat. It is called *hibernation*. During hibernation, the biological activities of these animals remain almost suspended. They are males as well as females. These creatures lay eggs. The frogs have reverse tongues. They catch their prey by thrashing their tongues out, and the insects and worms get stuck to their tongues. Thus they get their food.

REPTILES

Snakes, lizards, crocodiles, tortoises and turtles are all reptiles. They belong to the Reptilia group of animals. All the animals of this group crawl and, therefore, are also called *crawling animals*. Today, there are around 5,000 species of the reptiles. They came into existence some 500 million years ago. The reptiles are divided into three categories – the **Chelonia**, the **Squamata** and the **Crocodilia**.

Reptiles are cold-blooded animals, that is, their body temperature rises or falls according to the temperature of their environment. Their skin is dry, rough and covered by *horn scales*. The skeleton is made of bones. Respiration in them takes place through lungs. Females lay eggs on land.

The reptiles of the Chelonia group have their body enclosed in a hard shell of bones. Their jaws are without any teeth. Various kinds of tortoises and turtles are included in this group, which are generally found in water and damp places. These creatures have teeth on both their jaws. Their hearts are divided into four parts. They have sexual breeding.

The reptiles of the Squamata group include various kinds of lizards. This group is divided

Tree-living snake

143

into two sub-categories – the Lacertilia and the Ophidia.

Various kinds of lizards constitute the Lacertilia sub-category. These lizards have five toes on their legs. There are claws at the ends of the toes. The toes have *lamellae* which facilitate the easy movement of lizards both on the smooth surface and on the walls. In the cavities below their eyes lie the *tympanic membrane*. They have *mobile eyelids*. These animals can shake off their

Giant Tortoise

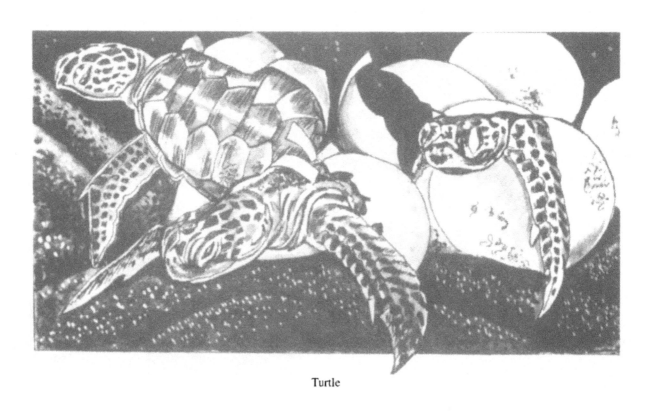

Turtle

144

tails. They can again have new tails after shaking them off.

All varieties of snakes come under the Ophidia sub-category. Snakes have a long, cylindrical body covered with dry scales. They do not have legs, backbones, bladder and tympanic membrane. They have only one lung for breathing. Their tongue is long and divided into two parts at the front end. It is through their tongue that snakes feel and smell. They have excellent eyesight. They do not have ears. Sounds are conveyed to these animals through the vibrations of the earth. Erex Jony is a two-headed snake. Python is the biggest snake.

Gavialis, crocodiles and alligators are included in the Crocodilia sub-category. They are carnivorous animals. Their thick skin is covered by hard scales.

Their fingers are connected by a membrane. The nostrils are situated at the tip of their mouth. Though these animals live in water, their females lay eggs in pits on land. These animals are founded in rivers and lakes. These are the largest of the existing reptiles. The extuarine species of the crocodile measures as long as 7.5 metres (25 feet). The largest crocodile, 'crocodilas' is the heaviest of all the crocodiles. It weighs 525 kg.

■■

Giant Commodore Dragon

Crocodile

BIRDS

The study of fossils indicates that the birds evolved from the reptiles long, long ago. The discovery of the Archaeopteryx, the earliest fossil bird, and the skeletons of the modern birds, as well as the recent researches have confirmed this belief.

Birds are called winged bipeds. They are vertebrate and warm-blooded creatures. It means that the body temperature of birds does not vary with the rise and fall in the temperature of the environment; it remains constant. The body temperature of the birds generally varies between 38° and 44°C, which is a little higher than that of mammals. The feathers covering the body of a bird are a bad conductor of heat. Therefore, they keep the bird's body temperature constant.

Feathers serve as a characteristic feature of birds. The feather structure of the birds of a particular group tells us about their lifestyle. The remarkable adaptability of the birds, which helps them withstand the hardships of extreme climatic conditions, is due to their feathers. Varying climates have little effect on birds, provided they are assured of adequate food. Then, it matters little for them whether they are in a desert of 60°C temperature, or in an icy region having 0° to 40°C temperature.

Birds have beaks in their mouths. They have no teeth. Their body is divided into four parts: head, neck, torso and tail. Their main food is insects, foodgrains and animal flesh. Their bones, though light and hollow, are strong. Like reptiles, birds too lay eggs. Their eggs are covered with a hard shell.

Birds are very efficient in flying in the air. They can stay in the air for a long time. But there, are some birds which cannot fly, for example, ostrich, cassowary, penguins etc. They are unable to fly because of their undeveloped feathers. For copulation, the male approaches and attracts the female. Apart from singing, birds produce different kinds of sounds to indicate anger, danger and hunger.

Birds have an adequately developed level of sight and hearing sensibility. They are, however, completely devoid of power of smell and have a poor taste sensibility. As compared to other animals, birds have a wonderful capability of adjusting their eyes very quickly – that is, they can shift the focus of their eyes instantaneously from a distant object to a nearby object.

Archaeopteryx

Ostrich

Birds build nests for laying eggs and for the care of their young ones. Almost all the birds, with some exceptions, hatch their eggs and look after their young ones till they are able to fend for and fly themselves.

There are about 9,000 species of birds. Their sizes vary from 5 cm to 7.5 metres. Ostrich is the largest bird. Its height is up to 2.5 metres (8 feet). It weighs upto 120kg. Though unable to fly, it can run very fast on its hind legs. It can run at a speed of 80 km per hour. Of all the birds, the wandering albatross has the largest expansion of wings. The expansion of its wings from one end to the other measures up to 3 metres (10 feet). Bee humming bird is the smallest in the world. It is found in Cuba and Pines Island. The male bird measures up to 57 mm and weighs only 1.6 grams. It flutters its wings about 90 times a minute. The fastest among the birds is swift: its speed is up to 170 km per hour. Ducks can also fly at a speed of 130 km per hour. Sooty tern is the bird which keeps on flying in the sky continuously for 3 to 4 years. It comes down to land only during the breeding season. Various birds migrate from one place to the other with the change in seasons. Beautiful colours and sweet songs of birds have always attracted human beings.

Penguin

Wandering Albatross

Bee Humming Bird – the smallest bird

147

MAMMALS

Mammals, the animals of the Mammalia group, are warm-blooded. They have a hairy skin. The sweat glands and sebaceous glands are located in their skin. Their bodies are divided into four parts: head, neck, torso and tail. Respiration takes place through the lungs. The word 'mammal' was given by the scientist named Linox in the year 1758. Both the males and the females have mammary glands. It is because of these glands that the animals of the Mammalia group are called mammals. The developed mammary glands of females secrete milk with which they suckle their young ones. All the animals of this group are *viviparous* – that is, the female does not lay eggs; instead it gives birth to its young ones. Only the mammals belonging to the Prototheria group lay eggs.

Mammals evolved from reptiles some 200 million years ago. Around 4,400 species of mammals exist today. Their size varies from 5 cm to 31 metres. Among all the animals, mammals are the most developed and intelligent group. Mammals have been divided into three sub-groups:

Giraffe

Spiny Anteater 'Echidna'

Duck-billed Platypus

Prototheria: This is a less developed sub-group of mammals. And the first mammals were evolved from the reptiles, therefore, the prototheria mammals still retain several features of reptiles. They lay eggs, although they have mammary glands. When the young ones are hatched out of the eggs, the mother suckles them.

These are cold-blooded animals. Today, in this group only six species of duckbilled platypus and five species of spiny anteater or Echidna are left. The mouth of the platypus is like the beak of a duck. Its body is covered with soft hair. The toes are bound together by a web or membrane. Adult platypus have no teeth. Echidna's body is covered with pointed thorns. It does not have teeth. These animals are found in Australia and Tasmania.

Kangaroo

149

Orangutan

Metatheria: It is also a sub-group of the early mammals. But today, the animals of this sub-group have reached a much higher state of development. The female gives birth to under-developed young ones. The young ones are brought up in the *marsupium* or the pouch-like fold of skin on the ventral side of the female. It is because of this that the mammals of this sub-group are called marsupials. They live on the mother's milk for months. They are found in Australia and South America. Kangaroo, Opossum, Koala, etc. are the animals of this sub-group.

Eutheria: The animals of this sub-group are highly developed and most intelligent. In females, the embryo and the foetus is nursed within the uterus through a tube called placenta. In this sub-group are included primates, that is, apes and human beings. Eutheria is sub-divided into the following orders.

150

Insectivora: These include shrew, hedgehog, mole, etc.

Chiroptera: These are *nocturnal* animals, that is, they are active at night. Various kinds of bats are included in this order.

Edentata: The animals of this order have undeveloped teeth. They have a sticky, thin, long tongue with the help of which they catch their prey. Dasypur, sloth, etc. belong to this order.

Pholidota: The animals of this order have a covering of hard plates over their body. Man is an example of this order.

Rodentia: These animals have one pair of chisel-shaped sharp incisors in each of their jaws. They gnaw their food using incisors. They are herbivores. Examples: rat, porcupine, squirrel, weaver, etc.

Logomorpha: These are gnawing animals. They too are herbivores. Examples: rabbit and hare.

Carnivora: This order includes flesh-eating animals. They have strong paws and sharp pointed strong carnives. Dog, hyena, weasel, bear, lion, cat, etc. belong to this order of animals.

Perissodactyla: The rhinoceros and horse belong to this order. They have hoofs in odd toes such as 3 or 5.

Artiodactyla: The animals of this order have hoofs in even number of toes such as 2 or 4. Giraffe, cow and goat belong to this order.

Proboscidea: Various kinds of giant elephants constitute this order.

Cetacead: The mammals of this order are found in the seas. Their bodies are adapted for living under water. Whale and dolphin belong to this order. Blue whale is the largest animal in the world.

Primates: The animals of this order have five fingers on their hands and feet. Their fingers have nails. They have eyes on the front side of the face. They have mammary glands. Monkey, gorilla, chimpanzee and human being belong to this order. Human beings are the most intelligent among the primates. They have achieved great heights in the fields of science and technology.

Blue Whale

Bull Walrus

NESTS

In birds, the breeding, hatching and upbringing of the young ones take place at a particular time of the year. It is called the *nesting season*. Birds build nests for the protection of their eggs and young ones. Generally, the nests are built some time before the birds lay eggs. Birds build different kinds of nests. But birds of a certain species always build a particular type of nest.

In case of human beings, intelligence and some acquired training are necessary for building anything. But the young birds have an inherent wisdom of making a specific type of nest at a particular time of the year, without having received any training from their parents. The inspiration for making a nest is instinctive in birds: it is hereditary and has been passed on from generation to generation. Birds build various types of nests, which are in harmony with their nature and fulfil their needs.

Weaver bird, sun bird and Tickell's flower-pecker build their nests hanging down from the branch of a tree. These large-sized sac-like nests hang vertically from the end of a twig. They have an opening on one side with a small hood over it. This opening is used as the entrance. The outer covering of the nest is made up of pieces of the bark of a tree, wickers, excreta of ants and shells of spider's egg. Weaver bird's woven nest is the most beautiful and a piece of art. It has its unique identity and can be seen on the trees.

Tailor bird and robin, etc. build funnel-shaped nests by putting together leaves of trees. Some birds of this category build purse-like nests, which are suspended from either long grass or stems of bushes. Tailor bird builds its nest by sewing leaves together. That is why it is named so.

Swift bird builds its nest in dark caves of rocks or in caves on islands. These nests resemble a half-cup and are made up of only saliva, or saliva mixed with straw, feathers etc.

Owl, mynah, woodpecker, horn-bill, etc. build their nests in the holes of tree trunks. The hole may be natural or drilled in some rotten

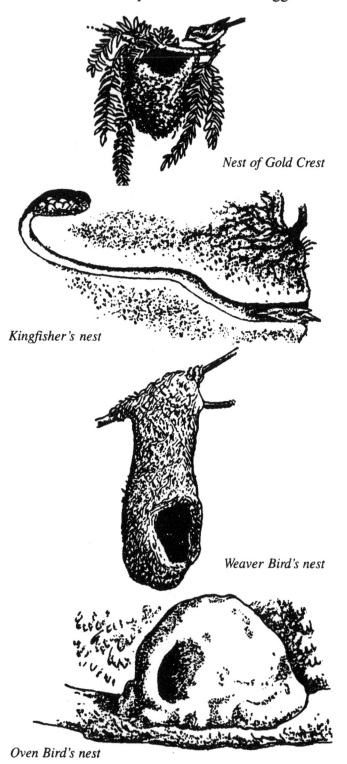

Nest of Gold Crest

Kingfisher's nest

Weaver Bird's nest

Oven Bird's nest

trunk. These birds fill their nests with soft material to prepare a thin cushion.

Kastura, swallow and martin birds bring wet soil from rainy ponds and mix it with their saliva to make their nest. Saliva does the work of cement. Meenrack and hoop make their nests on the projecting caves of houses or moulds. They dig the soil by their beaks and remove it with their toes to make a horizontal tunnel with an opening towards a water source. These tunnels are several feet in length. They are cylindrical like an egg chamber with an opening at the end.

Quail, tern and wild hen etc. make their nests on the ground by spreading grass or leaves on rags, straw, husk or peels. Sparrows use straw and leaves to build their nests on the trees and terraces of houses. The nests are built in such a way that they are comfortable and the birds can hatch their eggs.

Tailor Bird's nest

Swift's nest

Lapwing's nest

Nest of Dhanesh

Nest of Brush Turkey

Nests of different birds

153

MIGRATION OF BIRDS

Every year as winter sets in, the birds of some cold regions travel thousands of miles to migrate to warmer places of the world. They are called migratory birds. Their journey starts at a specific time every year, and they are so punctual that the possible dates of their arrival and departure can be foretold. These journeys to far-off lands take place in such a systematic way that it is almost as if the programme were controlled by a computer.

Birds migrate because of the non-availability of food in extremely cold weather conditions. They also lay eggs and bring up their young ones in warmer regions. These birds live at two different places during the two different seasons to ensure favourable weather conditions and availability of food. Birds, nevertheless, breed and build their nests at the places of their original home. Thus, their breeding takes place in the Arctic or sub-tropical region in the northern hemisphere, while they spend the winter near the Equator. Just the reverse takes place in the southern hemisphere. Though some birds migrate from the east to west and vice versa, the migration of birds is mostly from the north to south and vice versa. The most favourite places of their migration are North America, Europe and some regions in Asia.

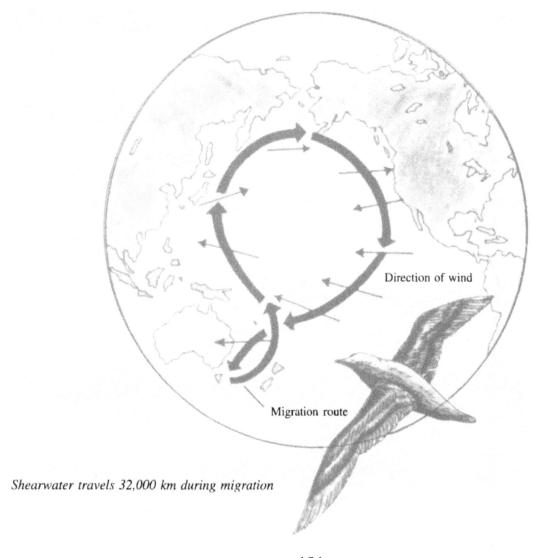

Direction of wind

Migration route

Shearwater travels 32,000 km during migration

154

Other animals also migrate, but the migration of birds is quite regular and systematic. The birds have to encounter great dangers and endure hardships during their journeys. Surprisingly, they still reach their destination without being strayed or any mishap. Moreover, they migrate without any previous experience or training. It is believed that they have an instinctive knowledge of the routes and destinations. This instinct is hereditary.

During the journeys of their migration, the birds fly at different heights, which may be usually from 900 metres to 9,000 metres.

The longest journey among the birds is

.... Migration route
● Breeding regions
▲ Cold regions

The Arctic Tern's migration route and regions

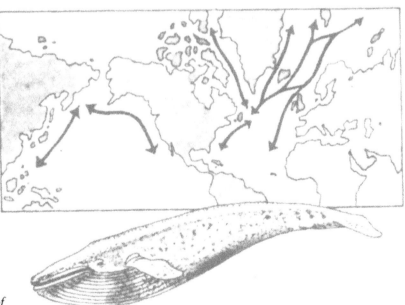

Grey Whale performs a journey of 9,000 km during migration

155

performed by the Arctic tern, which travels across the earth from the wintry Arctic to the Antarctica. After having spent the summer in Antarctica, it flies back to the Arctic. It travels some 36,000 km during its two-way journey.

Besides birds, mammals, insects, reptiles and amphibians etc. too migrate.

Migration records of some animals:

- **Caribou**—It undertakes a journey of 1,100 km from Arctic, northern America to the south.

- **Grey Whale**—It travels a distance of 9,000 km every year.

- **Eel fish**—It travels 8,300 km from Sargasso Sea to Black Sea. The journey takes several years to complete.

- **Painted Lady Butterfly**—It travels 6,400 km from northern Africa to Iceland.

- **Alaska Seal**—It undertakes a journey of 9,600 km.

- **Bat**—It covers a distance of 2,300 km during migration.

- **Green turtle**—It travels 5,900 km from southern America to Africa.

- **Toad**—It travels up to 3 km.

- **Shearwater birds**—It migrates, travelling 32,000 km.

Apart from this, there is a long list of migratory birds that migrate to the warmer regions during winters.

Caribou performs a journey of 1,100 km

156

SPEEDS OF ANIMALS

Scientists have conducted many experiments to ascertain the maximum speed of various animals in running, swimming and flying. These experiments have proved that animals can cover a distance of only 2-5 km with a fast speed. The records of speed of some animals are as under:

- **Tuna fish**—69 km per hour
- **Sailfish**—110 km per hour
- **Swordfish**—92 km per hour
- **Flying fish**—64 km per hour
- **Ostrich**—80.450 km per hour
- **Spine-tailed swift**—171 km per hour
- **Hawkmoth**—53 km per hour
- **Black Mamba**—11 km per hour
- **Black Bug**—95 km per hour
- **Lion**—88 km per hour
- **Grass snake**—8 km per hour
- **Leather backturtle**—35 km per hour

- **Kangaroo**—25 km per hour
- **Elephant**—35 km per hour
- **Gazelle**—92 km per hour
- **Pronghorn antelope**—89 km per hour
- **Zebra**—64 km per hour
- **Hyena**—45 km per hour
- **Cheetah**—101 km per hour
- **Man**—43 km per hour

Some of the slow-speed animals are given below:

- **Common garden snail**—0.83 m per minute
- **Giant tortoise**—4.57 m per minute
- **Sloth**—2.10 m per minute
- **Deer**—75 km per hour
- **Fox**—75 km per hour
- **Buffalo**—55 km per hour

Sailfish

Cheetah

MAXIMUM LIFESPANS OF ANIMALS

All those who are born on this earth must die one day. But each living being has a lifespan that is common to its species. Some bacteria buried under snow or salt may live for more than a million years. Some plants too have a long span of life. It is among the animals alone that some live for very short periods, while others have a long span of life.

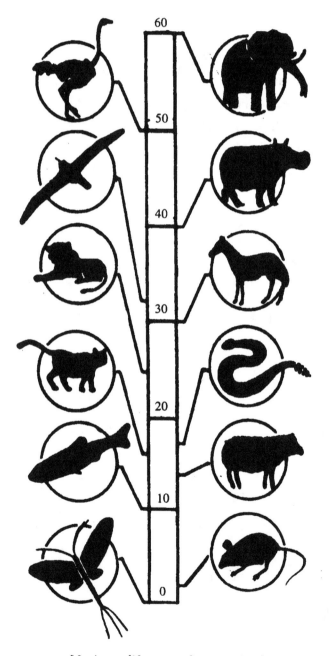

Maximum lifespans of some animals

Records of the maximum lifespan of some animals are as under:

- **Mayfly**—1 day
- **Rat**—2-3 years
- **Sheep**—10-15 years
- **Cat**—13-17 years
- **Rattlesnake**—18 years
- **Japanese Salamander**—55 years
- **Lake Sturgeon Fish**—82 years
- **Jellyfish**—1 year
- **Leech**—20 years
- **Lobster**—Over 50 years
- **Spider**—28 years
- **Beetle**—30 years
- **Tapeworm Echinococcus**—56 years
- **Clam Snail**—150 years
- **Tortoise**—152 years
- **Albatross**—33 years
- **Ostrich**—50 years
- **Andean Condor**—70 years
- **White Pelican**—51 years
- **Hippopotamus**—40 years
- **Elephant**—60 years
- **Lion**—25 years
- **Rhinoceros**—40 years
- **Horse**—30 years
- **Bear**—34 years
- **Monkey**—20 years
- **Dog**—22 years
- **Human being**—118 years
- **Parrot**—140 years
- **Sparrow**—23 years
- **Carp Fish**—25 years
- **Cat Fish**—60 years
- **Eel**—50 years

06

THE HUMAN BODY

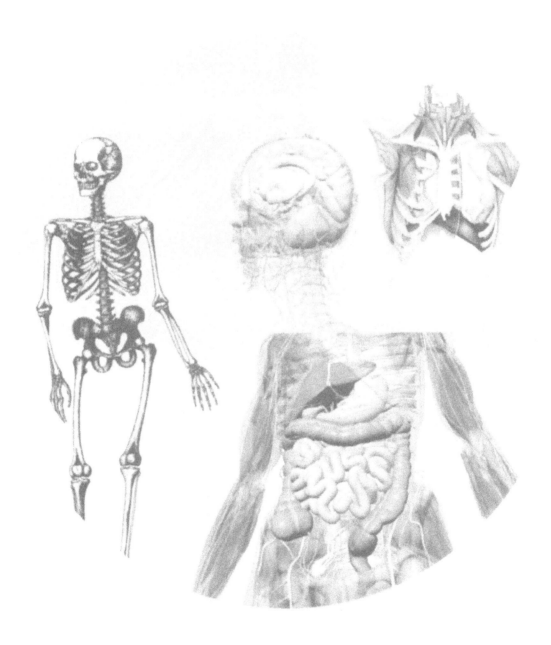

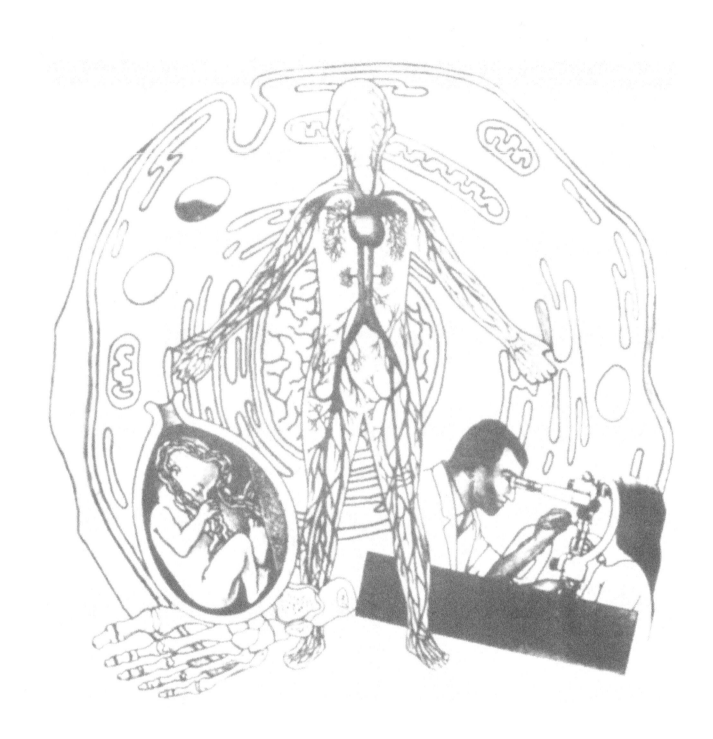

LIVING CELLS

A cell is the building block from which all living things are made. A human body is made of countless cells. Similar cells combine to form tissues and different tissues join together to make the various organs of a body. The organs work together to make different systems of the body such as nervous system, digestive system, circulatory system etc.

The cells in these systems vary in shapes and sizes and have their own specific functions to perform. A cell is made up of cell membrane, cytoplasm, lysosomes, nucleus, endoplasmic reticulum, mitochondria and golgi complex.

A cell is surrounded by membrane which forms its outer wall. It is a very thin layer and allows food and oxygen to pass into the cell and substances made by the cell or waste products to pass out. Membrane separates cells from one another and gives shape to each cell.

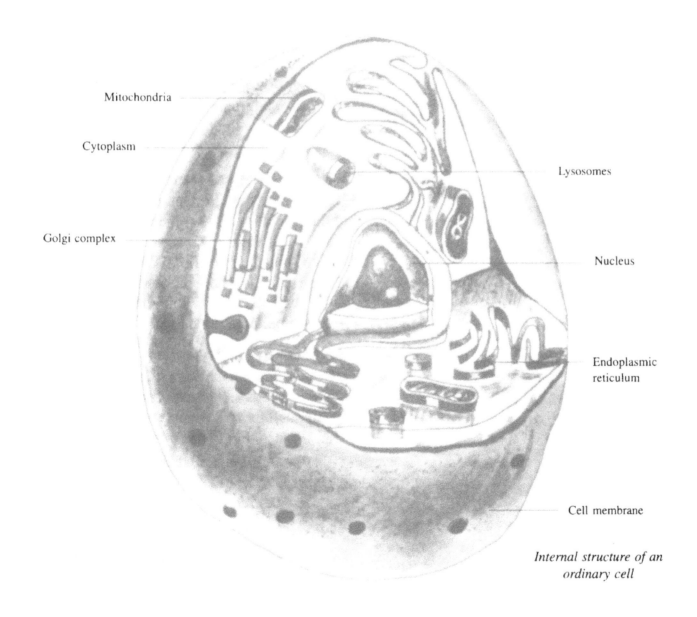

Mitochondria

Cytoplasm

Golgi complex

Lysosomes

Nucleus

Endoplasmic reticulum

Cell membrane

Internal structure of an ordinary cell

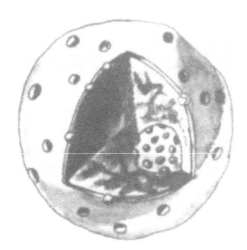

Nucleus

Endoplasmic reticulum

Mitochondria

Golgi complex

Cytoplasm is a watery, jelly-like material which surrounds the nucleus of the cell. It is living, colourless, translucent and granulated. All the parts of the cell float in it.

Lysosomes are circular tiny packets of digestive chemicals which can destroy harmful substances or worn out parts of the cell. They also play an important part in cell division.

Nucleus is a compact body. It is the control centre of a cell. It contains the chromosomes and DNA of a cell. DNA keeps the cell alive doing its own special job.

Endoplasmic reticulum is the folded, membrane like structure in the cytoplasm. Their main function is to transport substances throughout the cell.

Mitochondria are the cell's powerhouses. The energy needed for cellular activities is released here from food (glucose) and oxygen.

The golgi complex is the flat arrangement of pockets where some substances are stored and others are prepared for use outside the cell.

Different types of cells

Human body has different types of cells. Some of the smallest cells are in the brain, measuring only 0.005 mm. The largest are the ovum or egg cells with a diameter of 0.2 mm. Cells differ in shapes. Some of them are round and some are flat. Muscle cells are long and cylindrical in shape upto 60 mm in length.

Tissues and Organs

Despite being independent units, the cells often work together. The cells performing similar function, form a tissue. Every tissue has its special function to perform.

They are structured differently according to the task they perform. Muscles, bones and nerves are three different types of tissues.

Life of Cells

Living cells grow and many of them reproduce themselves so that a body can grow larger and replace the damaged or dead cells. The life span

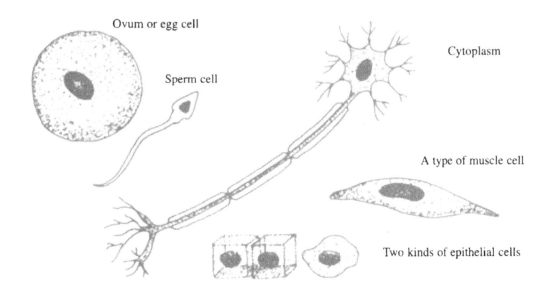

Different types of cells

of some cells is only a few days while some others can live for weeks, months or years. Bone cells last for about 15-20 years, while white blood cells only for four months. Skin cells do not live more than 3 weeks. Nerve cells do not have any reproductive property, therefore, they end with the life itself.

Through cell division, the cells multiply.

163

THE HUMAN SKELETON

The human skeleton is made of bones, cartilage and tough fibrous tissues. It gives support and a form to our body. It protects the organs and helps in movement. At birth, a baby has more than 300 separate bones. With the growth of body, many of these bones fuse together and thus an adult skeleton possesses only about 206 bones. The skeleton can be divided into two main parts: the axial skeleton (head and trunk) and the appendicular skeleton (arms and legs).

Human skull is made up of 29 plates of bones, most of which fuse together after birth.

The backbone or spine is made up of 33 vertebrae. The chest is enclosed by 12 pairs of ribs. Most of the ribs are connected to the backbone and to the breastbone in the middle of the chest.

Each hand is made up of 27 small bones. The bone in the upper leg, called femur, is the longest bone in the body. It is about 48 cm long. It is joined to the pelvic girdle in a ball and socket joint. There are 52 bones in the ankles and feet.

Our skeleton protects the vital organs like lungs, heart and brain and provides support to

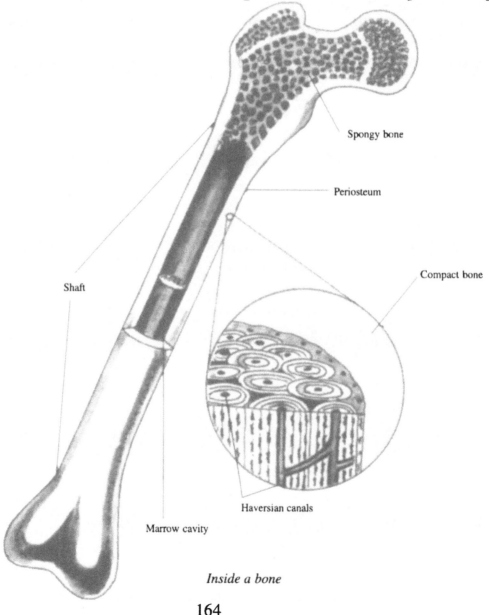

Spongy bone

Periosteum

Compact bone

Shaft

Haversian canals

Marrow cavity

Inside a bone

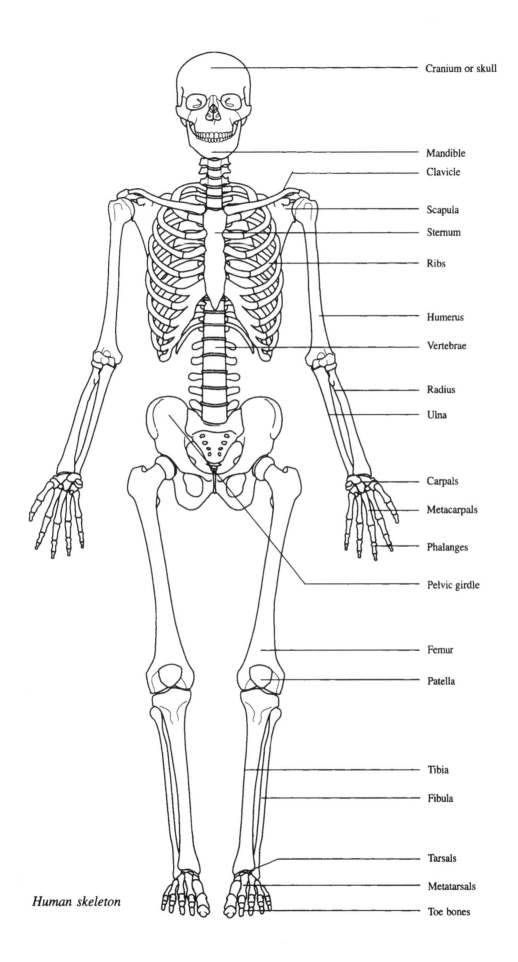

Cranium or skull

Mandible

Clavicle

Scapula

Sternum

Ribs

Humerus

Vertebrae

Radius

Ulna

Carpals

Metacarpals

Phalanges

Pelvic girdle

Femur

Patella

Tibia

Fibula

Tarsals

Metatarsals

Toe bones

Human skeleton

the muscles. It is also helpful in constituting the respiratory and hearing systems.

The bones are surrounded by a membrane called periosteum. All bones are hollow containing blood vessels which provide the essential nutrients to the bones. Most of the bones are covered with tough fibrous tissues. The ends of the bones in moving joints are covered with cartilage. This cartilage acts as a cushion in the joints. Bones are attached to other bones by the special connective tissues called ligaments. Muscles are attached to bones by connective tissues called tendons. Some bones are long while some others are flat but they all are very strong. Most of the long bones begin with cartilage while most of the flat ones with membranes. Bones are made up of calcium and phosphorus. They also contain collagen which is a gluey kind of protein.

Section through a long bone.

- Epiphysis
- Metaphysis
- Periosteum
- Hard, dense bone
- Spongy bone
- Marrow cavity
- Cartilage

Some important human bones:

- Cranium or skull
- Mandible or jawbone
- Clavicle or collar bone
- Scapula or shoulder blade
- Sternum or breast bone
- Ribs
- Humerus
- Vertebrae—they constitute spinal cord
- Radius
- Ulna
- Carpals or wrist bones
- Metacarpals
- Phalanges or finger bones and toe bones
- Pelvis or pelvic girdle
- Femur or thighbone
- Patella or knee cap
- Tibia or shinbone
- Fibula
- Tarsals
- Metatarsals

Joints

A joint is a junction between two or more than two bones. There are many different kinds of joints in the human body. Most of the joints provide a kind of movement to the body but there

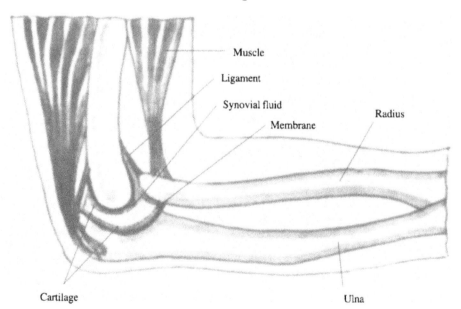

The elbow joint

Muscle

Ligament

Synovial fluid

Membrane

Radius

Cartilage

Ulna

166

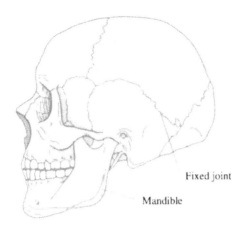

The immovable joints of skull

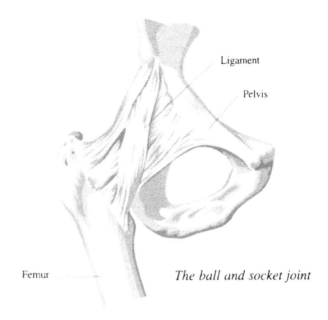

The ball and socket joint

are some immovable joints also i.e. those that do not move. Mainly there are three types of joints— movable joint, gliding joint and immovable joint.

Movable joints: These joints provide free movement to the body. These are found between the bones of the arms and legs and of hip and shoulder. The ends of the bones of such joints are covered with caps of tough cartilage. Cartilage does not wear out easily. Ball and socket joints, hinge joints, angular joints, pivot joints etc. are all movable joints.

Gliding joints: At these joints, the bones do not bend but simply slide over each other. The joints of the spine are of this kind. The joints where the ribs meet the breastbone are also partially movable joints.

Immovable joints: In an immovable joint, the bones are held tightly together. There is no cartilage between the joint. The joints between the bones of the skull are immovable joints. The plates of the bones fit together something like the pieces of a jigsaw puzzle. The joints where the teeth fit in the jaw bones are also immovable joints. The lower jaw is the only part of the skull that moves. These joints are also known as fibrous joints.

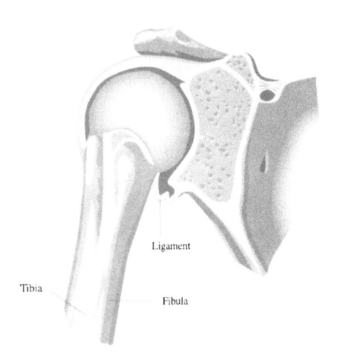

The hinge joint

MUSCLES

Muscles are the fleshy tissues which provide motion to the body. After receiving a signal from the nerve, the muscle contracts. As the muscle contracts, it pulls on the bone to which it is attached. This moves the bone into a new position. It also changes the shape of the body. These changes in shape help us to move from place to place.

In the human body, there are about 650 muscles. Muscles make up about 42% of a male body weight while 36% of a female body weight.

Muscles are of three types—unstriped or involuntary muscles, striped or voluntary muscles and cardiac muscles. Most of the muscles are made of tissues.

Unstriped or Involuntary muscles: Unstriped or involuntary muscles are made up of fibres. These muscles contract on their own i.e. involuntarily. They are controlled

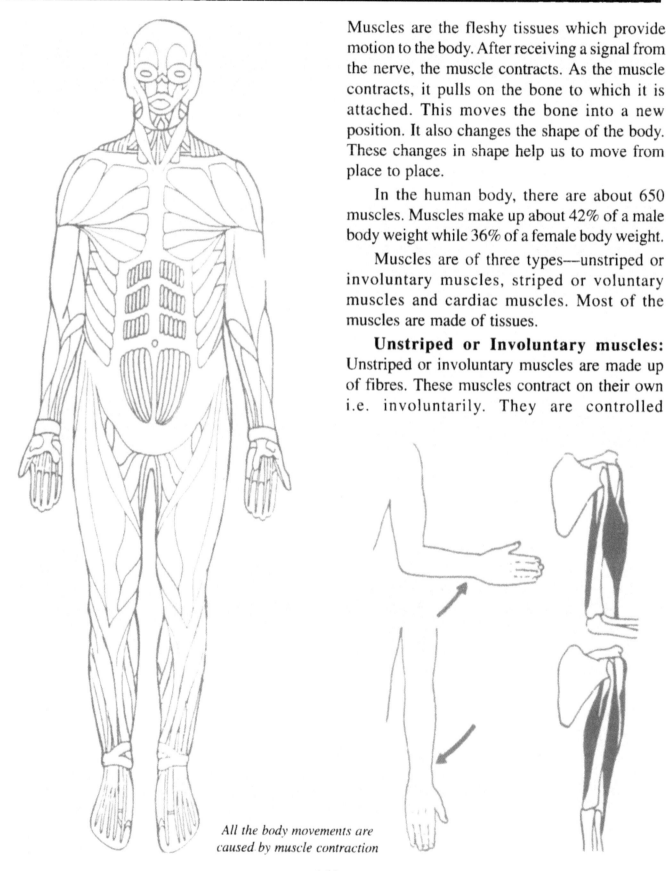

All the body movements are caused by muscle contraction

168

automatically by autonomic nervous system. The walls of stomach, intestines and urator are made up of involuntary muscles. The walls of the blood vessels are also made up of unstriped muscles. These are also called smooth muscles.

Striped or Voluntary Muscles: These muscles remain under our conscious control. Our arms and legs are moved by voluntary muscles. Most striped muscles are attached to bones and are made up of many fibres. These muscles can contract quickly and powerfully. They are also called striated muscles.

Cardiac Muscles: This muscle is found in the heart. Like striped muscles, it has fibres which can contract powerfully and rapidly. Like involuntary muscles, it is controlled automatically and does not get tired easily. Thus, this muscle has features of both voluntary and involuntary muscles. This muscle works the entire life without any halt. Cardiac muscle is the only kind of muscle that continues contracting even if removed from the body.

Muscles cannot push, they can only pull. They work only when they contract. Muscles often work in pairs, one of them moves a bone while the other brings the bone back to its normal position.

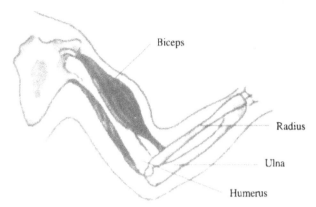

Tendons attach the muscles to the bones

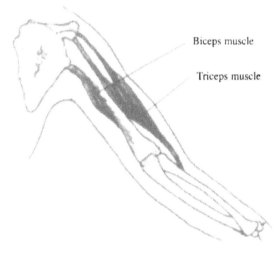

The working of muscles

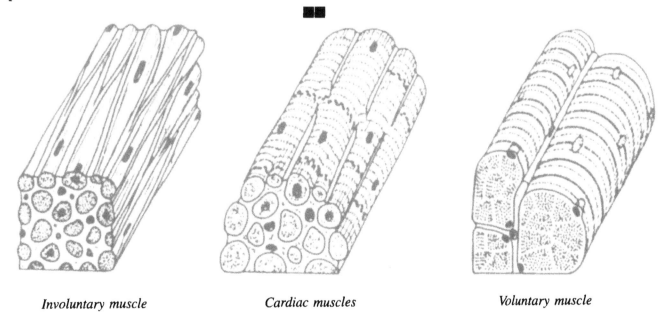

Involuntary muscle   *Cardiac muscles*   *Voluntary muscle*

169

THE BLOOD SYSTEM

Blood is the red fluid that supplies the cells with the food and oxygen they need for work and growth. A healthy person has about 5 litres of blood in his body. The human blood is made up of plasma and blood corspuscles. Plasma is a light yellow-coloured fluid which contains 92% water and 8% proteins, sugar, salts and other minerals. Blood contains three kinds of corpuscles, namely, red blood corpuscles, white blood corpuscles and platelets.

Red blood corpuscles or erythrocytes are disk-shaped, flat and biconcave. They do not contain nuclei. Red blood corpuscles contain a red pigment called haemoglobin which makes the blood look red. New red cells form in the bone-marrow. There are about 5000 million red cells in 1 cc of blood. Red blood cells live from 50 to 120 days. Their size is about 0.007 mm.

White blood cells or leucocytes are mostly formed in the bone marrow while some of them in the spleen and lymph glands. White cells are larger than the red cells. Their average size is .007 to .012 mm in diameter. There are about 11 million white cells in 1 cc of blood. White cells have a nucleus. They are actually transparent and white as they do not contain haemoglobin. White cells protect the body against diseases and help in fighting infection. Lymphocytes are white cells that control immunity.

Platelets or thrombocytes are tiny disks with a diameter of .002 to .004 mm. There are 150,000 to 400,000 platelets in one cubic millimeter of blood. They have no nucleus. If a small blood vessel is cut, platelets stick to the damaged edges and to each other. As they pile up, they form a temporary seal over the injury. At the same time, they release a substance that starts the process of blood clotting. Blood clotting prevents the loss of blood. People having a less number of platelets in their blood bleed excessively before the blood clotting takes place.

Blood Groups

Whenever a person bleeds profusely due to some injury or during an operation, the blood is transfused into his body. But it cannot be transfused without matching the blood groups.

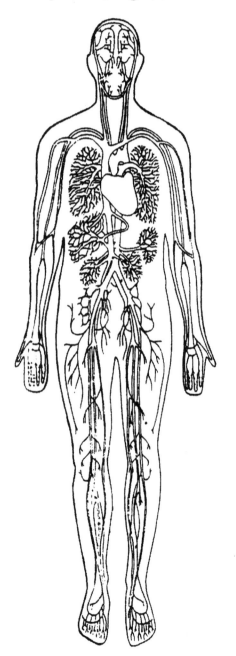

A human body has about 5 litres of blood

Karl Land Steainer in 1931 classified the human blood into three groups. Later, De-Castello and Stulri identified one more group. Human blood is divided into four different groups based on the presence or absence of certain antigens in red blood cells. These blood groups are A, B, AB and O.

Antigens are special proteins that stimulate the production of antibodies. If antibodies of one blood group are mixed with another group, they react with antigens and cause the red blood cells to clump together. Clumping can block small blood vessels and result in serious illness or death. The distribution of antibodies is shown below:

Blood Group	Antigen in RBC	Antibodies in Plasma
A	Only A	Only B
B	Only B	Only A
AB	Both A, B	None
O	None	Both A, B

Doctors prefer to use similar ABO blood types during transfusion to avoid any possibility of clumping. Type O is the universal donor and can be transfused into anyone. Type AB is the universal recipient and can receive blood of any ABO type.

In 1940, an additional rhesus (Rh) factor was discovered, which must be considered before the blood transfusion.

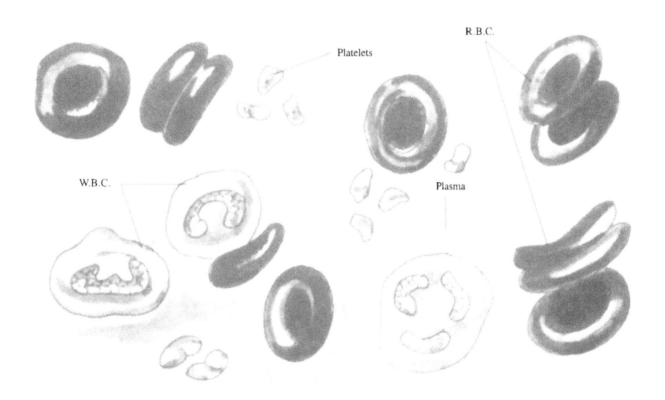

The red and white blood cells and platelets float in the plasma

171

CIRCULATORY SYSTEM

The organs involved in the circulation of blood in the body form the circulatory system. Heart and blood vessels are the most important parts of our circulatory system. Heart is a muscular organ and it weighs about 300 gms. It works like a pump and sends blood through a network of blood vessels known as arteries to different parts of the body. The blood returns to the heart for purification through veins. In this way, blood reaches every cell in the body to provide it with food and oxygen and to carry away waste products.

Our circulatory system is a fine network of arteries, arterioles, capillaries and veins.

Arteries and Veins: An artery is a large vessel that carries blood from the heart to the different parts of the body. It branches into the

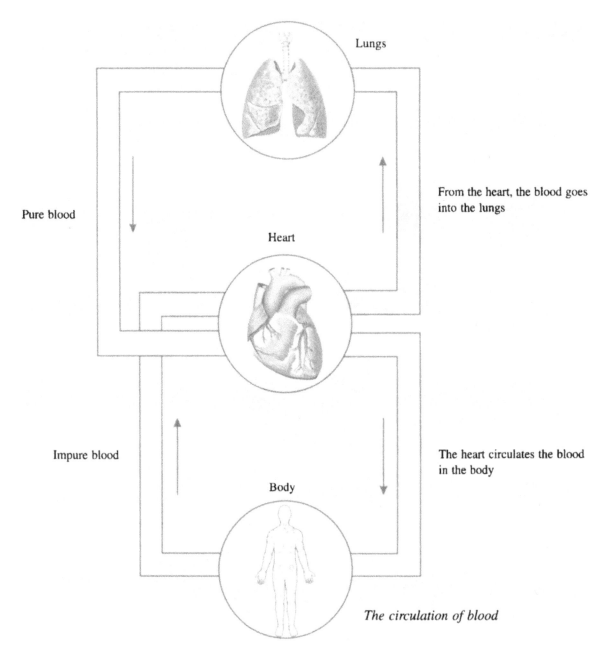

Lungs

Pure blood

From the heart, the blood goes into the lungs

Heart

Impure blood

The heart circulates the blood in the body

Body

The circulation of blood

smaller arterioles which further divide into very small vessels called capillaries. Capillaries carry blood to the cells. Capillaries change to venal capillaries, venules and finally veins. Veins carry the blood back to the heart. Scientists guess there are about 100,000 kms of blood vessels in our bodies.

Circulation: Blood is pumped from the right ventricle of the heart into the pulmonary artery. This carries the blood into lungs, where it absorbs oxygen and releases carbon dioxide. The blood returns to the left auricle or atrium of the heart through the pulmonary vein. The left auricle pumps the oxygenated blood into the left ventricle of the heart which pumps it into the aorta. The aorta, being the largest artery, carries the blood to other arteries and arterioles. The blood absorbs food when it passes near the small intestine. Waste from the cells is removed from the blood

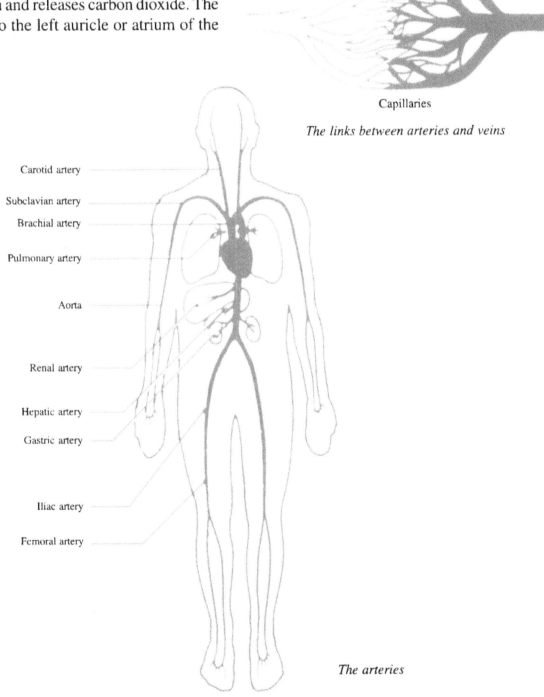

Arteries   Veins

Capillaries

The links between arteries and veins

Carotid artery

Subclavian artery

Brachial artery

Pulmonary artery

Aorta

Renal artery

Hepatic artery

Gastric artery

Iliac artery

Femoral artery

The arteries

when it passes through the kidneys. After the blood passes through the cells of the body, delivering food and oxygen and removing waste, it returns to the heart through the vena cava, the largest vein of the body. The impure blood enters the right auricle which returns it to the right ventricle. It is again pumped to the lungs to receive a fresh supply of oxygen to make another trip. In this way, this cycle goes on for ever.

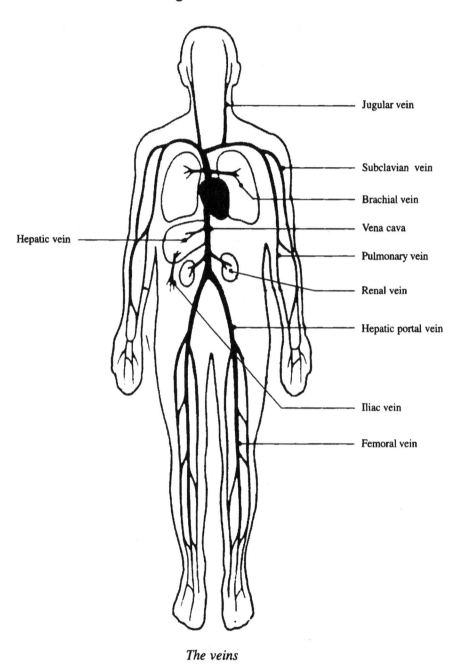

Jugular vein

Subclavian vein

Brachial vein

Vena cava

Pulmonary vein

Renal vein

Hepatic portal vein

Iliac vein

Femoral vein

Hepatic vein

The veins

HEART

Heart is one of the most important organs of our body which lies a little to the left side in the chest between the lungs. It is about the size of a closed fist and weighs about 300 grams. It is covered with a membrane called pericardium whose inner side remains in touch with the heart. It is, in fact, a muscle which is filled with blood. This is called myocardium and the layer which remains in contact with the blood is called endocardium.

The heart is a hollow organ divided into four chambers. Two chambers are on the right side and two on the left, separated by a wall called septum which does not let the blood of the two sides mix. The upper two chambers of the heart are called left and right atrium or auricle while the two lower chambers are called left and right

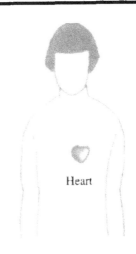

Heart is about the size of a closed fist

ventricles. Valves are fitted between atriums and ventricles. These valves open only in one direction. The blood can go from atrium to ventricle but not from ventricle to atrium.

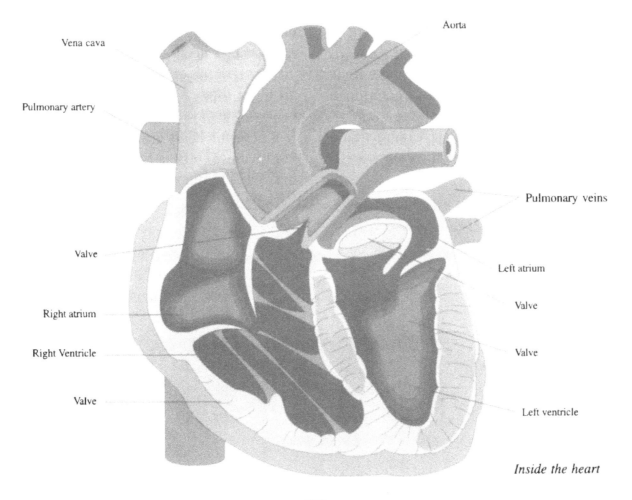

Inside the heart

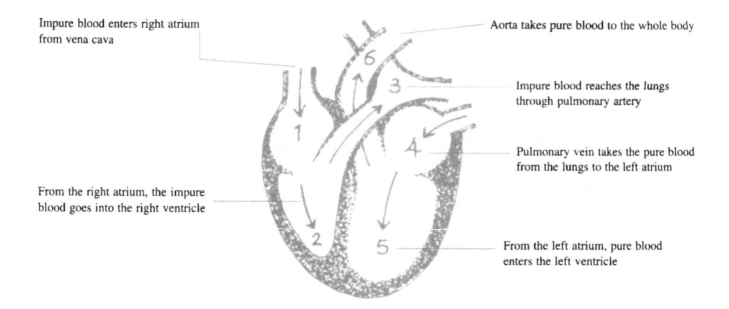

Impure blood enters right atrium from vena cava

Aorta takes pure blood to the whole body

Impure blood reaches the lungs through pulmonary artery

Pulmonary vein takes the pure blood from the lungs to the left atrium

From the right atrium, the impure blood goes into the right ventricle

From the left atrium, pure blood enters the left ventricle

The heart supplies pure blood to the body

Heart works as a powerful pump and sends blood to the whole body. Its two sides work together. They pump blood through the two chambers at the same time. The right side pumps impure blood to the lungs. The left side pumps purified blood to all other parts of the body.

Heart Beat: The heart beats automatically. It does not need to be controlled by the brain. Heart beat corresponds to the circulation of blood. Every heart beat has three stages— Contraction of the two Atriums, Contraction of the two Ventricles and the Rest period.

One heart beat completes one cardiac cycle. Heart beats can be heard by putting the ear onto the left side of the chest or with the help of a stethoscope. The count of heart beats of a new born child goes to about 140 per minute. A ten-year-old child's heart beats about 90 times a minute. In a healthy adult, it varies from 70 to 72 times per minute while a woman's heart beats 78 to 82 times a minute. During exercise, the beats rate goes as high as 140 to 180 per minute. If a person's heart-beat is not upto the proper count, an external device—pacemaker—is implanted to regularise it.

SKIN

Skin is the outer layer of our body. It is the largest organ of the body. It performs many necessary functions. Besides protecting internal organs from injury and preventing infection, skin relates the sense of touch and plays an important role in excretion and temperature regulation also. Skin makes vitamin D. An important function of the skin is to control the loss of the body fluids. It also excretes body wastes in the form of sweat.

Structure of Skin

Epidermis: This is the outer layer of the skin. Much of this layer is made up of dead, flattened, horny cells. These cells are constantly being worn away and replaced by new cells from underneath. It does not contain blood vessels.

Dermis: Next to the epidermis is dermis. It contains blood vessels together with sweat and sebaceous glands, fat cells and nerve endings that form receptors for the sense of touch, heat, cold, pressure and pain. The outer surface of the dermis is covered with a pattern of tiny elevations called papillae. The amount of blood flowing through the blood vessels of the dermis is controlled automatically by the nervous system. Blood circulation and sweat control the body temperature.

Sebaceous glands: These glands produce sebum which keeps the hair root lubricated.

Subcutaneous fat: It protects the skin.

Erector muscle: This keeps the hair erect.

Blood vessel: These blood vessels take oxygen through skin.

Sweat glands: These glands produce sweat to control the body temperature and excrete waste materials from the body.

Malphigian layer: It produces a pigment called melanin. This pigment is responsible for the colour of the skin.

The thickness of the human skin varies from about .05 mm to .65 cm. Skin is the thinnest on the eyelids and the thickest on the sole of the feet.

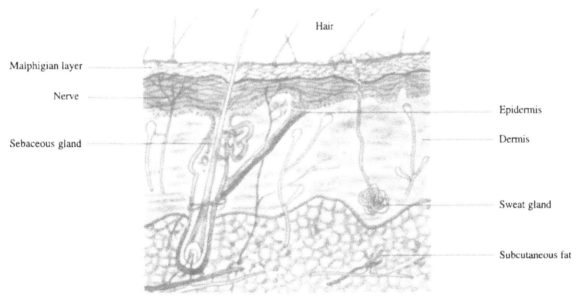

Under the skin

BRAIN

Human brain is a unique thing in the animal world. In fact, this is the central control system of our body. The brain coordinates all the activities of the body, such as seeing, hearing, thinking, feeling, and remembering the events. Brain is connected to the central nervous system, Nerves carry the messages to and from the brain and the body with the speed of about 400 km/hr. The human nervous system has about 13 billion nerve cells, of which about 10 billion are in the brain. The left side of the brain controls the right side of the body while the right side of the brain controls the left side of the body.

The human brain is safely placed between the thick bones of the skull. An adult brain weighs about 1.4 kg.

The human brain is divided into three main parts: the cerebrum, cerebellum and medulla oblongata.

Cerebrum: The cerebrum is the largest part of the brain. It is divided into two halves called the cerebral hemispheres. The outer surface of the cerebral hemisphere is the cortex. It contains the cell bodies of the nerves. The surface of these hemispheres is made up of grey matter. Below the grey matter is the white matter. Different regions of the cerebral hemispheres are called lobes. The frontal lobes are just behind the forehead where the processes of thinking, judgement and reasoning take place. Behind the frontal lobes are cells that control the movement of the whole body. The parietal lobes on each

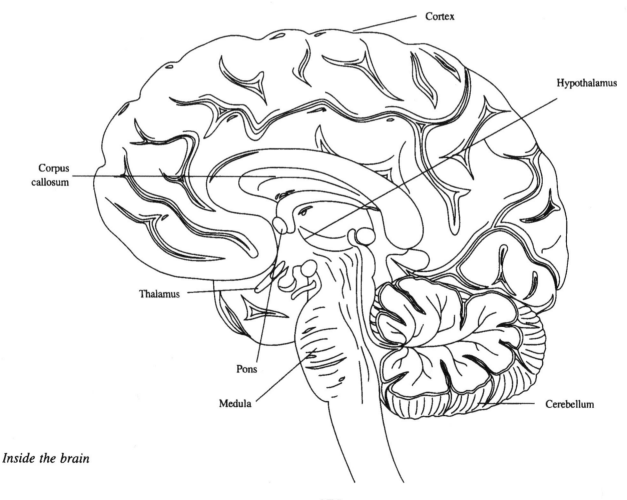

Inside the brain

178

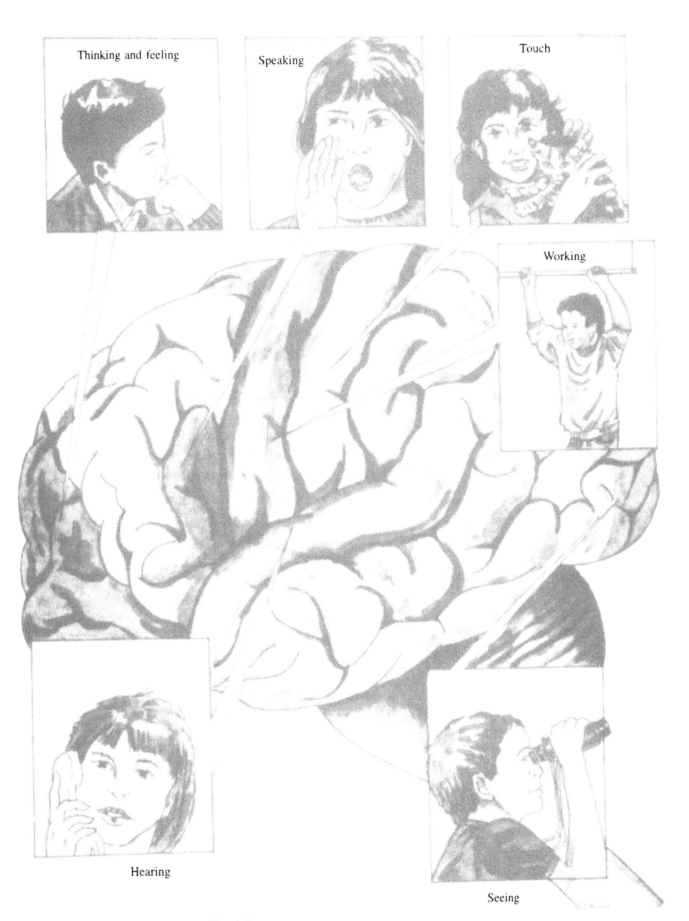

The different parts of brain and their functions

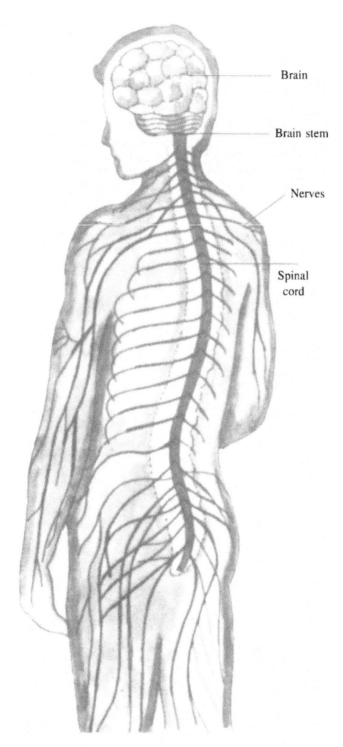

The nervous system

side of the brain receive the message connected with sensation of touch, body position, pain and temperature. The temporal lobes are the centres of hearing whereas the occipital lobes are the centres for the sense of light. One hemisphere is stronger than the other. If an individual has a dominant right eye and is right-handed, then he is likely to have a dominant left hemisphere.

Cerebellum: The cerebellum is the second largest part of the brain and lies towards the rear of the cerebrum. It is concerned with balance, position of the body and coordination of movements. This part of the brain keeps our body in equilibrium.

Medulla oblongata: The third part of the brain is the brain stem which connects cerebrum and cerebellum to the spine. It contains the important hypothalamus. The cells in hypothalamus regulate the movement of the intestines, reactions to emotions and sleep. Medulla oblongata regulates breathing, heart beat and the contraction of blood vessels.

The nervous system: The nervous system is made up of brain, spinal cord and nerves. Most nerves join together to form the spinal cord which runs through the backbone. The brain is connected to the rest of the body through the spinal cord. Any damage or injury to the spinal cord may stop the nerve signals from getting through.

DIGESTIVE SYSTEM

Alimentary Canal

- Mouth
- Pharynx or throat
- Oesophagus or gullet
- Stomach
- Small intestine
- Large intestine
- Rectum
- Anus

Food to a body is like fuel to the machines. Food is needed to maintain body tissues and is a source of energy for all body functions. A

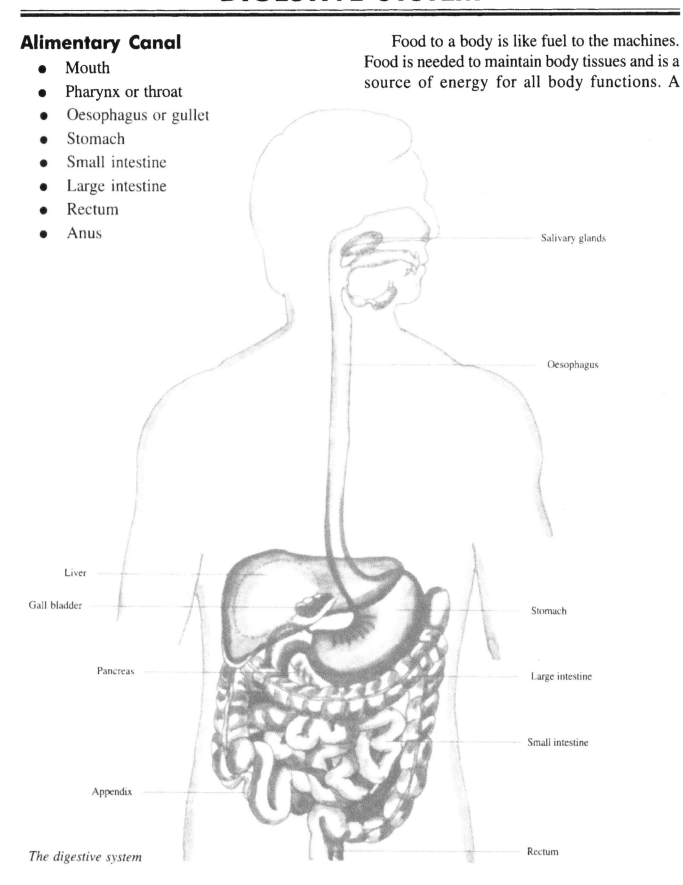

The digestive system

complete food contains all the essential nutrients that are vital for the growth and working of the body. Food, normally, is in the solid state and it is made soluble and absorbable by the chemical reactions of enzymes and digestive juices. The place where food is digested is called digestive tract and from where chemical substances are released, is called digestive gland. They together make the digestive system.

Taste: The pleasure of life lies in the different tastes of food. There are granular lumps on the upper surface of our tongue. These are called taste buds. Different taste buds are connected to the brain by nerve cells which carry the information of taste to the brain. Taste buds are of four kinds which identify sweet, bitter, sour and salty tastes. In the figure below are shown different taste buds and their locations on the tongue.

How food gets digested: Digestion of food begins in the mouth itself. Chewing makes food soft and breaks it into small pieces. The saliva in the mouth helps break down some carbohydrates into sugar. Saliva also makes the food slippery so that it can slide down the throat or oesophagus easily. A wave-like muscle contraction called peristalsis pushes the food down to the stomach. Churning of the stomach changes the food into a semi-liquid mass called chyme. At this time gastric juices and acids such as hydrochloric acid start digesting the food and help to kill the bacteria in the food. Liquid food now goes into the small intestine. Here some more enzymes are added to the food. Most of the food is broken down in the small intestine into its simplest forms—sugar, amino acids and fats. It takes about five hours. The nutrients of the food are absorbed by millions of tiny finger-like projections called villi. From there the nutrients are absorbed into the blood and are transported to every cell.

From the small intestine food goes to the large intestine. The small intestine is about 6.5 m long while large intestine is 1.8 m long. As it passes into the large intestine, water is removed and the waste product is passed out of the body as faeces through the anus. The faeces contain undigested food, cells, bile, salts and spleen salts.

Salty   Sour   Sweet   Bitter

The taste-buds on different parts of tongue respond to different tastes. The back of the tongue is sensitive to bitter tastes, the sides respond to sour taste and the front of the tongue responds to salty and sweet tastes. There are about 3,000 taste buds on our tongue.

EXCRETORY SYSTEM

The waste products and unwanted substances are removed from the body through excretory system. During the digestion of food, certain substances like ammonia, urea and uric acid are produced. Metabolic activities of the body produce wastes such as carbon dioxide and water. Since these waste products are poisonous in nature, they should be removed from the body.

Water vapour and carbon dioxide come out of the body in the process of respiration. About 0.2 litre of carbon dioxide is given out every minute by a person. Ammonia, urea and uric acid reach the liver where ammonia is converted into urea. Urea and uric acid get mixed in the blood stream and go to the kidneys from the liver. The kidneys work as filters and separate out these substances from the blood. About 120 ml of blood is filtered by the kidneys every minute. The whole blood is filtered by the kidneys about 30 times a day.

The filtered substances do not contain urea and uric acid only but some other harmful substances also. All these substances are excreted from the body as urine. A person passes out about one litre of urine in a day.

Our skin is also an important excretory organ. Water and salts come out from the skin in the form of sweat. About 0.7 litre of sweat along with small amount of salt is excreted from the human body each day through perspiration.

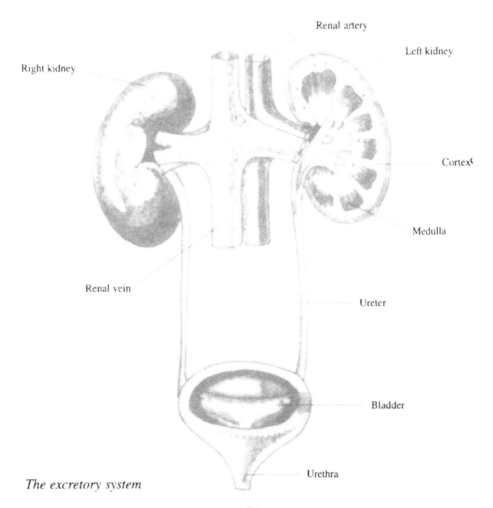

The excretory system

183

RESPIRATORY SYSTEM

Respiration is a natural process in most living things. In this process, living beings change energy that is locked in molecules of digested food into a form that can be used by the cells. In the process of respiration, water and carbon dioxide are produced. The body continuously takes in oxygen and releases carbon dioxide by this process. Intake of air is called inhalation and its expelling out is called exhalation. On an average, an adult breathes 15 to 17 times every minute.

Human beings breathe in and out through nose and mouth. When air enters the nose, it is warmed and moistened before it goes to the lungs. In addition, the nose helps remove small particles of dirt and dust. This air goes into the lungs through the wind pipe. During the process of respiration, bulging of chest is a muscular action. This action is performed by both voluntary and involuntary muscles. In the process of respiration, mainly intercostal muscles and diaphragm take part. In deep breathing, shoulders, neck and muscles of stomach also take part.

Lungs are the most important organs of respiratory system. There are two lungs, both of which lie in the air tight chest or thoracic cavity. The chest cavity and the lungs are covered with a membrane called pleura. Each lung is made of millions of alveoli. Lungs purify the blood coming from the heart by the oxygen which is inhaled and then the carbon dioxide of the blood is exhaled. After being purified, the blood goes back to the heart.

A person's breathing is automatically controlled by the respiratory centre in the brain. This centre is sensitive to the amount of carbon dioxide in the blood. If there is an increase in carbon dioxide, such as during strenuous exercise, the respiratory centre sends more nerve signals to the muscles that control breathing. As a result, the person breathes faster.

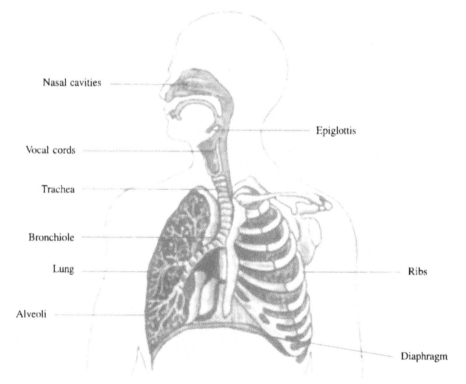

The respiratory system

184

EYE

The eyes are a part of the sense organs that give human beings and most other animals the most accurate and detailed information about their surroundings. Our eye is a sphere measuring about 2.5 cm in diameter. It rests in bony socket in the skull and can move in all directions due to the actions of the six ocular muscles.

Eyes work on the principle of a photographic camera. The light rays from the object enter the pupil and are focused by the lens on the retina. Before the light enters the pupil, it passes through the cornea and aqueous humour. Lens forms an inverted reduced image of the object on the retina. From the image part of the eye, signals are sent to the brain by the optic nerves. The brain makes the correct and real image of the object and perceives depth and distance. The sensitive cells of the eyes are called rods and cones. The number of these cells is about 130 million. Rods are sensitive to light while cones are sensitive to colours. The two images produced by two eyes are combined by the brain as one. The ability of the brain to form one image from the images of the two eyes is called binocular vision.

Main Parts of the Eye

- **Lachrymal:** This is the tear gland of the eye which produces tears. The tears clean the eyes.
- **Iris:** It makes the pupil contract or dilate to control the amount of light that enters the eye.
- **Suspensory Ligaments:** Lens is covered with a thin capsule which is connected to ciliary body by a soft ligament.
- **Lens:** This is a soft and transparent convex lens which forms the image of the object onto the retina.
- **Vitreous Humour:** It is a jelly type transparent fluid.
- **Aqueous Humour:** Water like liquid.
- **Cornea:** Transparent part in the centre of the eye ball.
- **Choroid Layer:** Black coloured soft layer which stops the spread of light.
- **Retina:** Light sensitive screen on which image of the object is formed.
- **Optic Nerve:** This carries the information of the image to the brain.

Nature has provided us two eyes which give the correct estimate of distance, depth and solid nature of the objects.

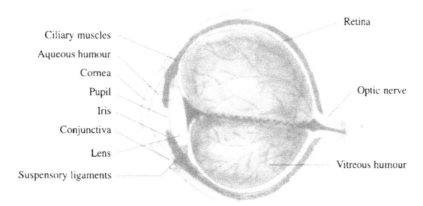

Parts of an eye

185

EARS

We not only hear the sounds with our ears but they also help us in maintaining the balance of the body. The ear is divided into three compartments—the outer ear, the middle ear and the inner ear.

Sounds are vibrations in the air. When a sound is made, its vibrations or sound waves travel all around. These waves strike with the outer ear and reach to the middle ear through a pipe. The ear drum or tympanic membrane is made to vibrate by these sound waves. Just behind the ear drum are three small bones, joined together. They are—hammer, anvil and stirrup. The three bones transmit these vibrations to cochlea. Cochlea is filled with a fluid in and contains thousands of the nerve cells. As cochlea vibrates, the fluid also vibrates and the nerve endings get excited. The impulses of the nerve endings are transmitted to the brain by auditory nerve. Brain analyses them and we hear the sound. Our ears can hear both feeble and loud noises. To keep the ears healthy, it is essential to get them cleaned from time to time. ■■

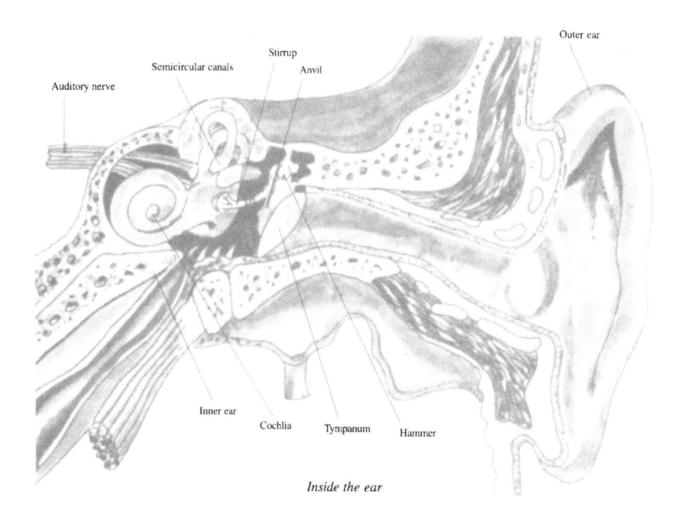

Inner ear   Cochlia   Tympanum   Hammer

Inside the ear

TEETH

Human beings have two sets of teeth during their life time: primary teeth and permanent teeth. Primary teeth are also called deciduous or milk or baby teeth and begin to break through the gums after the age of six months. All the 20 primary teeth are in place by the time a child is three years old. The permanent teeth begin forcing out and replacing the primary teeth when the child is about six years old. All the 32 permanent teeth are in place usually when the person is in his/her early 20s.

The 32 permanent teeth are arranged in pairs on each side of the upper and lower jaws. There are four types of teeth: incisors, canines, bicuspids or premolars and molars. The incisors are the eight front teeth, four in each jaw. The canines are four pointed teeth, one on each side of the incisors, two in each jaw. Next to the canines are two bicuspids, four in each jaw. Next to the bicuspids are six molars, three in each jaw.

All teeth have two parts: The crown and the root. The crown is the part of the tooth that is above the gum line. The root is below the gum. The part of the tooth at the gum line, where the crown and the root meet, is called the neck. The root is held light in the jawbone by a layer of bone tissue called cement.

All true teeth are made up of three layers— the outer hardest layer is called enamel, inside the enamel layer is dentine and the innermost layer is pulp. Dentine is like bone and its inside is called the pulp cavity that contains tiny blood vessels and nerves. The dentine is nourished by the pulp.

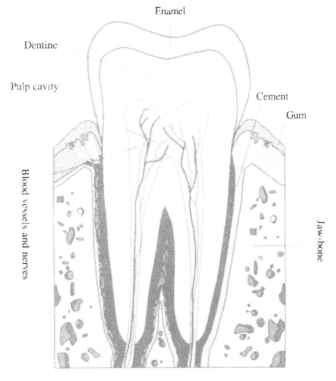

Inside the tooth

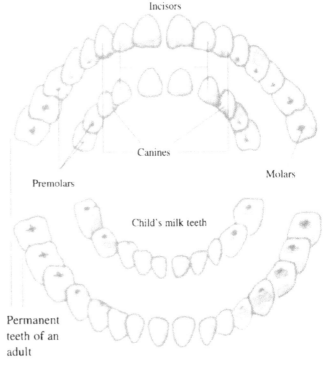

A child's and an adult's teeth

187

NEW LIFE

Every new life starts as a single cell. After the sexual intercourse, when a sperm from the father's semen enters the ovum of the mother, fertilization takes place. Pregnancy begins when a fertilized egg attaches itself to the lining of the uterus.

During the first two months of pregnancy, the developing baby is called embryo. After this period until the birth (9 months), the developing baby is called a foetus. After two months, the foetus, though only about 2.5 cm in length, can move its head, mouth, arms and legs. The child develops in about nine months in the womb. After nine months, a new child is born on earth.

■■

Egg

Sperm

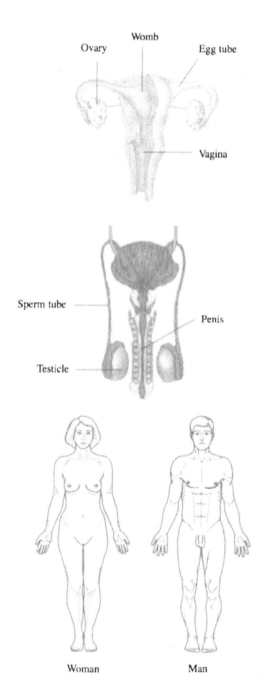

Ovary

Womb

Egg tube

Vagina

Sperm tube

Penis

Testicle

Woman

Man

ECOLOGY

Human civilisation has developed very slowly. In the ancient times man was totally dependent on nature for his needs but as he learnt about agriculture he started interfering with the laws of nature. He shaved off the forests for agriculture and tamed many wild animals for his food and clothing. He left the caves and started living in houses. With the development in man's lifestyle, several civilisations of Sumaria, Egypt and Indus Valley spread over the vast areas of the world. It is a general belief that man has conquered the nature but still he cannot control natural disasters.

All the living beings on the earth utilise solar energy and get essential elements from atmosphere, lithosphere and hydrosphere. Animals and plants either depend on plants or

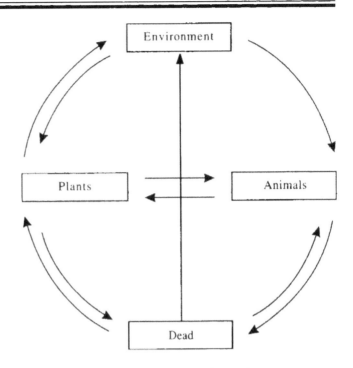

The relationship between the environment, plants and animals

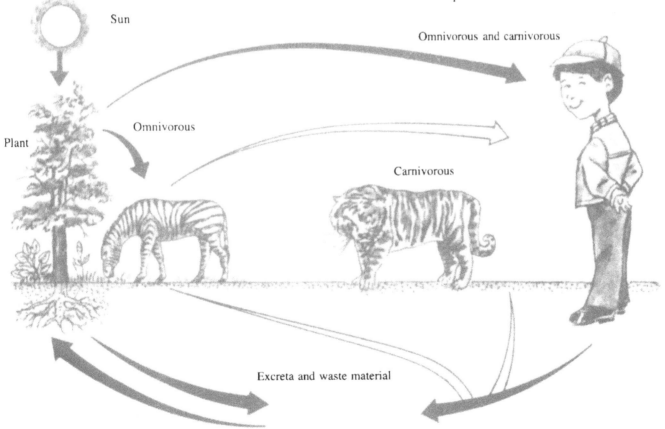

Eating relations in a community

191

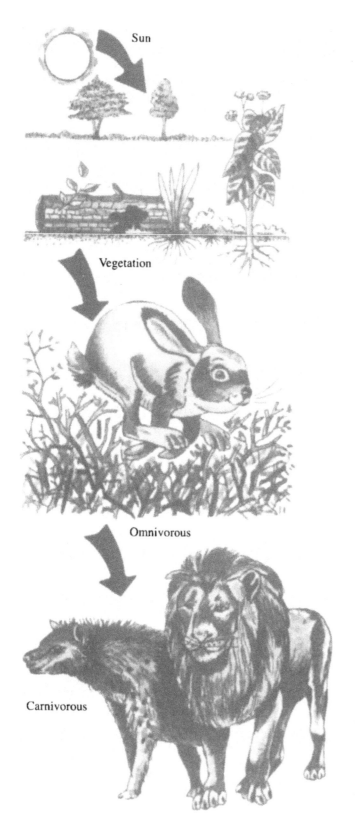

Sun

Vegetation

Omnivorous

Carnivorous

on other animals for their food. A balance exists between living and non-living things in the nature. Whenever any element of nature increases or decreases in large proportions, ecological balance gets disturbed. The study which deals with the relationships of life forms with each other and with their surroundings is called ecology. A group of animals living in a particular physical environment is called Ecosystem.

Today rapid industrialisation and ever increasing human requirements have disturbed these natural elements and created an ecological imbalance. Increasing population, emission of carbon dioxide and other toxic gases by factories and vehicles, occurrence of acid rain, nuclear tests and poisonous chemicals have posed a serious threat to the ecological balance. The entire earth and its inhabitants are affected by pollution. Life on earth is endangered. Now, the man has realised the need of ecological conservation and safety of environment. Public awareness is being created for the protection of life on the earth. Many programmes for a better environment and conservation of natural resources are being carried out all over the world. The scientific conservation programmes are mainly based on the principles of ecology. We can get a better environment only when every-body, specially the children, becomes aware about it.

To study the ecosystem, it is necessary for us to acquire deep knowledge of the trees, plants, ponds, rivers, lakes; air, water and land; crops and creatures; marine animals, etc. This way we will be able to make our ecosystem even better. It is our duty as well as responsibility to ensure the safety of our food chains.

THE LIVING PLANET

Among the nine planets of the solar system, earth is the only one where life exists. Plants and animals exist on earth because it has all the favourable conditions required for survival of life.

Even on earth life is not found everywhere because survival requires a particular type of environment. That part of the earth where plants and animals exist is called biosphere. Biosphere is further divided into three parts: Lithosphere, Hydrosphere and Atmosphere. Lithosphere is that part of earth where rocks, sand, soil, plants, animals, etc. are found. That part of the earth where water is present is called Hydrosphere. All the animals and plants need water to live. The blanket of air which surrounds the earth up to a height of about 1000 km is called atmosphere. Atmosphere is a balanced mixture of several gases like nitrogen, oxygen and carbon

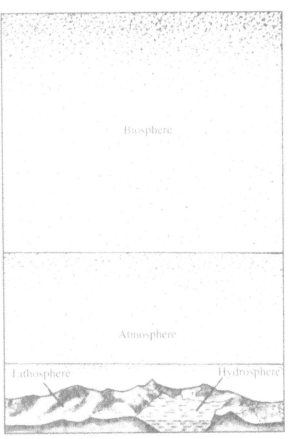

Biosphere: Atmosphere, Lithosphere and Hydrosphere

The earth has life

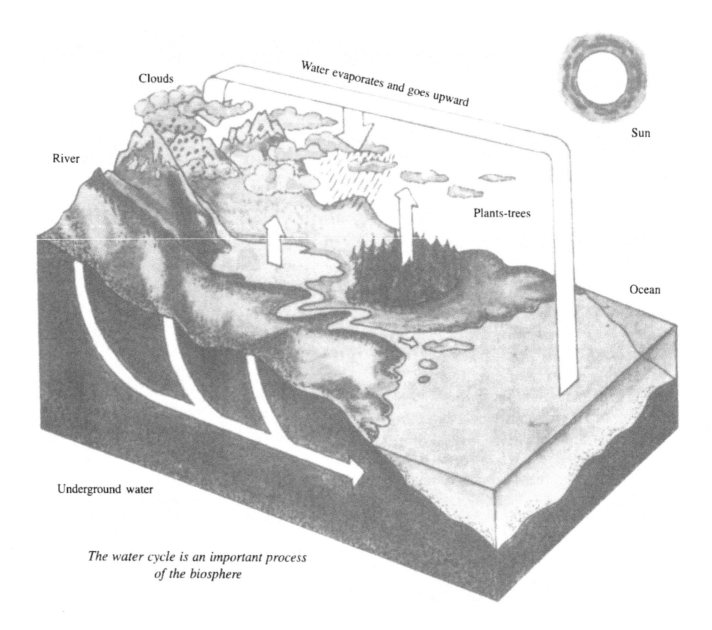

Clouds

Water evaporates and goes upward

Sun

River

Plants-trees

Ocean

Underground water

*The water cycle is an important process
of the biosphere*

dioxide. The major atmosphere extends upto 320 km above lithosphere and hydrosphere. About 10 km below and 10 km above the surface of the earth is biosphere. In reality 90 per cent living beings live in the area between 1 km below and 1 km above the earth's surface.

Biosphere or ecological framework always remains in a dynamic equilibrium. In this many biological and physical elements are present in a balanced state. Air, water, soil, animals, plants and bacteria are all linked to each other or to their surroundings by some means or the other. This system is called environment. Environment includes all those conditions which affect our life. Plants, animals, water, air, land are important parts of our environment because living being cannot survive without them. Our houses, villages, towns etc. all are included in the environment. Man has created an imbalance in the environment by its haphazard exploitation to fulfil his needs. The trees are being cut mercilessly, water and air are becoming highly polluted. If environmental pollution keeps on increasing with the present rate, the very existence of the life on earth is endangered.

■■

CYCLES OF CARBON, NITROGEN, OXYGEN AND HYDROGEN IN ECOSYSTEM

Innumerable kinds of plants and animals are found on earth which are different from each other in shape, size and structure. All these are basically made up of carbon, nitrogen, oxygen and hydrogen. Animals get these elements from nature only. The processes like formation and decay, growth and decomposition normally occur in nature. Various cycles are going on in the nature as a result of these processes. Due to these cycles-plants, animals and their environment are linked to each other and a balance is established between them.

Carbon cycle

Carbon is an essential element found in all living beings. Plants and animals get this element from carbon dioxide present in the atmosphere. Plants take carbon dioxide from air and in the presence of water and sunlight they make their food by the process of photosynthesis. Plants are consumed by other animals as food. In this way they get compounds of carbon. Energy is produced by digestion of these carbon compounds. Carnivores get carbon from flesh of herbivores. Omnivores like man get carbon

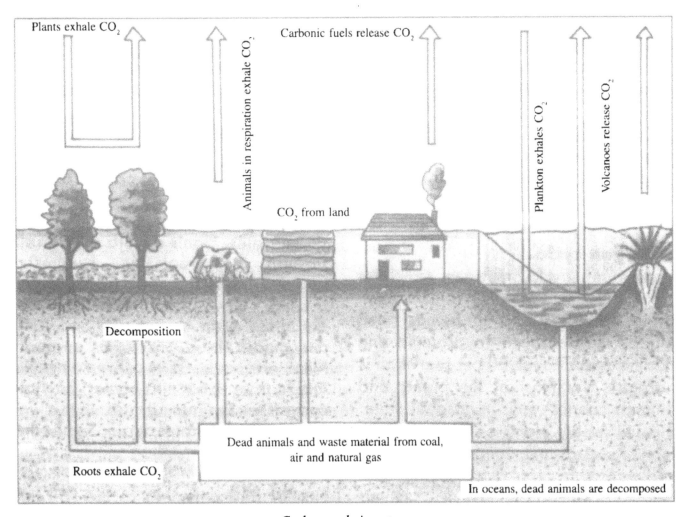

Carbon cycle in nature

195

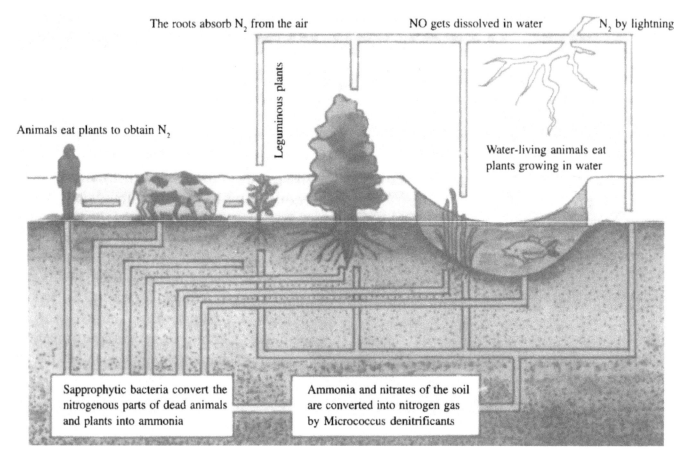

The roots absorb N$_2$ from the air

NO gets dissolved in water

N$_2$ by lightning

Leguminous plants

Animals eat plants to obtain N$_2$

Water-living animals eat plants growing in water

Sapprophytic bacteria convert the nitrogenous parts of dead animals and plants into ammonia

Ammonia and nitrates of the soil are converted into nitrogen gas by Micrococcus denitrificants

The nitrogen cycle in nature

from both plants and animals. Carbon cycle shows that all the animals absorb carbon dioxide directly from atmosphere or from the plants. Animals convert carbon into carbon dioxide during respiration or oxidation and again this carbon dioxide is released into the air. Carbon dioxide is also liberated during decay of plants and animals and burning of fuels which enters into the atmosphere and completes carbon cycle.

Nitrogen cycle

In nitrogen cycle, nitrogen circulates continuously through soil, water, air and living beings. All living beings need nitrogen. It is found in proteins and nucleic acid. Though 80 per cent of air is nitrogen, yet most of the plants and animals cannot use it in gaseous state. It is utilised by plants or animals in the form of several compounds like ammonia or amino acids. Many bacteria of soil like Azotobacter and Clostridium convert the atmospheric nitrogen into nitrogenous compounds which can be absorbed by the plants. Many small nodules are present in the roots of leguminous plants which contain bacteria named Rhizobium leguminosarum. These bacteria convert atmospheric nitrogen into nitrates. Plants absorb nitrate salts through their roots and synthesize proteins in their cells. Nitrogen enters into the body of the animals when they eat these plants. Bacteria and fungi decompose the dead bodies on earth. Nitrosomonas bacteria convert ammonia into nitrites and nitrobacter bacteria convert nitrites into nitrates. In this way plants absorb nitrates again. Some other bacteria like Thiobacillus denitrificans and Micrococcus denitrificans convert nitrates and ammonia of the soil into nitrogen gas. Apart from this, a fraction of nitrogen in atmosphere changes into nitric acid during lightning which forms nitrates with rain water in soil that is absorbed by the plants. In this way atmospheric nitrogen moves through many food chains and again comes into

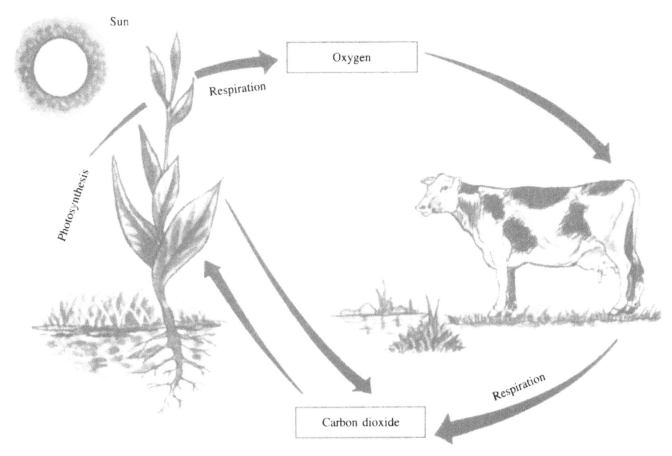

The oxygen cycle in nature

the air through a cycle. Though this process takes billions of years but each atom of nitrogen comes back into the air.

Oxygen cycle

Life is not possible without oxygen. About 20 per cent of air is oxygen. All the living beings on earth need oxygen for respiration. Oxygen is essential for burning. You will think that the constant use of oxygen from the atmosphere will bring its end some day. But in nature the oxygen cycle is constantly going on and no disturbance has occurred to the balance of this gas. This cycle like nitrogen cycle is a continuous one. Green plants absorb carbon dioxide gas from the atmosphere and convert it into oxygen by the process of photosynthesis. In the daytime, all the trees and plants give out oxygen and take in carbon dioxide. At night, they give out carbon dioxide. This way, the quantity of oxygen in the atmosphere remains stable.

Hydrogen cycle

Water is very important for all plants and animals. The hydrosphere gets its most abundant element, the hydrogen, from water only. Water is formed by the combination of oxygen and hydrogen. During photosynthesis hydrogen is produced by the dissociation of water which forms glucose after combining with carbon dioxide. Plants provide food for herbivores and these animals get glucose and proteins from plants only. Hydrogen forms carbohydrates which is an important source of energy for living beings. We take these carbohydrates in the form of food. These help our life cycle to continue.

THE FOOD CHAIN

All the animals, directly or indirectly, depend on plants for their food. Herbivores get their food directly from plants. Carnivores eat small and weaker herbivores. The inter-relationship between plants and animals is very interesting and is an important subject in ecological studies. This inter-relationship is called food chain. Food chains can be observed in places like grazing lands, deciduous forests, rivers, seas, deserts etc. which are inhabited by plants and animals.

Insects eat plants, and insects are themselves eaten by shrew and shrew is eaten up by the owl. In this way a food chain (Plant-Insect-Shrew-Owl) is formed. Zebra eats grass in African

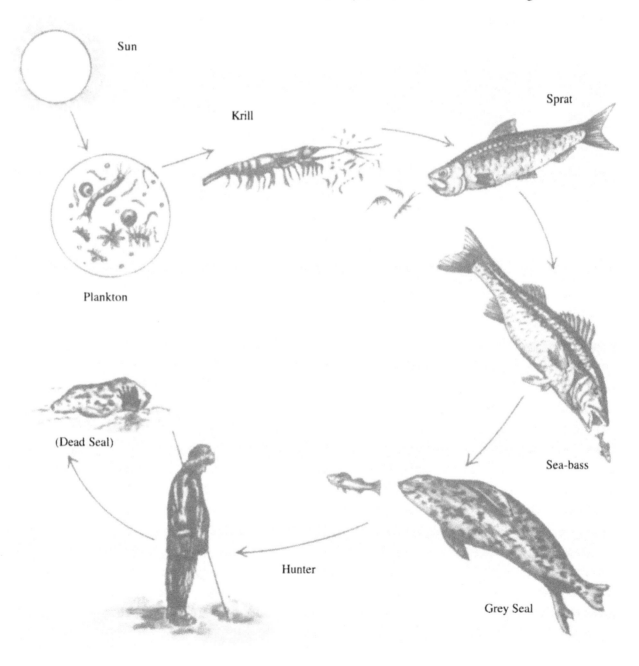

A food chain of the ocean

198

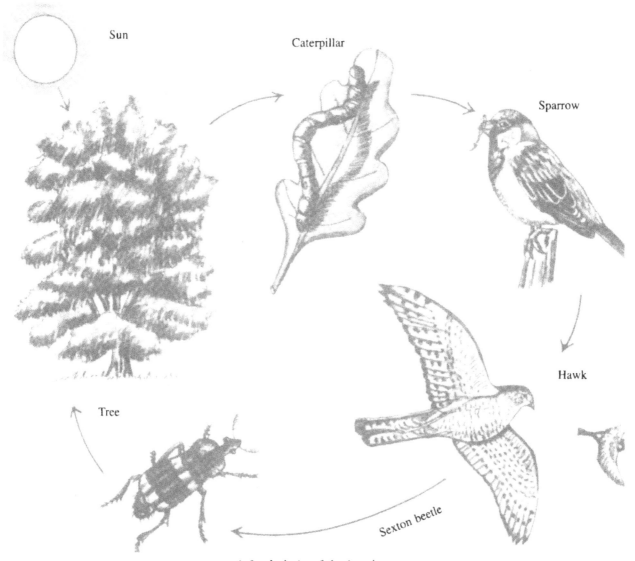

A food-chain of the jungle

Savana and lions consume zebra (Grass-Zebra-Lion food chain). Similar food chains exist in sea also. For example, small fish eats algae and it is eaten by bigger fish, bigger fish is eaten by gull and gull is further consumed by seal. Similarly, deer eats grass and lion eats deer. There are many food chains present in the environment. This cycle does not end here. When many food chains combine then a food web is formed. Most of the food chains start from plants.

Food chains are very delicate and they can be broken easily. If one relationship in the food chain is broken it becomes dangerous for the whole chain. If the farmer uses pesticides to kill the pests then shrews will also die of starvation because pests are their food. Owls, which eat shrews, will also die in want of their food. At the Peru coast, extensive hunting of anchobi fish led to a decrease in the number of carmorent birds which ate anchobi fishes only. Vanishing carmorent birds caused lack of their faeces which were consumed by plankton. And decreasing planktons caused food deficiency for anchobi fishes which led to the reduction of anchobi fishes.

Man is also an important member of several food chains because he eats plants (vegetables) as well as meat, fishes, eggs etc. For the animal conservation we should have the knowledge about food habits and food webs of the animals.

SAVING THE SOIL

The uppermost layer of the earth is called soil. It is made up of rocks, organic substances, minerals, water and air. Soil is essential for life on earth. Plants grow in soil which are eaten by other animals. Various insects, worms and other organisms breathe in the air present inside the soil.

About 300 million tiny creatures live in every one hectare of soil. Rodents, insects, millipedes, mites, sea bugs, snakes, spiders, earthworms and other types of worms, reptiles and protozoa are commonly found in soil. They grow due to base exchange capacity of soil in the presence of potassium and phosphorus. These creatures produce organic substances for soil by decay of their dead bodies and waste products.

When man learnt about importance of soil for the production of food, he shaved off jungles, built dams for water and worked against laws of nature. In this way, he progressed a lot but forgot that excessive crop cultivation, cutting of forests

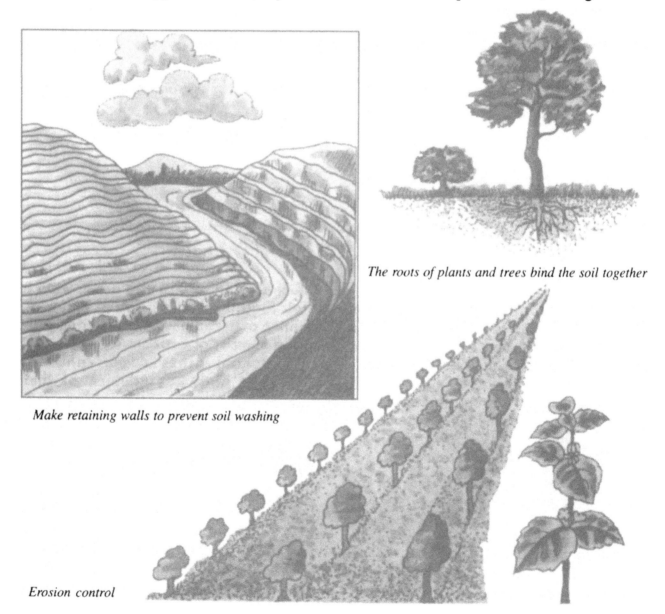

Make retaining walls to prevent soil washing

The roots of plants and trees bind the soil together

Erosion control

and overgrazing will cause soil erosion, and consequently turn fertile land into deserts. The areas, where ancient civilisations of Nile Valley, Mesopotamia, Sindhu Valley and Vi-ho Valley of China were flourished, have been converted into barren lands today. Only 20% area of the old Iraq civilization is under cultivation now. Today, even after knowing the importance of the soil, we are destroying more than 3000 tons of soil every second. For good growth of crops we have to conserve the soil and this can be done by controlling erosion. Soil erosion is caused by wind and water. Erosion is also caused by overgrazing, forest fires and growing same crops repeatedly. Erosion makes the soil waterless and causes the death of wild animals. It increases the possibility of floods. Erosion can be controlled by planting trees, bushes and grass. It can also be stopped by controlling soil flow with the help of dams and retaining walls. Contour ploughing, deep ploughing, step farming and mixing of organic matter in the soil is also helpful in controlling soil erosion. The repeated use of chemical fertilizers also reduces the fertility of the soil. It can be controlled to some extent by the use of compost and home made manures.

Soil erosion

Briefly, we can do soil conservation in the following ways:

- By planting grass and plants
- By constructing dams
- By doing contour farming
- By prohibiting animal feeding
- By prohibiting cutting of forests
- Alternate cultivation of crops

We can prevent soil erosion by applying these methods.

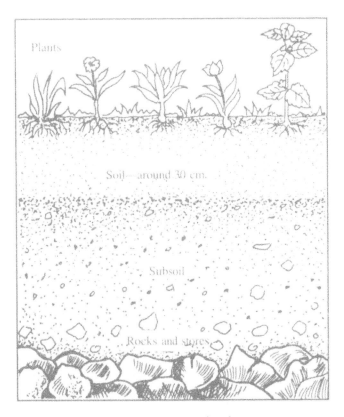

The formation of soil

201

ACID RAIN

The term acid rain refers to the fall of acids mixed with rain and snow. The main cause of acid rain are the gases emitted by the factories and motor vehicles which pollute the atmosphere. Mainly sulphur dioxide is responsible for acid rain which is produced in the burning of fossil fuels. This gas is also produced by natural phenomena like volcanic eruptions, forest fires etc. Sulphur dioxide reacts with water vapour and forms sulphuric acid in the presence of sunlight. This acid gets mixed with rain water or snow flakes and falls on to the earth. Similarly, motor vehicles burn petrol and diesel and produce oxides of nitrogen. These oxides produce nitric acid with rain water. In 1872, Robert Angus Smith was the first person to discover acid rain.

The acid rain may occur thousands of kilometres away from the source of gas emission. During the Gulf war many oilfields caught fire and the smoke produced was so much that it reached even up to the Kashmir valley and turned the snow into black. Acid rain badly affects the

Acid rain has reduced the sparkle of the Taj Mahal

marine life. The reproduction capability of fish gets reduced. Trees defoliate before time and the crops having edible roots get destroyed. The Taj Mahal at Agra and New York's 'Statue of Liberty' have lost their lustre due to acid rain. Water lily plant gets most affected by acid rain.

■■

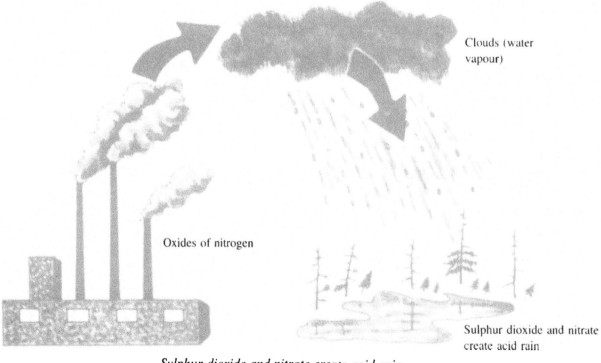

Clouds (water vapour)

Oxides of nitrogen

Sulphur dioxide and nitrate create acid rain

Sulphur dioxide and nitrate create acid rain

POLLUTION

The term pollution refers to the contamination of the environment—air, water and land—with harmful wastes resulted from human activities. Environmental pollution has become a major worldwide problem. It has badly affected the whole biosphere.

Pollution has been caused by man. Man has been polluting the air with smoke and poisonous gases; rivers, lakes and oceans with sewage and harmful chemicals. During loading and unloading of crude petroleum in huge tankers millions of tons of oil spills in the oceans adversely affecting the aquatic life. You can observe the dead bodies of birds and fishes near sea shores or floating on the water.

Increasing human population of the world and industrialisation are the major causes of environmental pollution. The jungles have been shaved off to accommodate the increasing population. This has considerably reduced the vegetation and forest cover over land. During the last 50 years industries have multiplied manifold. The harmful chemical discharges of factories are polluting the rivers and oceans. The smoke produced by automobiles is increasing the air pollution and the noise produced by the engines of motor vehicles is aggravating the noise pollution. Modern means of agriculture have increased the use of pesticides which have also added to land and water pollution.

The marine life in the big lakes like Ari has vanished due to pollution. The sewage water, industrial wastes and the water from agricultural farms containing extensive amount of fertilizers had filled this lake with too much of phosphates and nitrates. These chemicals have permanently distorted the biological balance of the lake. All the fishes of the lake have died. The water of Ganga and Yamuna rivers is also becoming badly polluted. The Dal Lake of Srinagàr is in worse condition because of pollution. Pollution has affected mankind adversely. For the improvement of degradation of biosphere only man is responsible. And protection of flora and fauna also depends on the ways and means adopted by the human beings for their livelihood.

Industries cause pollution

203

AIR POLLUTION

Air not only supports the human life but is also essential for the survival of plants and other animals. Polluted air can cause many diseases such as cancer and lung malfunctioning. The main pollutants of air are the smoke, dust, ash, pollengrains, poisonous gases and tiny particles of materials. The air pollution is mainly caused by burning fossil fuels like coal, petrol, natural gas etc. Industrial plants, electric plants and automobile engines also add to air pollution. They produce oxides of nitrogen, sulphur dioxide, carbon monoxide particulate matter etc. and pollute the air. The smoke emitted by factories, automobiles, trains, ships etc. also release poisonous particles in the air. The smoke of jet planes pollutes the upper atmosphere. Air pollution causes many diseases like allergy, cough, cold, heart troubles asthma and many types of cancers. Green leaves turn yellow due to air pollution. The growth of the plants is slowed down. The paint of the buildings gets faded and the colours of various materials and clothes get bleached due to air pollution.

The increasing quantity of carbon dioxide has posed a serious threat to the environment. Prior to industrial revolution the amount of carbon dioxide was 275 to 285 parts per million in the atmosphere. By 1980s this increased to 338 parts per million.

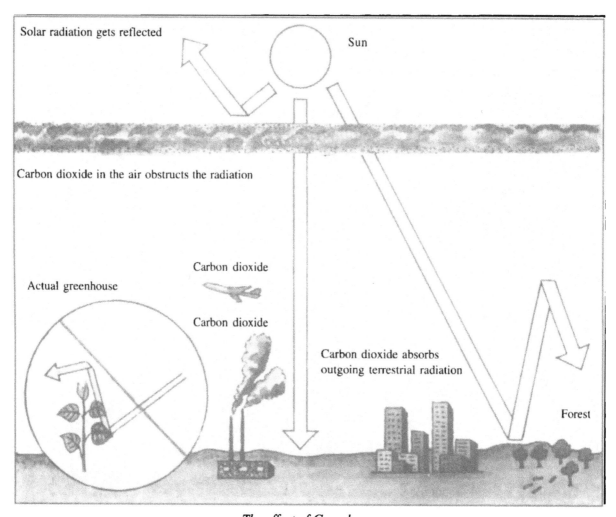

The effect of Greenhouse

Environment is our heritage

In the beginning of the 21st century, this has further increased to 400 parts. The heat energy radiated from the earth is in the form of long waves. About 90 per cent of this heat is absorbed by the atmosphere. Atmosphere heats up mainly due to this terrestrial radiation. The complete absorption of terrestrial radiation and noncapturance of solar radiation by the atmosphere is called Green House Effect. The Green House Effect was explained for the first time by the Swedish chemist Arthenius in 1986. In a green house, the glass walls and roof allow the sunlight to reach the earth but they do not allow the terrestrial radiation to go out. This causes the temperature to rise.

On the global scale, carbon dioxide in the atmosphere acts as a green house. The carbon dioxide is largely transparent to incoming light energy but outgoing infrared radiation from the earth is absorbed by carbon dioxide leading to a warming of atmosphere. If the amount of carbon dioxide continues to increase in the atmosphere then the atmospheric temperature will become so high that ice of polar regions will melt and water level in oceans will rise, leading to flooding of coastal areas. The air pollution also affects direction of wind, temperature, climate and weather. Air pollution is adversely affecting the historical monuments. The Taj Mahal of Agra is being adversely affected by the smoke emitted by the petroleum refinery situated in Mathura. The capital city of Japan, Tokyo, is the world's most polluted city.

Air pollution can be controlled to a large extent by planting more and more trees and by fitting electrostatic filters in the chimneys of the factories that emit the smoke hazardous to the environment.

OZONE DEPLETION

According to meteorological researches the atmosphere is divided into five thermal zones—the troposphere—10 km above the sea level; stratosphere 10 to 40 km; mesosphere 40 to 70 km; ionosphere 70 to 400 km; and exosphere above 400 km.

Stratosphere contains negligible amount of water vapour and dust. Strong jet streams blow in this layer which affect the weather of the earth. For convenience this layer may be divided into two parts called stratospheric layer and ozonosphere. Stratospheric layer exists only up to 25 km height and its temperature remains almost constant. Ozonosphere contains ozone and it exists between 25 km to 40 km. Ozone is essential to life on earth because it acts as a safety shield for the earth. Ozone forms a hot layer by absorbing about 99% of short ultraviolet rays of sun which are harmful for the life on earth.

In nature ozone is formed from oxygen by ultraviolet radiations of the sun. These two gases remain in equilibrium or balanced condition. But recently oxygen-ozone balance has been upset by the excessive presence of chemical pollutants in the atmosphere. Scientists believe that ozone layer is becoming thin over Antarctica.

Man-made chemicals specially chlorofluorocarbons adversely affect the ozonosphere. Freon gas is also a chlorofluorocarbon which is used in refrigerators. These chemicals are also used in airconditioners, aerosol spray and plastic foams. Free chlorine atoms of these compounds have disturbed the ozone equilibrium. Ozone depletion will cause dreadful diseases like skin cancer and skin allergies. It may also have hazardous effect on plants and other animals. To save our earth we have to discover substitutes for chlorofluorocarbons which do not cause ozone depletion in atmosphere.

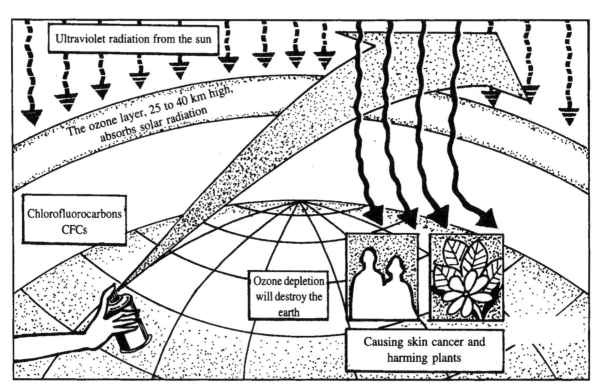

Chlorofluorocarbons (CFCs) adversely affect the ozonosphere and mankind.

WATER POLLUTION

The earth, from a spacecraft, looks blue, because its two-third part is covered with water. This part is called hydrosphere ('hydro' means water and 'sphere' means region). Water is elixir of life. Neither plants nor animals can survive without it.

These days rapidly increasing population and industrial units have been polluting the water. The water coming from the factories and homes contains liquid and solid wastes. This is called sewage. The river sources and sea coasts all are polluted by sewage. This polluted water contains organic compounds of mercury and lead which enter into the body of aquatic animals. Eating such fishes is harmful for the eyes and brain of the man. In Minamata Bay of Japan, mercury was deposited in the body of the sea animals through industrial wastes. These animals were eaten by local people. It is horrifying to know that 10 thousand people were poisoned by mercury, and 100 people died mysteriously. In muddy polluted water sun rays cannot penetrate and light does not reach the aquatic plants. This halts the process of photosynthesis

Industrial units and liquid and solid wastes from homes pollute water

207

Aquatic animals die in large numbers due to water pollution

and plants die. Consequently, the fishes which are dependent on such plants also die.

Oil pollution is very harmful for the ecological balance of the oceans. The spilled oil spreads widely on the water surface and badly affects the aquatic life. About 30,000 aquatic birds can die due to the oil spill from a big tanker. Fishes do not get sufficient oxygen due to oil pollution and they start dying in large numbers. The oil spill during the Gulf war has been the worst in the history of mankind.

Agricultural fertilizers and insecticides are also adding to water pollution. Heat of nuclear power plants too poses a danger to aquatic life. The river water in large quantity is used to control the nuclear reactions in nuclear power plants. The river water gets heated to very high temperature and is again fed into the river after use. This waste water increases the temperature of river water at the point where it goes into the river. The hot water kills many aquatic plants and animals.

Every nation is striving to provide clear drinking water to its citizens. This can only be done by controlling water pollution. Presently, only 1.6% of the total water content is potable, i.e., fit for drinking. Water cycle in nature takes around 1000 years to complete itself.

NOISE POLLUTION

Today, noise pollution has become a serious health hazard in the big cities. Automobiles, trucks, motors, factory machines and even the means of entertainment like rock music, DJs, etc. produce loud, irritating sounds. Man feels disturbed by such sounds. The harmful effects of intense sounds are referred to as noise pollution.

The intensity of sound is measured in decibels. It is based on a logarithmic scale which means that ten decibels is ten times more powerful than one decibel and 20 decibels are 100 times more powerful, and so on. Man's ears can tolerate the sound of upto 80 decibels. The sounds of higher intensity than this are harmful for ears. Normal human voice is of about 60 decibels. People living in noisy places become accustomed to high intensity sounds but their ears lose the sensitivity and capacity to send sound waves to the brain. This leads to deafness.

Studies reveal that in India ten per cent people living in cities and seven per cent in the villages are suffering from auditory defects. Abnormally high noise disturbs the sleep and

The clamour of motors, trucks and automobiles

increases heart beat. Noise pollution causes gastric problems, dialation of pupil of eyes, ulcers in stomach, weakening of teeth and defects in kidney. It is also harmful for the nerves of the womb and adversely affects the embryo of a pregnant woman. Noise pollution can only be checked by the people who are aware of its ill-effects.

The music systems producing very loud sounds can be banned to check noise pollution.

Noise is a hazard

209

LOSING FOREST

Forests are the homes for various plants and animals. There was a time when about 40 per cent of land was covered with forests. Today only one-third part of land has forests. Twenty-two per cent of the world forests are present in the republics which formerly comprised the Soviet Union, 22 per cent in South America, 20 per cent in North America, 17 per cent in Africa, 15 per cent in Asia and the Pacific area and 4 per cent in Europe. The total area covered by the forests

Presently, every minute, 40 hectares of forests are destroyed worldwide

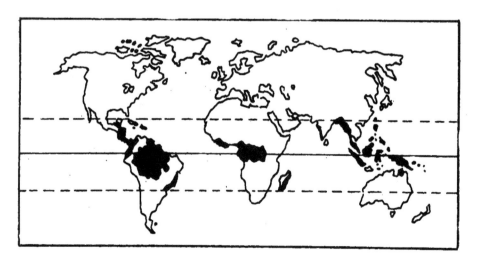

in the world is 40,500 lakh hectares.

Man has been destroying the forests to meet his needs. According to some surveys about 40 hectares of forests are shaved off per minute. Temperate forests have suffered immense destruction. About 42 per cent of tropical rain forests have been destroyed by man. Such forests are mainly destroyed in Africa where they have reduced to half. In South America 37 per cent and in Asia about 41 per cent of forests had been cut down. Destruction of forests will not only result in vanishing of greenery on earth, but the whole ecological balance will be disturbed.

Forests provide us with wood, raw materials and protect our environment against pollution. Forty-one per cent land of Soviet Union, 32 per cent of America, 50 per cent of Brazil, and 69 per cent land of Japan is covered with forests. These forests are mainly used for furniture wood, pulp and fuel. Forests are useful in many other ways also. They prevent floods by controlling the flow of rain water. This increases underground water and soil erosion is reduced. Forests prevent fast evaporation of water from land and increase the humidity of air by transpiration.

Forest fires occur due to man's carelessness and natural causes

The destruction of forests by man, if not stopped, will one day lead to a catastrophe. The forests can be protected from destruction by planting more and more trees, preventing cutting and burning of forests.

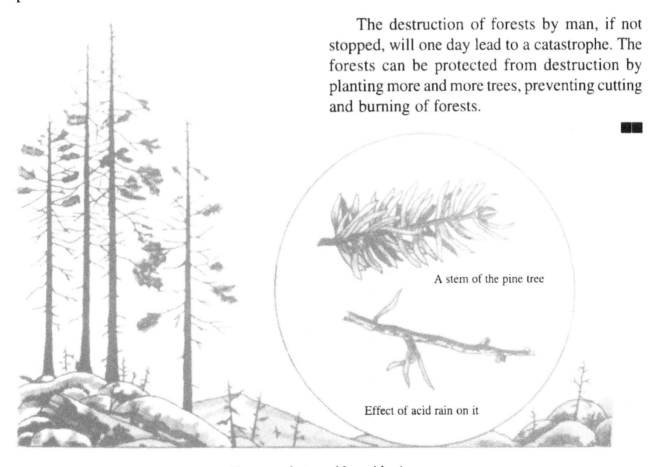

A stem of the pine tree

Effect of acid rain on it

Trees are destroyed by acid rain

NEW SPECIES

The beauty of earth lies in plants, animals and greenery. Grasslands, fragrant pine forests, evergreen forests and shady deciduous trees contribute to this beauty. Various kinds of animals are also found with plants. The natural groups of plants and animals found in big areas are known as biomes. Biomes are found at those places where environment is favourable for vegetation. The natural vegetation areas and the biomes found in them are interdependent on each other.

The vegetation of any place depends on the climate of that place. The animals found in any area adjust with the climate and vegetation of that place. Today bison grass is not found in Great Plain of North America. So the antelope, the animal dependent on this grass, had also become extinct. Hilly goats and sheeps are also lost from conifer biomes. White pine and red wood of California have become extinct in the same way. The unnatural environmental changes brought about by man's lifestyle are responsible

Rain forests are dense and high

212

for this destruction. Environmental activities all over the world are campaigning to stop all such human activities which are dangerous for biomes.

Rain Forests

Scientists are now concerned about rain forests which are found in Amazon basin of South America and coastal regions of Brazil, coastal parts and western islands of central America, Congo basin in Africa, Gini coast and eastern Madagascar, New Guinea island of Asia, Sri Lanka, India's Western coast and north-eastern states, countries of South-east Asia and southern China.

Rain forests are dense and high. Epiphytes like ferns, archids, algae, lichen etc. are

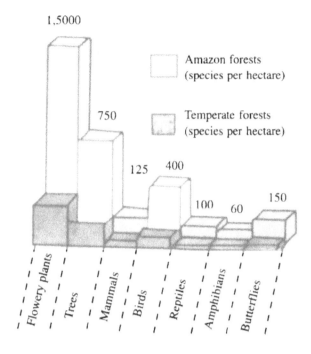

A variety of creatures inhabits the Amazon rain forests

213

The Jojoba plant

Rose Periwinkle

exist on the earth. Scientists know about 1,150 million species and they are working to discover more. About many species of plants and insects we are still ignorant. Rain forests are main source of new species. If these forests are destroyed, then probably all these species may be lost forever.

The number of mammals is very small in rain forests because mammals cannot easily roam in dense forests. Monkeys, bears, jaguars are commonly found in these forests. The number of crocodiles, *gharial* and rhinoceros in rivers is also very small. Mainly reptiles, insects and birds are found here in large numbers. Besides animals, there are many hidden treasures of natural resources in these forests which can be very useful to mankind.

Today about 40% medicines are obtained from plants. A new fascinating plant Rosy Periwinkle has been discovered from Amazon rain forests which can cure a kind of cancer. Still there are many undiscovered plants in these forests which can be of great help to medical science.

Another plant known as petrol plant is also discovered from Amazon rain forest. The juice of this plant can be used in place of petrol in car engines. Jojoba plant is a similar type of plant recently discovered which can yield good quality oil. Its oil can be used in place of whale oil in many industries. It is expected that in future many more useful products can be obtained from these forests. Presently, many farm houses containing various kinds of trees and plants are being built in the world which use the products obtained from these trees and plants for some or the other use. In our country, Forest Research Institute, situated in Dehradun, is doing a commendable job in this regard.

commonly found among them. These plants remain attached to the stems, leaves and branches of trees. A unique feature of rain forests is that different varieties of trees in large numbers grow at one place. Surprisingly more than 1,200 kinds of trees can be found in just one square kilometre area of a rain forest. We do not know exactly how many species of plants and animals

214

WILDLIFE UNDER THREAT

Extensive hunting made the Woolly Mammoth extinct

The destruction of forests, pollution, climatic changes and extensive hunting have resulted in the extinction of some species of plants and animals, and many more are approaching towards extinction. Man has realised today that destruction of wildlife means an end to his own survival.

The Aak bird of northern polar region, which was commonly seen some 100 years ago, has been so extensively hunted for meat and feathers that its whole genus has vanished. Some 300 years ago, the Moa bird of New Zealand became extinct. Dodo was a bird found in the Mauritius island which became extinct about 300 years ago as a result of merciless hunting. Thylacine, an animal with pouch, which was also known as Wolf of Tasmania, is reported to have become extinct some 300 to 400 years ago. Indian tiger, Indian rhino, pink-headed duck are the animals which have become extinct in our country during the present century. The number of rare and

The extinct bird Dodo

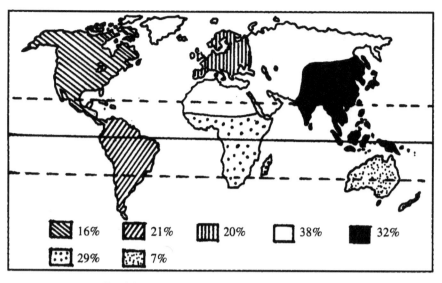

Worldwide fear of creatures getting extinct

16%	21%	20%	38%	32%
29%	7%			

Whale

Snakehead Fritillary

Big Blue Butterfly

endangered animals has been increasing with time.

The destruction rate witnessed during the last few years was never seen before. In a period of 400 years from 1550 to 1950, on an average, one species of animals have become extinct every year due to ignorance of man. By 1985, this rate increased to one species per day. If such trends continue, then by the end of the present century, 50,000 species will become extinct. Studies show that during 1980 to 2000, about 5,00,000 species of plants and animals will face extinction i.e. 10% of total plants and animals will become extinct.

Wildlife has been under threat of extinction all over the world. Nothing but man's modern lifestyle is responsible for this situation. In India, this is to a large extent. Today about 1,000 types of animal and 20,000 types of plant species are endangered. Some rare vertebrates like Gila

Trout, Houston Toad, *Gharial*, Californian Condor, Javan Rhinoceros, Lion, Puma, Polar Bear, Antelope, humpback Whale and Panda have reached the stage of extinction.

Among the plants species, Monkey Orchid, Alpine Catchfly and Snakes Head Fritillary are under threat of extinction. The species of following Vertebrates are also facing the danger of extinction:

Fishes	190
Amphibians	40
Reptiles	150
Birds	430
Mammals	320

Californian

Monk Seal

Pygmy Hog

Panda

217

To control pollution, the effort is on all over the world to bring new areas under forest cover. Cutting of forests is banned to prevent further destruction of forests. The wild animals are saved by keeping them in natural environment. In this direction, Yellow Stone National Park was set up in 1872 in USA, which was the first Conservation Area of the world. Canada's Wood Buffalo National Park, Alberta is the world's biggest conservation park, with an area of 45,480 square km. It was established in 1922. Today, there are 1200 such national parks and animal reserves in the world. India has 59 national parks. Now, zoological parks are also used as conservation areas. About 500 zoos all over the world have saved many animal breeds from extinction.

International Union for the Conservation of Nature and Natural Resources (IUCN) has done pioneering work for the cause of healthy environment and World Wildlife Fund (WWF). This society has helped a lot to save wildlife. The wildlife can be saved from extinction through the following methods:

- Prohibiting hunting of animals.
- Prohibiting the use of leather obtained from the skin of animals.
- Controlling pollution.
- Planting more and more trees and preventing cutting of forests.
- Construction of national parks and reserves.
- Construction of zoos.

By adopting these methods, the wildlife, flora and fauna can be saved from getting extinct.

White Rhinoceros

Lion

218

TRANSPORT

INVENTION OF WHEEL

The wheel has played a very important role in the development of human civilisation. In fact, the wheel has become the pivot of the whole human civilisation. If the wheel had not been invented, there would have been no car, no scooter, no cycle, no train and no aeroplane. The wheel has marked the beginning of scientific and industrial age. Almost all machines make use of the wheel in some form or the other.

Exactly when and where the wheel was invented is not known to us. Man did not have any means of transport about 15,000 years ago. At that time, man's biggest worry was food and shelter. He used to go into the forest and kill some animal for food using his primitive weapons. After killing a large animal, he had no means of transport to bring that animal to his cave. He used to drag or lift the animal to his cave. It must have been a very tiring job for him.

It is said that necessity is the mother of invention. He might have put some wooden planks under the hunted animal and would have made his first sledge. This would have proved a better way of bringing the large animals to the cave and, perhaps, this was his first means of transport. After this, a modified sledge was developed by putting leather belts between the two wooden planks. This type of sledge proved very useful in snow-bound areas.

It is believed that the Pyramids of Egypt and Stonehenge were constructed by using rollers for transportation. When were the rollers for transportation invented is not known. It is supposed that the idea of rollers might have come from the round stem of a tree with which carrying of rock pieces would have been easier for the ancient man. He might have started using rollers of tree stems to carry heavy things to the desired places.

The sledge-cart of primitive man: the first means of transport

221

*The idea of wheels might have come from
the round stem of a tree*

*The Pyramids of Egypt: a miraculous example of
the use of rollers for transport*

Wooden rollers were used by man as a means of transport during the middle Stone Age. After a long span of time, perhaps, the man might have cut a disc from the stem of a tree and made a hole in its centre. He might have used it as a primitive wheel. After putting two such wheels on one axle and a wooden plank on it, he might have used it as the first cart. In this cart, both the wheels and axle rotated together. After this, the potter's wheel was invented.

According to some evidences, the wheel was invented in Syria and Sumeria in about 4000-3000 BC. Around 3000 BC, the wheel was being used in Mesopotamia. In about 2500 BC, the wheel came to be used in the Indus valley also.

For a long time, the wheel remained crudely shaped, because it was made with the help of stone tools. Only after the development of metal tools, a better shape to the wheel could be given.

Egyptians could know about the wheel after several hundred years. Spoked wheels were invented in about 1800 BC in Egypt. The

Transportation through rollers

Egyptians made the spoked wheel by joining its inner and outer circles with small bars i.e. spokes. Instead of placing a wooden plank on the wheels, the Egyptians made a box-like cart of it. One or two persons could sit in this cart with some of their belongings.

This two-wheeled carriage was also adopted by the Greeks and the Romans. In the beginning, bullocks were used to pull the cart. Later on, horses were used for the same purpose. The Romans started using a four-wheeled cart which was called a chariot. With the passage of time, many improvements were made in these carts and wheels.

The wheeled carriage marked the beginning of road transport. This gave birth to bullock carts, horse carts, buses, cars, trucks, scooters, bicycles etc. which are being used all over the world for transportation purposes. Roads were first constructed during the Roman empire, but with

Stonehenge: The famous monument of Britain

223

The wheel was being used in Mesopotamia in about 3000 BC.

its fall, the condition of these roads worsened. The industrial revolution in 18th century introduced the cemented and coaltar roads. Germany was the first country to construct national highways before the Second World War. Hitler made about 3200 km long roads in Germany. At present, USA has the longest network of roads with the length of 6.4 million kilometres. India has a 16,75,000-kilometre long network of roads. About 40% of the total load is carried from one place to another by roads.

The main roads which connect a country from one end to another are called 'highways'. The roads which connect the capitals of two countries are known as 'international highways'.

The road that connects Kolkata (in India) to Peshawar (in Pakistan) is the oldest highway of our country, which is functional even today. The biggest highway in the world is the Pan-American Highway.

BICYCLE

The bicycle is a popular vehicle of our time. It can be seen almost in every house. The shortage of petrol has made it even more important.

A modern bicycle consists of two wheels fitted in a frame. The rear wheel is rotated on a chain and a toothed wheel with the help of pedals. A handle turns the bicycle in a desired direction. Karl Von Drais invented the bicycle in 1818. It consisted of two wheels fitted in a wooden frame and a seat in the centre for sitting.

One day, people saw Drais driving his bicycle on the roads of Mahaim in Germany. He wore a green coat and a high hat and was pushing the bicycle by his feet. The ride on this strange vehicle became a subject of laughter for the people.

Drais named his bicycle as Draisine. The invention of this jocular vehicle by a high official like Drais made his senior officers unhappy and, as a result, his services were terminated. Instead of praising his invention, everyone criticised his bicycle. He died in poverty in 1851.

Later on, Draisine was further improved in England, France and USA. After 20 years i.e. in 1839, Kirkpatrick Macmillan, a blacksmith in Courthiel, invented a lever-driven bicycle. It was named as velocipede. After 10 years, a German mechanic, Philip Heinrich Fisher, fitted a paddle in the front wheel of Macmillan's bicycle. In 1876, J. Lawson fitted a toothed wheel and paddles between the two wheels. Swis Hans Ronald invented the roller chain. After this, wheels with wire spokes, springed seat, ball bearing, gear, gear shift and freewheel were fitted in the bicycle.

In 1874, James Starley of England and H.J. Lawson developed the modern bicycle. There are about 24 to 40 wires in its tyres that work as a spoke. They can tolerate the jerks of the uneven roads. To make the bicycle lighter and faster, John Dunlop invented pneumatic tyres. This invention made the bicycle a popular vehicle. In 1902, the Sturmey-Archer Company of Britain established the speed gear system in the rear wheel hub.

In 1905, India started to import bicycles from England. India started making its own bicycles in 1938. In China, about 80 million people use bicycles today. Every year a cycle-racing competition is organised in France, which is called 'Tour de France'. It was started in 1903. This race is completed in 20 days and covers 4,000 kms.

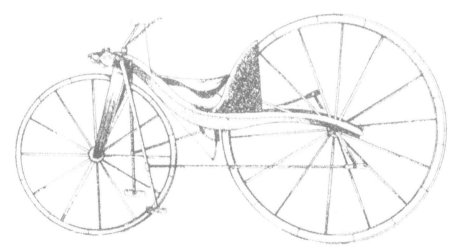

The Bicycle of Karl Von Draise

RAILWAYS

Nowadays, trains are being extensively used for carrying goods and passengers from one place to another. The real development of railways took place during the last 150 years only. In fact, the railways is an outcome of industrial revolution. The spread of railways is linked with the power of steam. The first steam engine was made in 1776 by James Watt of Britain. After this, many modifications and improvements were made in the steam engines, which led to the development of railways.

The first automatic steam locomotive was made in 1803 by Richard Trevithic. The steam locomotive could not move successfully on roads, and it was concluded that steam locomotive cannot travel on roads. That is why, the steam locomotive was run on rails.

James Watt (1736-1800)

The first mechanically propelled vehicle was 'The Rocket', designed by a Britisher named George Stephenson in 1814. He became famous all over the world as the inventor of railway engines. He was born in 1781 in a village near New Castle. He was very fond of engines right from his childhood. He did not study upto the age of 18, after which he joined a night school, where he could get some education. He used to work in a coal-mine. There he was known as the 'Engine Doctor' because he knew about engines more than even the qualified engineers. His master was very happy with him. He helped Stephenson in the development of steam engines.

After two years' hard work, the rail engine, named Blucher, was made. It was capable of pulling 8 carriages, loaded with 30 tons of coal, with a speed of 7km/hr. This engine gave Stephenson confidence that trains could be used in future as a public transport system.

Stephenson presented a plan of a rail line before the Westminsters Parliament house. The Parliament house, with a great difficulty, sanctioned a rail line, from Stockton to Darlington. This rail line was meant to carry both goods and passengers. Stephenson was given the contract of making rail engines.

On September 27, 1825, a ten-mile long railway line was inaugurated. The engine was named 'Active'. Stephenson himself was its driver. This train consisted of 33 bogies, out of which 4 bogies were loaded with coal and flour,

George Stephenson
(1781-1848)

227

In 1829, George Stephenson's 'Rocket' won a reward of 500 pounds

one bogy was for the people of the company, 22 bogies for general passengers and 6 bogies at the end were also loaded with coal. This train with 600 passengers travelled a distance of 8 kilometers in one hour.

In 1826, when another proposal was sanctioned, Stephenson made his second engine for the railway line between Liverpool and Manchester. At that time the government took a policy decision and declared that other builders of railway engines should also get a chance of making steam engines. To give the contract of making railway engines, a competition was arranged with a declaration that whichever engine will be the most successful, the government will purchase it for 500 pounds. In this competition, four engines took part.

First of all ran Stephenson's engine 'The Rocket'. Other three engines could not compete with it and 'The Rocket' was declared successful. This engine could pull a load of 13 tons for a distance of 19 kilometers in 65 minutes. This exercise was repeated 20 times. Seven such engines made by Stephenson were used to inaugurate Liverpool-Manchester line on September 15, 1830.

Stephenson's efforts made trains a grand success. This made him a rich person. He died in 1848.

A railway line was laid from Liverpool to Manchester

Soon after England, railway lines were laid in other countries also. In 1863, the first underground railway was made in England. In 1877, vacuum brakes were introduced in trains which reduced the number of accidents to a great extent.

After this, many new revolutions came in the train industry. In 1884, electric locomotives were introduced which reduced air pollution to a great extent. In 1925, diesel locomotives were developed in Canada. In 1964, bullet trains were introduced between Tokyo and Osaka in Japan. This train had a speed of 163 km/hr. By 1976, a fast diesel train having a speed of 230 km/hr was already in use in Great Britain. After this, Meglev train, monorail etc. were also developed.

The diesel and electric locomotives are now gradually replacing the steam engines

The first train in India was started between Bombay and Thane on April 16, 1853. The biggest railway line in the world is between Moscow and Vladivostok. This railway line is 9,438 kms long. Its name is Soviet Trans-Siberian Line. Development of modern technology has made it possible to manufacture tube trains, channel trains and trains without drivers. Now, trains having a speed of more than 270 kms per hour have also been manufactured.

229

SHIPS

Most of the ancient civilisations emerged at the banks of the rivers. At that time, crossing the rivers was a big problem to man. When and how did the idea of making a boat come in the mind of man is not known but the primitive boat would have been a log of wood. He might have made the first boat by hollowing a log of wood. After that, he would have made the front part slightly sharp and made the real boat.

According to known facts, perhaps boat-making started in about 4000 B.C. Masts and sails were developed around 3500 B.C. In the beginning, the use of boats was confined to Dazala, Farat and Nile rivers. Egyptians were the first who used boats in the sea. The oldest boat which could be found and is safe even today was constructed in about 2500 B.C. This boat is 43.4 metres long and weighs about 40 tons. It was found buried in Egypt near the Khufu Pyramid. The oldest wrecked ship is assumed

The idea of making a boat might have originated by seeing a floating log of wood

A Phoenician mast-boat

A Portuguese warship

to have been constructed in about 2700 to 2250 B.C.

Ancient Phoenicians used to make trips around Africa by boats. These were the first Mediterraneans who traded with Britain by sea and identified sea-routes to India in about 600 B.C. Harbours were also made by them. Egyptians, Greeks, Romans, Indians and Arabs carried out trade on a large scale with each other. Greeks, Romans and Vikings also made ships. Vikings used to call them as long ships.

After the invention of magnetic compass, bigger ships were made. The ship-making technology could be fully developed only in the 14th century. In this field Italy was the first, and then the lead was taken by England. The Portuguese and Spanish were their rivals. For about 100 years, mast ships could not replace sail ships. These were developed by the Britishers along with their naval forces. In 1620, the submarine was developed. In the 17th and the 18th centuries, sailing ships acquired larger speeds and became more capable.

In the 18th century started the construction of steam-driven ships. Britain was at the top in making such ships. A ship named 'East India Main' was constructed in England. For faster movement 'Frigate' boats were made. The Cliper ship was also developed which was used to carry tea from South-East Asia to England. By the year 1863, wood was replaced by iron in the ship-building industry.

The first steam-driven Pyroscaphe steamer was made in 1783. The first successful steam-driven ship Charlotte Dundas was made in 1801-1802. Paddle and wheels were fitted in its rear part. Screw propeller was invented in 1836.

An Aircraft Carrier Enterprise

Bartha is the oldest mechanical ship which was 50 ft. long and its capacity was 48 tons. It was designed in 1844.

A British engineer, Charles Parsons carried out the first experiment of steam turbine in a ship. Nine diesel engines were used in 1911 in ship building. The American submarine, Nautilus, driven by nuclear power, was made in 1955. Today there are many submarines that are driven by nuclear power and they are capable of staying inside the water for months. The first nuclear-powered ship named Sevana was made in U.S.A. in 1961.

In modern ships the size of the hull, engine, propeller, rudder etc. is very large. In ships, today, we have the facilities of electric generation, bedrooms, bathrooms, kitchens, recreation etc. These ships have turbine engines.

A Missile Launching Cruiser

A ship floats in water because the weight of the ship is less than the weight of the water displaced by it. For stabilising the motion of the ship, stabilisers are used. Ships may be several hundred meters long and can carry 100 millions of tons of goods. They can carry upto 2500 passengers.

Today, we have different kinds of ships for different purposes, such as, bulk carriers, oil tankers, container ships, dry cargo ships, lighters, oceanographic research ships, fish processing ships, war ships, petrol ships, frigates, destroyers, passenger ships, aircraft carriers etc.

Sonars, earthquake-detecting and ice-breaking equipment are fitted in the modern ships. The ice-breaking Russian ship, Sibir, is driven by nuclear power. It is of 75,000-horse power and has the capability of breaking 4-metre thick ice.

Some super tankers are as big as 450 metres in length and they can carry 5,00,000 tonnes of crude oil at a time. In 1960, a bathescaphe was constructed. It was named Troste. It could go 10,917 metres deep into the sea.

The American-built aircraft-carrying ship, Enterprise is 335 metres long. Eight nuclear engines are fitted into this ship.

A submarine driven through nuclear power can carry 16 directional missiles, each fitted with 10 war-heads.

The ship named Elvin was used to take out the remains of the Titanic ship. It went 3,800 metres deep into the sea.

A Submarine

HOVERCRAFT

A hovercraft or aircushion vehicle is such an amazing means of transport that hovers on a cushion of air. It can travel over any fairly flat surface. In this vehicle, a fan sends air into the pipes fitted at its bottom. This air comes out from the holes and makes a cushion of air. A hovercraft can move on land, water, marshy places, snow etc.

The credit of developing the hovercraft goes to a British engineer Sir Christopher Cockerell. The idea of developing a hovercraft came in his mind in 1954. He got his idea patented on December 12, 1955. His idea was that a lot of power is wasted in a ship, which can be saved by making it travel on the surface of water on a cushion of air. It will not only save energy, but larger speeds can also be attained in less power. For this purpose he carried out many experiments.

In a hovercraft, a fan feeds air to the underside of the craft and the openings at the sides provide for the escape of air currents. The air is kept inside the cushion by a plenum. This allows the vehicle to hover two metres above the water or snow or land.

When the first trial of a hovercraft was carried on May 30, 1959 at Cowes in England, it impressed people so much that its news spread far and wide. When a hovercraft of 4 ton weight travelling over the surface of water came to the bank and from there on to the ground, people were very astonished to see it.

If a big-sized hovercraft is made and allowed to travel at larger speeds on the waves of water, the passenger does not feel the pitch and the rolling motion. A hovercraft can carry passengers and freight between ports. A hovercraft can travel with a speed of 150km/hr.

After the invention of a hovercraft by Cockerell, many modifications have been made in it. The first public service of the hovercraft began in July 1962 by V.A. 3, with a capacity for 24 passengers. It was able to travel with a speed of 60 knots (111 km/hr.). This service was across Dee Estuary. The biggest hovercraft of the world was made in Britain which weighs 305 tons and is 56.69 metres long. Its name is SRN 4 MK III. It can accommodate 418 passengers and 60 cars. Its maximum speed can go upto 65 knots (120.5 km/hr). Presently, there are many hovercrafts of all sizes. These are known as ferries. Some of the hovercrafts are so big that even cars, buses, trucks etc. can also go straight into them.

A Hovercraft

MOTOR CAR

A motor car is a petrol- or diesel-driven four-wheeled passenger vehicle. As a road transport, it is very popular among people. Its development is a result of continuous efforts of the scientists.

In 1769, Nicolas-Joseph Cugnot developed two steam-driven tractors. The first one of these two was the first passenger vehicle. Its speed was 3.6 km/hr. The first steam-driven motor car in Britain was developed by Richard Trevithick in 1801. The internal combustion engine was invented in 1876 by the German engineer, August Nikolaus Otto. The petrol-driven car

The first successful steam-driven motor car was made by Richard Trevithick

Karl Beng (1844-1929)

Karl Beng made the first successful car with internal combustion engine

The use of petrol in passenger vehicles was started in about 1885

The model T Ford 'Tin Lizzie' (1908-1927)

The Lanchester, 1895

with internal combustion engine was called a 'Motor Wagon'. It was developed by Karl Friedrich Beng of Karlsruhe in 1885. This three-wheeled vehicle weighed 250 kg and its speed was 13 to 16 km/hr. Its single cylinder four-stroke engine produced 0.85 horsepower by 200 rotations per minute. In 1885, he made two more vehicles, one of them is kept still in working condition at the Deutsches Museum, Munich.

Karl Beng, a German engineer, made a three-wheeled car in his small workshop by using the internal combustion engine of Otto. It had a four-stroke petrol-driven engine. Beng developed electric ignition system for this car. He also used water to keep the car cool. It had a seat for the driver and the passenger. A small wheel was fitted on a stick as a steering wheel to change the direction of the car. This car, developed by Karl Beng, was exhibited in an exhibition in Paris in 1887, but nobody gave any attention to this car. Beng started driving his car in Munich. People became astonished seeing his car and he started getting orders for motor cars. In fact, Karl Beng marked the beginning of the modern car industry.

The motor car industry flourished very fast. Henry Ford of USA manufactured his first car in 1896 in Detroit. This car had a four horsepower engine with two cylinders. Henry Ford had the capability of improving the already invented cars. He could understand the shortcomings of the European cars. He

The Rolls Royce, 1905

The Mercedes Sports Car, 1928

developed a low cost, strong and stable car. His model T Ford or Tin Lizzie became so popular all over the world that he became famous and very rich. He opened a new car factory in Detroit, which was 300-metre long. He made assembly lines in his factory so that several cars could be made at a time. From 1908 to 1927, about 18 million Tin Lizzie cars were sold. After this, Ford made several beautiful models of cars. Even today, the Ford cars are too expensive.

The Union C Car, 1936

Today, several kinds of cars are being made in different factories all over the world. C.S. Rolls and Henry Royce made costly cars. After this, a four-seater baby car named Austen Saven was made, which was liked by the people.

A modern car has an engine, transmission, drivetrain, steering, brake, suspension, chasis etc. as its parts. All cars have either petrol or diesel engines. When a car is started, a spark of fire from the spark plug burns the fuel. The hot gases produced by the fuel make the car move. Today, scientists have developed computer-controlled and noise-controlled cars. Cars have become an essential part of our lives. Every year, car production goes up by 15%. The motor car, in a short span of time, has become a very successful transport system for us. Presently, many big cities of India are flooded with cars. Cars fitted with A.C. are favourite amongst many people. Jeeps are available for comfortable driving on uneven roads. Sports cars and racing cars are also available in the automobile markets. Computers have played a very vital role in the modern car technology.

The Penhard, 1892

The Duria Racing Model, 1895

MOTOR CYCLE

A motor cycle is a two-wheeled petrol-driven means of transport. The first motor cycle with an internal combustion engine was made by Gottlieb Daimler of Germany in 1885. Its frame was made of wood, and its speed was 19 km/hr. In 1898, two motor cycles were made in Britain. These were named as Holden Flat-Fore and Clide Singal.

Motor cycle industry developed only in the beginning of the present century. The first large scale manufacturing factory of motor cycles was established in West Germany at Munich by Heinrich, Wilhelm Hildebrend and Alois Wolfmuller. In this factory, more than 1,000 motor cycles were made in the first two years. All these motor cycles had four-stroke, two-cylinder, water-cooled, 1488 cc engines. They were capable of producing 2.5 brake horsepower (bhp) by 600 rotations per minute. Mopeds are cheaper and are capable of moving up on elevated roads. Their engines are of the capacity of 50 c.c. or less and have a speed limit of 50 km/hr. The engine of the scooter is also small in size.

A modern motor cycle is fitted with a two or four-stroke internal combustion engine. Some motor cycles also have rotary wenkel-type engines. These engines are either air-cooled or water-cooled. In an internal combustion engine, fuel mixed with air is burnt by the spark produced by a spark plug. The gases produced by the burning of the fuel push the piston in the cylinder. The piston is connected to the wheel. This makes the motor cycle move. The frames of modern motor cycles are made of steel.

Today, several kinds of motor cycles are available in the market. Motor cycles for general use can attain a speed upto 248 km/hr while those meant for racing can attain a speed of more than 300 km/hr. Their engine's capacity is 125 c.c. The engine capacity of big cars can be upto 1,000 c.c.

A motor cycle

AIRCRAFT

Man has been fascinated from ancient times by seeing the birds flying in the sky. To give a practical shape to this fascination, Leonardo Da Vinci made the design of a flying machine and parachute. In the 18th century, two brothers Joseph Michel and Jacques Etienne Montgolfier, made man fly in air with a hot air balloon.

They made a silk balloon and suspended a cage from it. In this, they sent a chicken, duck and sheep first. Hot air was supplied from below

The first balloon in which two men took a flight, was made by two brothers, Joseph Michel and Jacques Etienne

Wilbur Wright and Orville Wright

by straw fires and people watched it rise in the air. It stayed in the air for about 8 minutes.

Finally on November 21, 1783, two young Frenchmen, Pilatre de Rozier and Marquis d' Arlandes rose in a Montgolfier balloon and made a free flight for 25 minutes over Paris. They were the first human beings to fly.

The practice of inflating balloons with hot air soon came to an end and people started using hydrogen gas in balloon. After balloons, Wilbur Wright and Orville Wright of USA marked the beginning of gliders. These brothers were bicycle makers in Dayton, Ohio. In their gliders, the rear edges of the wings were made flexible and were worked by ropes in order to control the lateral or side-by-side position. The gliders of the Wright brothers proved to be very successful.

After several years of hard work, the Wright brothers succeeded in 1903 in making a powered plane, fitted with a petrol engine of 12 horsepower under the right side wing. On the left side, there was a seat for the pilot. The engine was connected with two propellers fitted at the rear of the aeroplane.

On December 17, 1903 at 10.35 a.m. at Kitty Hawk, North Carolina, Oriville got into his plane named Flyer. The motor was started and warmed

241

up while his assistants held the plane in position. The signal was given and the craft was released. The truck slid down the rail. The plane rose in the air, leaving the truck behind. It flew for a distance of 36.6 metres, remaining aloft for 12 seconds with a speed of 48 km/hr and then landed on its skids. It was the first powered flight in world history. This was a two-winged plane. This flight was witnessed by Orville's brother,

Wright Brothers made a demonstration flight

242

four others and one child. This plane is still lying safe in National Air and Space Museum at Washington D.C.

After this success, Orville Wright made several improvements in the plane and had a demonstration flight in 1908 in France. After a year of the first successful flight in 1903, Wilbur Wright died. Orville saw the development stages of aeroplanes. He lived upto 1948. By that time, aeroplanes capable of flying with a speed of 100 km/hr were already in use.

In a short span of time, many countries developed the technology of making airplanes. In 1909, Louis Bleriot of France crossed the English Channel. Two Britishers, Alcock and Brown, made a flight across the Atlantic Ocean

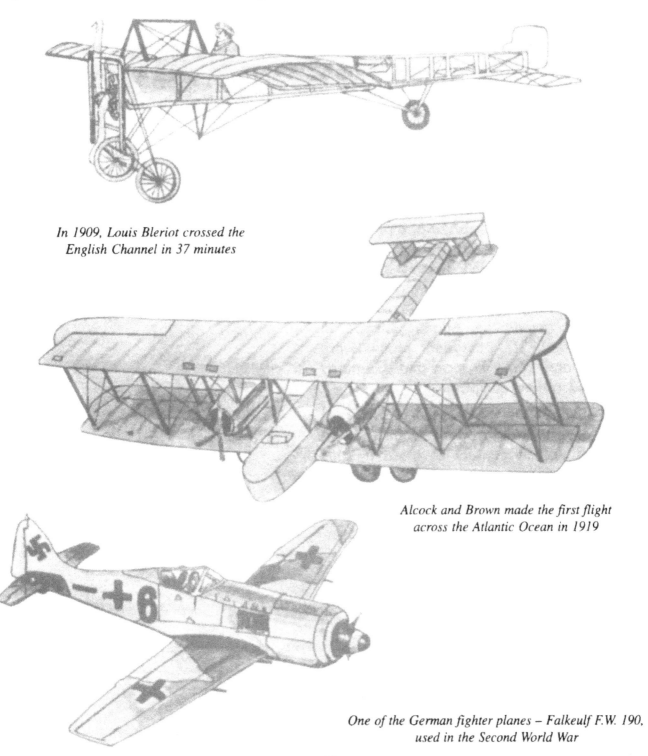

In 1909, Louis Bleriot crossed the English Channel in 37 minutes

Alcock and Brown made the first flight across the Atlantic Ocean in 1919

One of the German fighter planes – Falkeulf F.W. 190, used in the Second World War

243

The Boeing Jumbo Jet (USA, 1969)

in 1919. In 1920, passenger planes were made and USA, France and Germany opened airline companies.

Today, we have several types of planes. The most modern planes are jet planes, which can fly at great heights with speeds faster than the speed of sound. These planes mainly have three parts–body, wings and rear part assembly. The body has a cockpit for the pilot and seating arrangement for the passengers. At the tail, there is a rudder. The rear part gives the stability to the plane. During flight, a plane is under the action of four forces, namely, weight, lift, thrust, and drag. These four forces keep the plane in the air and make it fly.

Today, aircrafts having the speed more than that of sound have also been manufactured. There are also such aircrafts that can take off straight and even land in the same way, thus not requiring a runway for taking off and landing on the ground. The shape of their wings can also be changed.

244

JET PLANE

A jet plane works on the principle of Newton's third law of motion. This law states that for every action there is an equal and opposite reaction. In a jet engine, combustion of fuel is an action and exhaust of the gases from the nozzle is the reaction. The gases move at high speed and, as a result, the engine moves in the opposite direction of the exhaust gases. Jet engines are 80% more efficient as compared to piston engines.

A jet engine, while flying, leaves a white line of smoke. A jet plane makes use of jet propulsion. A liquid or gas when it comes out of a small hole is called a jet. When a jet comes out with a great speed, its reaction makes the thing move in the opposite direction of the jet. The jet engine of airplanes has a gas turbine in the rear portion and a compressor fan in the front. A jet engine uses oxygen in the air around it.

The idea of employing jet engines in airplanes was proposed for the first time by Captain Marcone of France, Henry Coanda of Rumania and Maxime Gillaume in 1921. In 1930, Frank Whittle got a patent of a jet-propelled airplane. The first experimental flight of a jet-propelled airplane was proposed by Frank Whittle, and constructed by the British Power Jets Ltd. which took place on April 12, 1937. The first flight of a turbojet-propelled airplane was materialised by Heinkel at Marien He in Germany on August 27, 1939. After this, different types of gas turbine-propelled engines came into existence.

A jet engine can fly at an altitude of 3,05,000 metres. Their speeds go higher than 3,500 km/hr. Jet planes have added a new chapter in aeroplane technology and have made the world smaller in size. Concord is a world-famous jet plane. Lockheed jet plane had broken the world record in 1976 by flying at a whopping speed of 3,529.56 km/hr. Columbia flew at a magnificent speed of 26,715 km/hr. Boeing 707 is also a world-famous aeroplane.

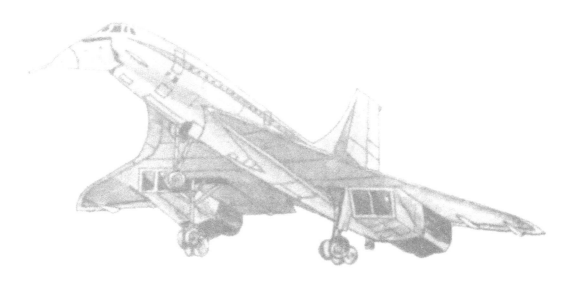

The Anglo-French Supersonic Cancord (1969)

HELICOPTER

A helicopter is an aircraft which can fly straight up or down in air without any runway. It can hover in one position like a humming bird. A helicopter can fly in any direction. It can fly upwards or downwards and even forward or backward.

Helicopter is a Greek word made up of 'heli' and 'copter', which means screw and fan. It was known to the Chinese about 500 BC as a flying toy. In about 1500, Leonardo Da Vinci made a design of a helicopter, based on Archimedes nut and screw principle. During the 19th century, people tried to develop helicopters but only in 1936 Focke-Wulf Company of Germany succeeded in developing a helicopter. This first helicopter in 1937 could fly at a height of 11,000 ft. reaching a speed of 70 miles per hour. After this Igor Sikorsky, a Russian engineer, made his first helicopter for the United States Army, which was named as XR-4. It was successfully tested in 1941. Since then many designs of helicopters have appeared in the world.

Nowadays, helicopters are used to evacuate sick persons from the sea, flood victims, and delivering food, medicines and other supplies to the army and stranded communities. These are also used to spray insecticides on crops and for survey. Every off-shore oil rig has its helicopter platform. They check remote power lines, monitor forest fires, hunt submarines and find a shipping path through ice-field. Helicopters with small tanks are used by army also. The uses of helicopters are multiplying day by day. The helicopters used in wars are of modern technology.

A Helicopter

SPACE SHUTTLE

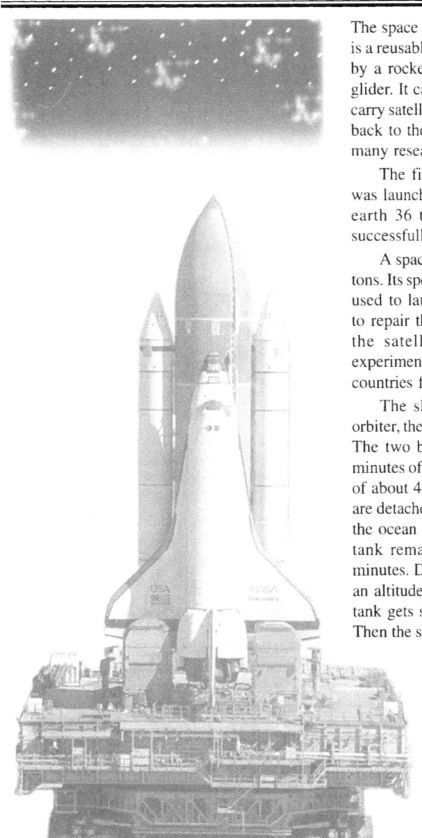

The space shuttle, developed by U.S. scientists, is a reusable spacecraft. It is launched into space by a rocket but it returns to earth like a huge glider. It can be reused for space travels. It can carry satellites into orbit and even bring satellites back to the earth for repair. It is being used in many researches in space.

The first space shuttle, named Columbia, was launched on April 12, 1981. It orbited the earth 36 times and came back to the earth successfully on April 14, 1981.

A space shuttle can carry a load of about 21 tons. Its speed is about 28,000 km/hr. It is mainly used to launch satellites and space telescopes, to repair the defective satellites, to bring back the satellites and to carry out scientific experiments in space. It can be hired by other countries for scientific missions.

The shuttle has mainly three parts—the orbiter, the external tank and the booster rockets. The two booster rockets give the shuttle two minutes of sturdy thrust and take it to an altitude of about 43 kms. After this the booster rockets are detached from the shuttle and parachute into the ocean for recovery and reuse. The external tank remains with the shuttle for about ten minutes. During this period the shuttle acquires an altitude of 200 kms. After this, the external tank gets separated and falls back to the earth. Then the shuttle is taken into its orbit by its two

The launching of space shuttle

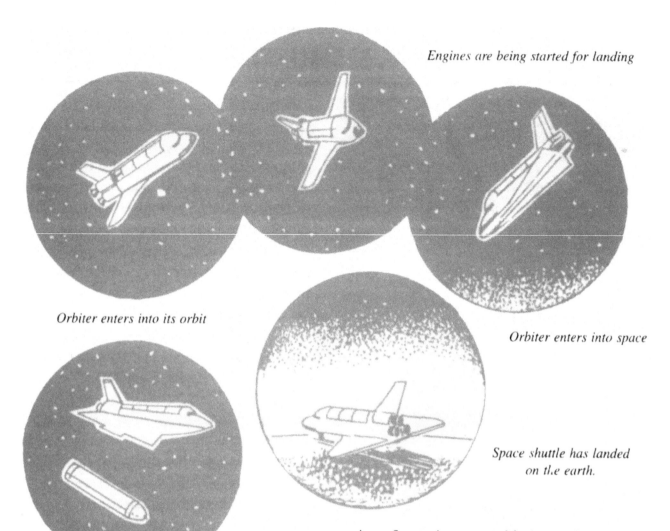

Engines are being started for landing

Orbiter enters into its orbit

Orbiter enters into space

Space shuttle has landed on the earth.

External tank being separated out

Booster rockets being separated out

engines. It continues to orbit the earth as long as it is desired. The engines are used again at the end of the mission to bring the orbiter back into the atmosphere. After re-entry, the orbiter glides down through the atmosphere and lands like an aircraft on a long run-way. Apart from the large fuel tank, all other parts of a space shuttle can be reused.

In addition to Columbia, the Discovery, Challenger and Atlantis space shuttles have made several flights. On January 26, 1986, the Challenger exploded just after the take off, killing all the crew members on board. This has been one of the biggest setbacks in the history of the space shuttle. Another space shuttle named Columbia exploded in the year 2003 (1st February, 2003) shortly before landing on the earth. Kalpana Chawla of the Indian origin and all other members of the crew died in this fatal accident. ■■

248

SOUND

WHAT IS SOUND?

Sound is a kind of energy which is detected by the ears of living beings. Sound travels in the form of waves through the material medium and reaches our ears. These waves make our eardrums vibrate. From the inner ear, nerve messages travel to the brain and we hear the sound. Sound cannot travel without a material medium. It is of great importance to us.

Sound is produced only by the vibrating bodies but our ears cannot detect all the vibrations. Human ears are sensitive to the vibrations ranging in frequencies from 20 Hz to 20,000 Hz. This range of frequencies is called the audible range. Beyond this range of vibrations, our ears are not sensitive. However,

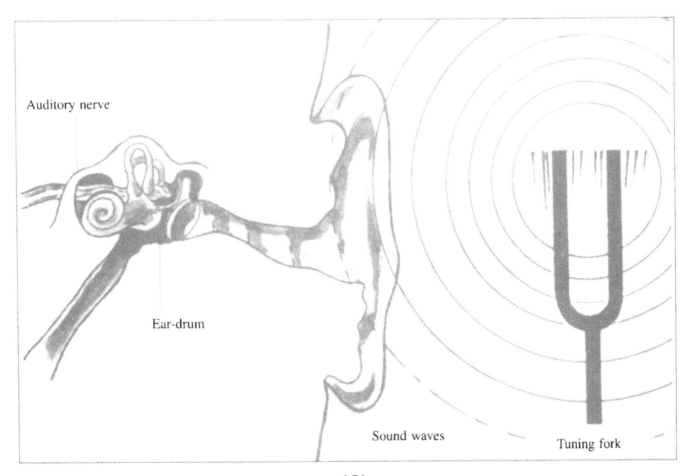

Auditory nerve

Ear-drum

Sound waves

Tuning fork

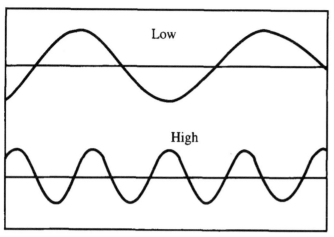

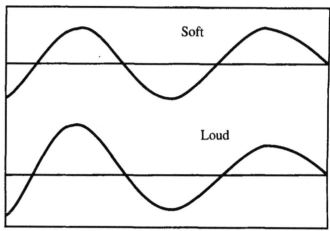

some animals like dogs can hear the frequencies even beyond this range. The word, **sound** is used only to those vibrations for which our ears are sensitive.

The range of audible frequencies differs slightly from person to person but the common range lies between 20 Hz to 20,000 Hz.

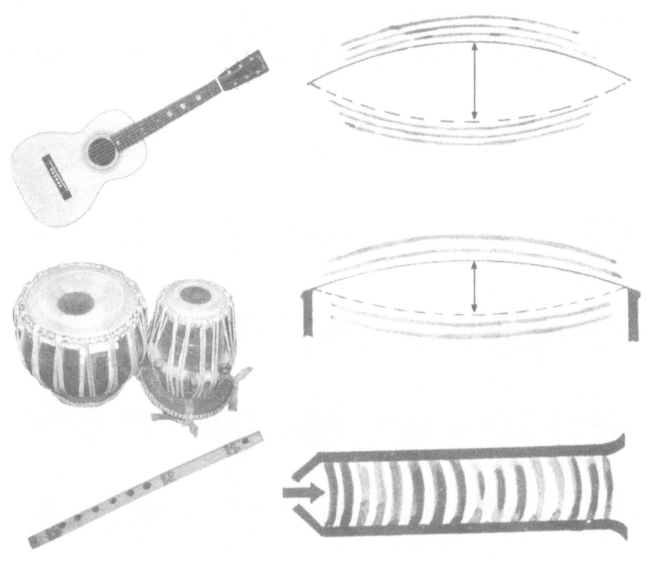

Various musical instruments produce different sound through vibrations

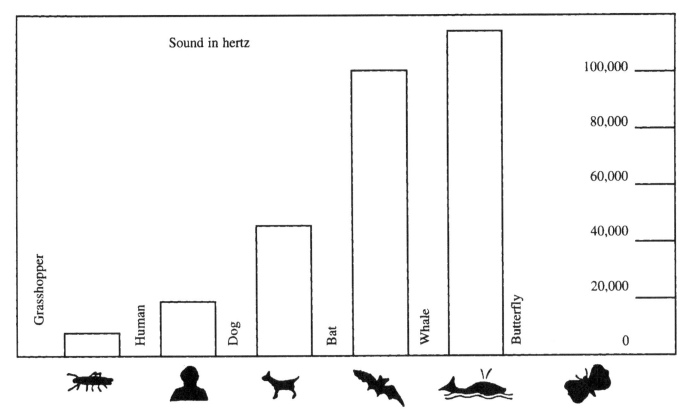

Most of the living beings produce different types of sound and can even hear them. Even though butterfly cannot produce any sound, it has the capacity to hear the sound of more Than 100000 Hertz frequency. The ears of Bat and Whales are very sensitive. Compared to other creatures, the hearing capacity of human being is less

More is the frequency of vibrations, the sharper is the sound. High frequency voices are called **high pitched sounds** while low frequency sounds are called **hoarse sounds**. Our ears are more sensitive to the frequencies from 1000 to 4000 Hz. We can distinguish some 400,000 kinds of different sounds.

The sounds having frequencies below 20 Hz are called **infrasonic** while those having frequencies above 20,000 Hz are called **ultrasonic**. Ultrasonic waves are very useful in scientific laboratories, diagnostic instruments and ultrasonic machines. These ultrasonic sounds can be heard by dogs, bats etc. Sound is produced through many sources. Infact, any vibratory thing can produce sound. Musical instruments also produce different kinds of sounds. The sound which is not pleasant to the ears is called noise.

■■

MECHANICAL WAVES

The wave concept in modern science is of upmost importance. A wave is a motion that carries energy from one place to another in a medium. A wave in water is probably the most familiar kind of wave but there are many other kinds of waves all around us. For example, voices travel to our ears in the form of longitudinal waves. Radio and television programmes reach our homes in the forms of electromagnetic waves. Light rays, X-rays and Gamma rays also travel as waves.

When a stone is thrown in the pond of water, the waves are created on the surface of water by the up and down movement of water molecules. The shape of the waves is usually like the shape of hills and valleys. The hills are called crests and the valleys are known as troughs. When a wave moves along a medium, there are two things to watch: The movement of the wave and the movement of the medium. The particles of the medium in which waves are produced vibrate to and fro about their mean position. The

When a stone is thrown on a still water pond, it creates waves all around it

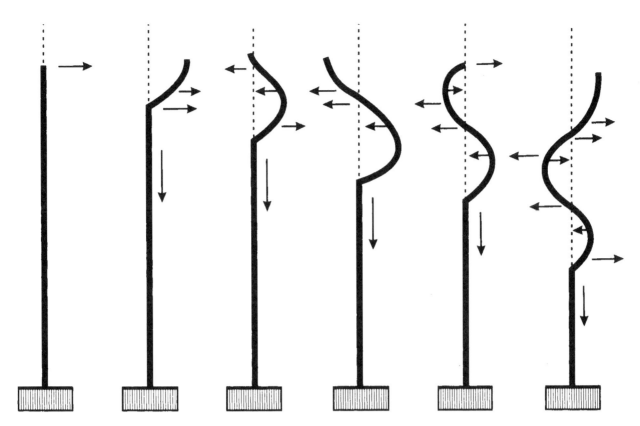

If one end of the rope is continuously shaken, you can find the crest moving towards the other end of the rope. The continuous shake will create innumerable waves. The distance between one crest to another crest is known as wave length

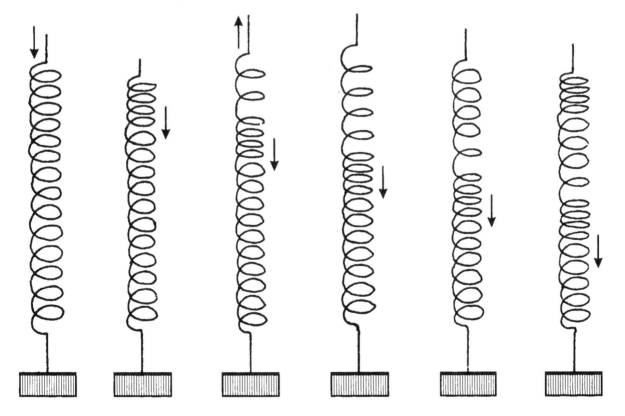

If one end of the spring is tied to a wall and the other end is pulled to and fro by your hand, you can hear innumerable vibration emittions. This type of vibration shows longitudinal waves

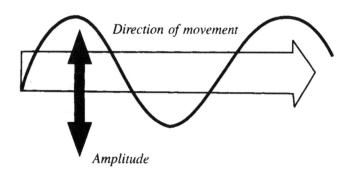

Direction of movement

Amplitude

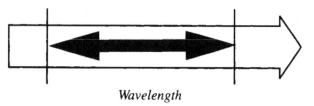

Wavelength

There are two distinctive characters of waves—amplitude and wavelength. The area covered by a unit (the amount of displacement) during its vibration is known as amplitude and the distance between two crests is known as wavelength

particles of the medium only vibrate, they do not travel with the wave. Wave in fact is a result of periodic motion of the particles.

For mechanical waves to travel, the medium should have the property of elasticity. This property makes the particles to come back to the original position. Medium should have the inertia but least force of friction. These properties are possessed by all materials such as air, water, steel etc.

Some kinds of waves do not require materials medium to travel. These are called electromagnetic waves. Light waves are electromagnetic waves and can travel even in space.

Mechanical waves are of two types namely transverse waves and longitudinal waves.

Transverse waves are those in which the particles of the medium vibrate to and fro about their mean position perpendicular to the direction of wave propagation. The waves generated on the surface of water or in strings are transverse in nature. Longitudinal waves are those in which the particles of the medium vibrate in the direction of wave propagation. These waves are created as a result of compressions and rarefactions. Sound waves are longitudinal in nature. Waves generated in the air are longitudinal in nature. Longitudinal waves can travel in solids, liquids and gases.

Some definitions

Frequency: Number of vibrations per second is called the frequency of the vibrating body. Frequency is expressed in cycles per second or Hertz. It is expressed as n.

Time Period: The time taken by a body to complete one vibration is called its time period. It is measured in seconds and expressed as T. The relation between time period and frequency is given by $T-1/n$.

Amplitude: The amplitude of a wave is the amount it rises or falls from its usual position.

Phase: Phase is defined as the instantaneous position and direction of motion of the wave.

Wavelength: It is defined as the distance between the two nearest points in the same phase or the distance between the two consecutive crests or troughs, or compressions or rarefactions. It is expressed by λ.

Speed of the wave: The distance travelled by the wave in one second is called its speed. If *via* is the speed of a wave having wavelength λ and frequency n then $v=n\lambda$.

Generally, waves travel in air with the speed of 340 metres per second. Generally, waves are measured in Angstrom units (A) or (AO). Actually, One Angstrom unit = 0.1 nanometres which is approximately one-third the size of a carbon atom. In fact, one Angstrom is equal to the diameter of a hydrogen atom, the smallest element on earth. However, the bigger waves are measured in metres, centimetres and millimetres.

Pitch: Sounds may be generally characterised by pitch, loudness and quality. The

256

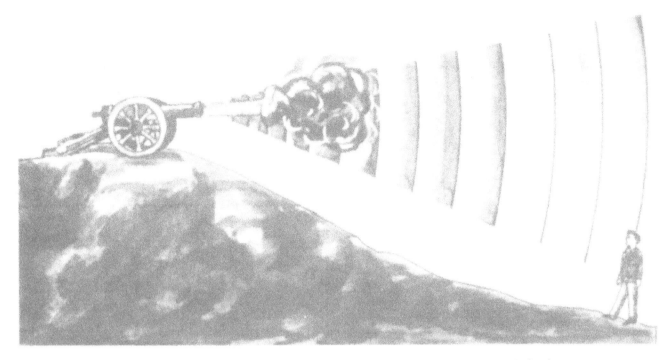

The speed of sound when compared to light is lesser. When a cannon is fired,
you see the light first and the sound later

perceived pitch of a sound is just the ear's response to frequency, i.e. for most practical purposes, the pitch is just the frequency. The pitch perception of the human ear is understood to operate basically by the place theory, with some sharpening mechanism necessary to explain the resolution of human pitch perception. The just noticeable difference in pitch is expressed in cents and the standard figure for the human ear is 5 cents.

SHOCK WAVES

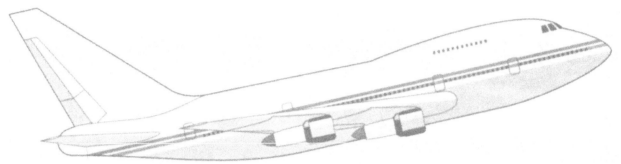

Sound barrier

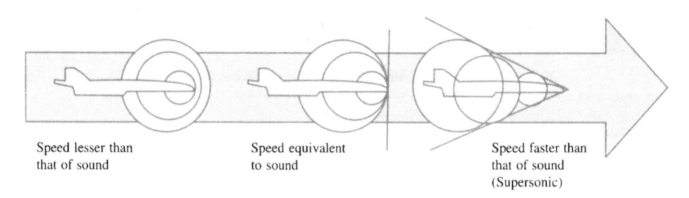

Speed lesser than
that of sound

Speed equivalent
to sound

Speed faster than
that of sound
(Supersonic)

Shock waves are created by supersonic planes

A shock wave is a wave of high pressure that forms in a gas or liquid as an object moves through it at or above the speed of sound or as the gas or liquid flows around the object. The shock waves are also called pressure waves and they move away from the object at the speed of sound. These waves contain high energy and are very dangerous. They are produced during cloud thunders, eruption of volcanoes and in lightening. Every year shock waves cause a big loss to many lives and property.

A shock wave is created in air when a supersonic airplane flies overhead at the speed of sound or faster. When the wave reaches the ground, a sonic boom is heard as the waves strikes the ears. A sonic boom can shatter the window panes.

Explosions also produce powerful shock waves which can damage constructions upto several kms. Shock waves can also be produced by the high speed of the bullet of a gun. When the speed of a motor boat in water becomes very high, shock waves are created on the surface of water. Earthquakes also cause shock waves. Explosions of atom and hydrogen bombs create very powerful shock waves which can cause deaths to large number of people. Shock waves are also produced due to the explosion of chemical bombs.

NOISE

A musical sound is the one which produces pleasing sensation on the ears while noise gives an unpleasant sensation to the ears. The increasing use of machines, automobiles, radio, television etc. has attracted the attention of man towards the physical and mental effects of noise. Scientists are studying the effects of noise, its measurement, and the ways to get rid of it.

The most important characteristic of noise is its loudness. The loudness of the two same intensity noises or same frequency noises may have different effect on the ears of two persons. The loudness felt by a person depends upon the sensitivity of the ears of the listener. The loudness of noise can be measured by several methods.

It may also increase blood pressure. Intense noises increase heartbeat and produce stomach ailments like ulcer. In fact, noise is proving very harmful to the health.

Loudness of noise is measured in Decibels -dB. Zero decibel sound cannot be heard by our ears. The level of normal talks is about 50dB. A crowded road has a noise level of 80dB. The noise of 140dB may permanently impair hearing. At the crossings of the busy roads, the sounds of the motor cars and vehicles are around 100 dB (decibles).

The different noise producing sources are car, scooter, bus, truck, train, aeroplane, machines in the factories, radio, television, loudspeaker, explosives, fireworks etc. The machines being used at construction sites in the big cities also produce noise.

The noise does not produce any physical harm but the long term exposure to noise may produce hard hearing. Noise of 1000Hz frequency with an intensity of 100dB may produce temporary hard hearing. The temporary dumbness of air force people is an example of this.

Noise also produce harmful effects on nervous system. According to Lord Horder it weakens the nervous system bywhich the

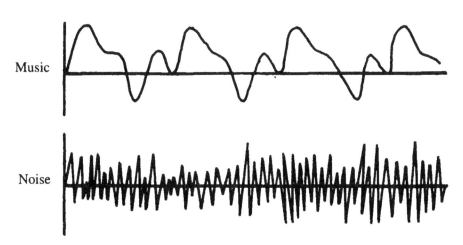

There is a difference between the sound waves of music and noise

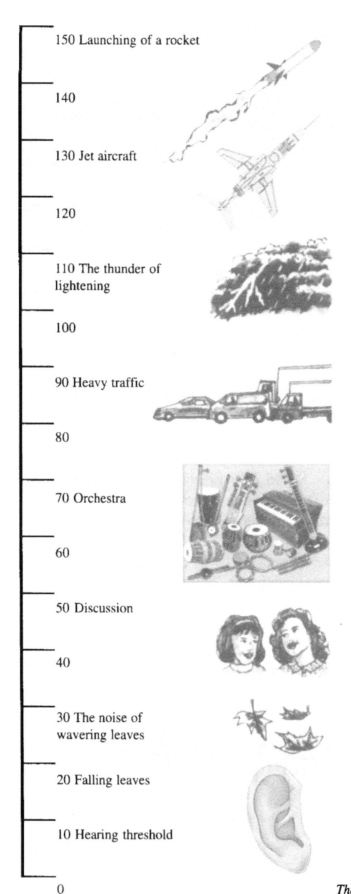

150 Launching of a rocket

140

130 Jet aircraft

120

110 The thunder of lightening

100

90 Heavy traffic

80

70 Orchestra

60

50 Discussion

40

30 The noise of wavering leaves

20 Falling leaves

10 Hearing threshold

0

resistance to diseases is reduced. Noise creates irritation, mental tension and anger.

The noise levels of some sound sources are given below:

Launching of a rocket	150dB
Jet airplane	130dB
Lightening	110dB
Heavy Traffic	90dB
Orchestra	70dB
Swinging leaves	20dB
Hearing threshold	10dB

A peaceful environment is good for health whereas a noisy atmosphere is harmful for the heart, stomach and increases the blood pressure giving rise to various ailments in human beings.

■■

The noise level of some sound sources (in decibels)

REFLECTION OF SOUND

Sound gets reflected from hills, buildings etc. just as light get reflected from polished surfaces. The sound waves reflected from the sides of a forest or mountain or a high building come to our ears as echo. When the voices uttered by a person are heard two-three times after the reflection from some object, they are called echoes.

If the area of the reflecting surface is large, the words spoken are heard as clear echo because the words become smaller and of larger frequency.

The effect of any voice, claping or bullet firing sound persists on car drum for a period of about 1/10th of a second. Any sound which comes as reflection to ears after 1/10th of a second travels a distance of about 34 meters. Therefore to hear echo, the barrier reflecting the sound should be at least at a distance of 17 meters.

The words spoken near a hill resounds and are heard as echo

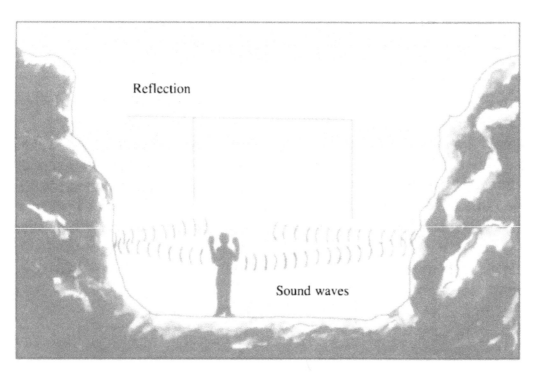

When a sound is made in between two huge blocks (hills), the sound gets repetitively reflected between these hills

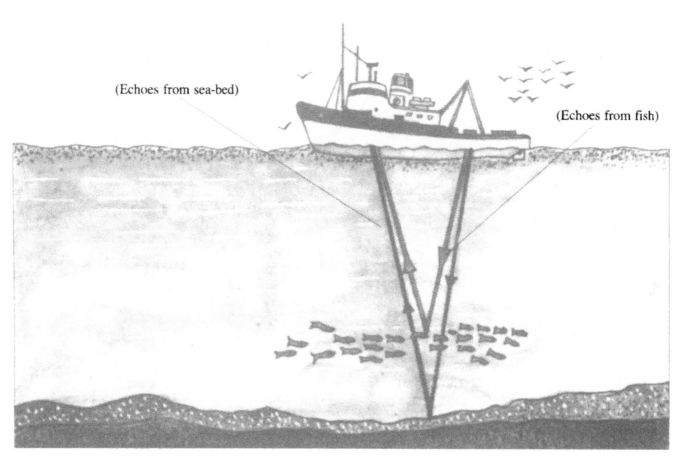

The reflection of sound is tapped for locating the shoal of fish and for measuring the depth of the sea/ocean

Multiple echoes are heard when the sound waves get reflected several times between the two parallel hills or high buildings. Such multiple echoes produce thundering sounds as produced by the clouds.

Principle of echo is used in determining the depth of ocean bed, in radars and for locating the submarines in oceans.

If you speak in an open place it is very difficult to hear the sound because the sound energy gets scattered widely but in large buildings the walls behave as sound reflectors. To make the sound clearly audible concave or parabolic reflectors are mounted at the back of stage and speaker delivers the talk from the focus of the reflector. The reflected voice reaches to the audience clearly.

The whispering galleries also work on the principle of reflection of sound. Normally the walls of dome shaped buildings and churches are made in round shape and on the sides of these walls are the places for sitting. In these galleries the sound after several reflections reaches to the audience. That is why these are called whispering galleries. In these galleries if a person speaks near the wall, the sound after many reflections can reach another person who after putting his ears near the wall wishes to listen the voice.

The reflection of sound or echoes is used by fishermen to detect the shoals of fish, position of submarines and by geologists to detect the minerals under the ground. Sonar (Sound Navigation and Ranging) also work on the

Multiple echoes produce thundering sounds as produced by the clouds

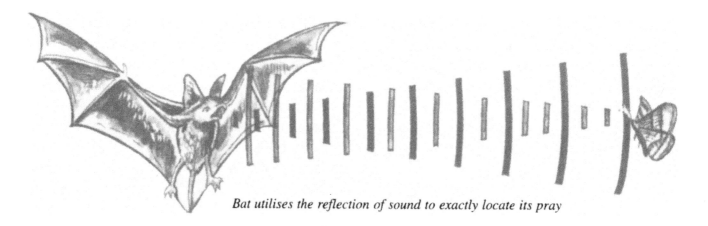

Bat utilises the reflection of sound to exactly locate its pray

principle of reflection of sound waves. In this instrument sound waves from a transmitter at the ship are transmitted into the sea water. These waves get reflected from the ocean bed. The reflected waves are picked by the receiver of sonar and the time taken from the transmitter to the bed and back to the receiver is measured. By using the speed of sound in the sea water and time, the depth of the bottom of the sea is measured.

Bats, Whales, Porpoises etc. make use of echoes to communicate with each other and can locate the position of their pray. These animals transmit waves of certain frequencies in all directions and receive the reflected waves and locate the pray on the basis of the direction of these waves.

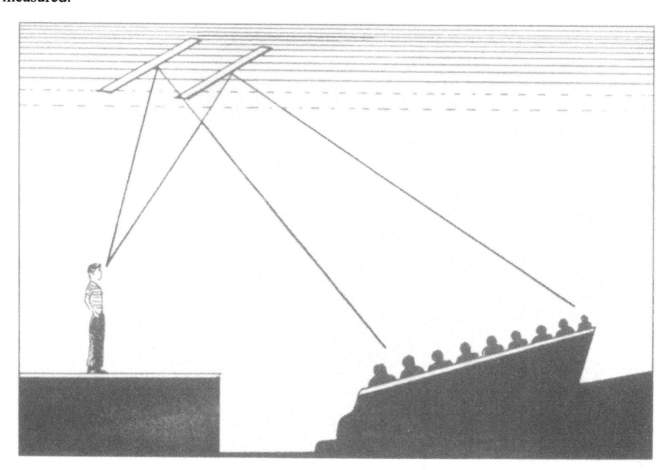

Whispering galleries have been made/built on the theory of sound reflection

DOPPLER'S EFFECT

The Doppler's effect is the apparent change in frequency of sound, light or radio waves caused by the relative motion of the source, medium or observer. For example, when an approaching train blows its whistle, the pitch of the whistle seems higher as the train comes toward you. The pitch seems to lower when the train passes and goes away from you. The cause of this effect was first studied in 1842 by Christian Doppler, an Austrian physicist and today known as Doppler's effect.

According to Doppler's effect whenever sound source moves towards or away from the observer, the pitch of the sound as heard by the observer is slightly different than the one emitted by the source. When the source of sound is approaching, each wave sent out by the source has a shorter distance to travel than the wave that was sent out earlier. Each wave reaches the listener a little sooner than it would have, if the source had not been moving. The waves seem to be more closely spaced. They have a higher frequency. As the source moves away each wave starts a little farther away. Each wave seem to be longer than it would ordinarily be. The pitch is lowered. When the source and observer are at rest or in motion in the same direction such that the speed of both is the same, the pitch of the sound does not change.

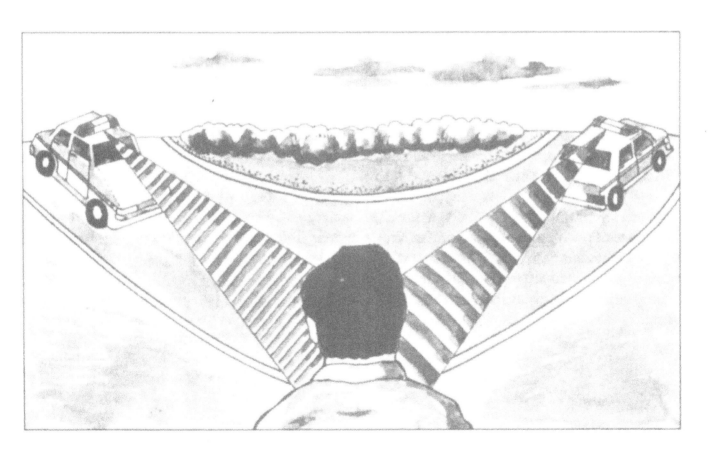

When we are standing on the road, the intensity of an approaching car's horn increases. Likewise the intensity decreases as the car moves away

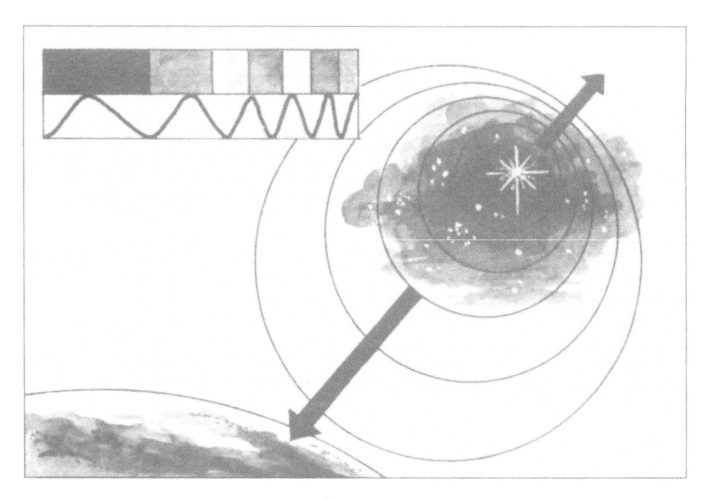

The speed of stars, galaxies and other bodies can be calculated on the basis of Doppler's effect

Doppler's effect is true in case of light waves also but since the velocity of light is very large as compared to sound, it can not be detected by simple methods. Astronomers study the motion of stars and galaxies by measuring the apparent change in frequencies of light due to motion. It has been observed by the scientists that galaxies are moving away from us. This has been concluded on the basis of **red shift** and on the basis of the **violet shift**, we can find out if any star is coming towards us in this vast galaxy of stars.

The rotation of the sun can be measured with the help of Doppler's effect. For this purpose the wavelength of light coming from the two edges of the sun is measured.

The speed of fighter planes and submarines is measured by using Doppler's effect in the battlefield and the Doppler's effect also helps to determine the position of the fighter planes and submarines, particularly, during a war. An instrument called the Laser Doppler Velocity meter has been developed with the help of Laser beams which is used basically to determine the velocity of fast moving objects.

INTERFERENCE

Hillock

The fog siren is not heard in a silent zone

When the two or more than two waves of same frequency and amplitude travelling in the same direction are superposed on each other the phenomenon of interference takes place.

The sound waves are longitudinal in nature, therefore compressions and rarefactions exist in the medium. When two compressions or two rarefactions in the same phase meet each other, they reinforce and the resultant wave has twice the amplitude of each wave producing maximum intensity. Similarly, when two compressions and rarefactions of opposite phase meet each other, the amplitude of the resultant wave is zero i.e. there exists a silence. This is known as the interference of sound waves.

In bad weather, when there is a fog, the blinking of light houses does not reach the ships clearly. In such a weather, fog sirens are used to guide the ships. The sound of fog sirens can be heard over long distances but in some areas silence zones exist. This is the result of interference of sound. The interference of sound waves takes place between the direct and reflected waves.

The interference is also produced between the light waves which produces bright and dark fringes. Light interference has been quite useful in measuring short distances.

BEATS

When two waves of slightly differing frequencies travelling in the same direction are superposed on each other the intensity of the resultant sound at any point alternately rises and falls many times in one second. This regular waxing and waning or rising and falling in the intensity of sound is called the phenomenon of

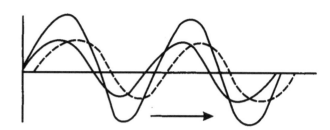

The displacement curve produced by two sound waves at one point but at different time

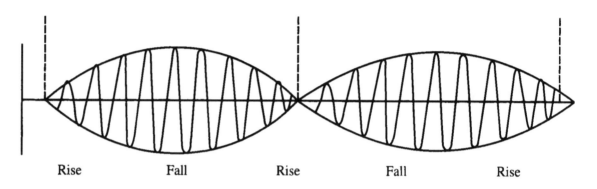

Rise   Fall   Rise   Fall   Rise

The waxing and waning of the intensity of sound is called Beats

beats. The production of beats is due to the phases of the two sets of waves being alternately the same and opposite. The number of beats is equal to the frequency difference of the two waves. In other words, **Beats** are the periodic and repeating fluctuations heard in the intensity of a sound when two sound waves of very similar frequencies interfere with one another.

By using the phenomenon of beats tuning of musical instruments is done. The two musical instruments are played at the same time. If they are not in tune with each other, beats are produced. In such a case the frequency of one instrument is changed such that no beat is produced. After achieving this condition, the two musical instruments are said to be tuned with each other.

The phenomenon of beats is also used in the heterodyne detection of radio reception. The incoming high frequency signals are mixed with some slightly differing frequencies at the receiving station. This gives rise to a pulsating frequency which lies in the audible range.

DIFFRACTION OF SOUND

Diffraction means the bending of waves round the corners. It is a matter of common experience that if some orchestra is being played at the back of a building it can be heard in front of the building but can not be seen. This is because of the fact that sound waves bend round the corners of the building and reach to our ears. In the similar way the sound of the common talks in a room, reaches to our ears through doors and windows by the phenomenon of diffraction. The bending of sound waves from holes, corners or obstacles is known as diffraction of sound.

The wavelength of the sound waves audible to us is less than a meter and in general the size of the obstacles is also less than a meter. For the diffraction of sound the size of the obstacles should be almost equal to the wavelength of sound.

As compared to the ordinary sound, the wavelength of the ultrasonic waves is much less, i.e., less than a centimetre and the size of the obstacles is quite big. That is why diffraction of ultrasonic waves is not easily observable.

As light is also a wave, it has the capability to bend around the corners. This is called as **diffraction of light**. Diffraction of light helps us to measure the diameter of very minute hole and extremely thin wires. When there is a need to separate light of different wavelengths with high resolution, then a diffraction grating is most ofter the best choice.

This **super-prism** aspect of the diffraction grating helps in measuring the atomic spectra in both telescopes and laboratory instruments. A prism can also be used for the above purpose.

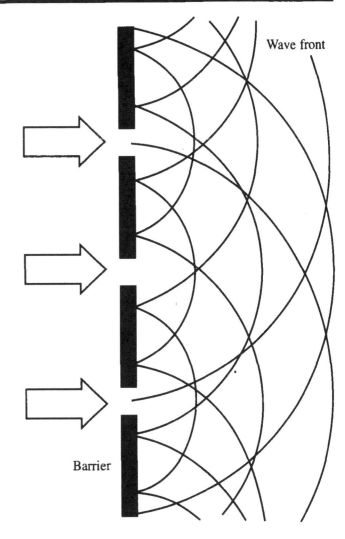

When there are obstacles or the sound waves have blocks on their way, they bend over and pass through the available space, this is known as diffraction

269

ULTRASONIC WAVES

The ultrasonic waves also known as ultrasound were first produced by Galton. The frequency of these waves is more than 20,000 Hz. In the beginning these waves were called as Galton's whistle. Now a days the ultrasonic waves of different frequencies can be produced by

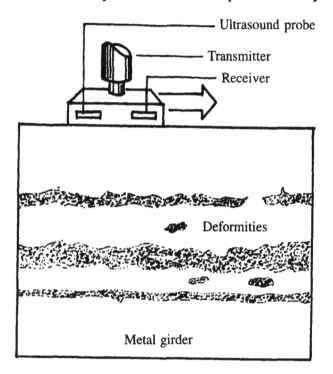

Metal girder

Screen

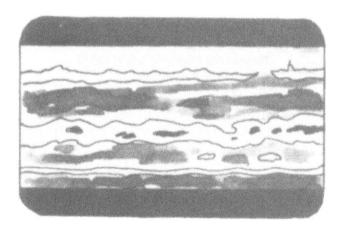

Ultrasound scanning helps in locating faults of thick covered objects

vibrating quartz crystals. The wavelength of ultrasonic waves in water is less than 1.6 cm.

The wavelength of the ultrasonic waves is quite small or the frequency of these waves is very high that is why the energy associated with them is very high. These waves can bring such miracles which ordinary sound waves can not do.

These waves are used to measure the depth of the ocean bed and to detect icebergs, underwater rocks, big fish and submarines. For this purpose signals of ultrasonic waves are transmitted into the ocean water which get reflected from these objects. The reflected waves are received and the time taken by the transmitted and the reflected waves is measured. Using the velocity of ultrasonic waves in water the distance of these objects is measured. This instrument is called the **Sonar** and it is of great use in the battlefield. These waves are also used to detect and locate the faults in thick metal sheets.

Ultrasonic waves are also being used to clear dust and coal particles on fogy days at airports. These waves are also used to clean the complex parts of the machines. They are also used to kill small insects.

Nowadays, ultrasonic waves are finding large scale applications in medicine and surgery. Tumors in the body can be located with the help of these waves without any operation. The sex of a child in the womb can be known by ultrasonic scan. These waves are also being used for the extraction of teeth. After three months of pregnancy, we can also detect whether the baby is a boy or girl with the help of ultrasonic scan or papularly known as **ultrasound**. Even various heart ailments, stones in the kidney or gall bladder, etc. can be detected from ultrasound. Hence, these ultrasonic waves are a boon for medical science. These waves are also

Ultrasound scanning helps in detecting the sex of the still born child (child in womb)

being used to diagnose the stomach ailments. They are not harmful to the body.

The valuable optical components are cleaned by ultrasonic waves. Ultrasonic machines are being used to drill holes in glass and ceramics. These waves, in fact are very useful for us.

BOUNDED MEDIUM

The well defined boundary of a medium is called bounded medium. Such medium vibrate only in definite frequencies. These frequencies are called the characteristic frequencies. The stretched wire of a sonometer or sitar, the air column of a flute, the stretched membrane of a Tabla etc. are the well known examples of bounded medium. When these musical instruments are played, they produce the tones of definite frequencies.

Bounded media are of two types namely free and rigid. For example; the open end of an organ pipe comes under the category of free boundary and the closed end comes under the category of rigid boundary. A wire attached between two fixed points constitute a rigid boundary. The particles of a free bounded medium are completely free to vibrate while the particles of the rigid bounded medium can not vibrate. The displacement at the rigid boundary is always zero while at the free boundary this is not the case. The study of bounded media has proved very useful in the development of musical instruments.

To produce any tune or musical sounds from a musical instrument, we require a bounded medium so that all the notes that are produced are in harmony with each other. In **electric guitar**, the message is transmitted from the amplifier to the loudspeaker and then the sound/note is produced which reach our ears. All the instruments that are used in **Rock Music** such as violin, saxophone, drums, trumpet, etc. are categorised and based upon the **Bounded medium principle.**

■■

Playing an instrument creates tones of definite frequencies

272

VIBRATIONS OF AIR COLUMNS

The melodious sounds produced by flute, whistle or clarinet etc. are generated as a result of vibrating air columns. When two longitudinal waves travel in opposite directions in an air column, they produce longitudinal stationary vibrations. The air column is called organ pipe. The organ pipes are classified as open pipes and closed pipes. Open pipe is open at both the ends while a closed pipe is closed at one end.

The diameter of the organ pipe is quite small as compared to its length. The walls of the pipe are rigid. The air column inside a pipe can vibrate

reflected from the closed and as a wave of compression and a wave of rarefaction is reflected from this end as a wave of rarefaction. When the wave of compression reaches the open end of the pipe where the layer of air is much more free than the layer inside the pipe, it is reflected as a wave of rarefaction. When the source of sound is such that the air column is set into resonance, nodes and antinodes are formed. These nodes and antinodes of stationary waves have fixed position within the tube. At the open end where the air has freedom of motion, an

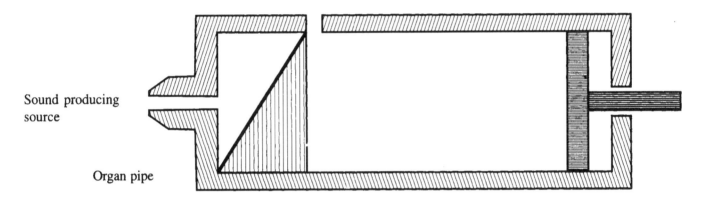

Sound producing source

Organ pipe

in a number of ways but there is always an antinode at the open end and a node at the closed end.

When air is blown across the mouth of a pipe, the enclosed column of air is set into resonant vibrations. The frequency of the note produced depends upon the length of the pipe. The closed end acts like a rigid wall and reflects back the wave which strikes it. Stationary waves are produced by the combined action of the direct and reflected waves. A wave compression is

antinode is formed. On the other hand, the layer of air in contact with the closed end can not move freely, hence a node is formed at this end. The different vibrations produce sweet and melodious sounds in the musical instruments based on the vibrations of air columns.

The study of air columns is an important branch of science because some of the vital musical instruments such as flute, trumpet, a clarionet, etc. are based on this principle.

■■

STATIONARY WAVES

When two waves of the same frequency and amplitude travelling in opposite directions are superposed on each other, the resulting wave so formed is called a stationary wave. The stationary wave does not appear to be moving i.e. it looks to be stationary. Stationary waves can be produced by the superposition of transverse or longitudinal waves. For example, the travelling longitudinal wave in an air column when gets reflected from one end and is superposed on the original wave, the stationary wave is formed in the air column. In the air columns of clarinet, flute, band etc. longitudinal stationary waves are formed. In the same way the transmitted and reflected transverse waves in stretched wires of sonometer, sitar, guitar, violin etc. produce transverse stationary waves.

In a stationary wave, the particles of the medium at different places vibrate in the same manner. For example particles of the medium at certain points remain stationary. These points are called nodes. On the other hand, the particles of the medium at some other points vibrate with maximum displacement. These points are called antinodes. The distance between the two adjacent nodes or antipodes is equal to the half of the wavelength.

The equation for stationary waves :

$$y = A \sin \frac{2\pi vt}{\lambda}$$

Where A = Amplitude.

Two similar transverse waves move equally in opposite direction in a bonded medium.

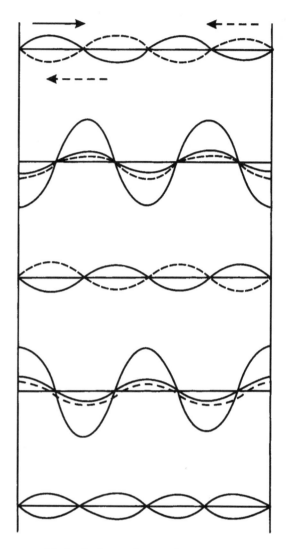

Method of creating stationary waves

■■

274

CHARACTERISTICS OF SOUND

Musical sounds differ from one another in three characteristics namely loudness, pitch and quality or timbre.

Loudness: It is the characteristic of all sounds whether musical or nonmusical. It refers to the magnitude of sensation produced on the ear. It is associated with the intensity of sounds which is measured by the energy of sound waves crossing per unit time through a unit area at right angles to the direction of waves. Loudness depends upon the amplitude of vibration, distance of the source, size of vibrating body, density and motion of the medium.

Pitch: The pitch of a note is that characteristic of a musical sound which enables us to distinguish between high and low, acute and grave, sharp and flat sounds.

The sensation of pitch depends on the frequency. The greater is the frequency the higher is the pitch. If the pitch is higher, the sound is said to be shrill. Pitch is almost independent of loudness and quality. A woman's voice is more melodious than a man's voice because of its high pitch.

(Low frequency)

(Low pitched sound)

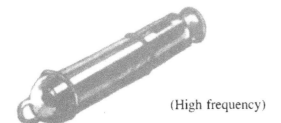

(High frequency)

(High pitched sound)

Pitch is related with frequency

275

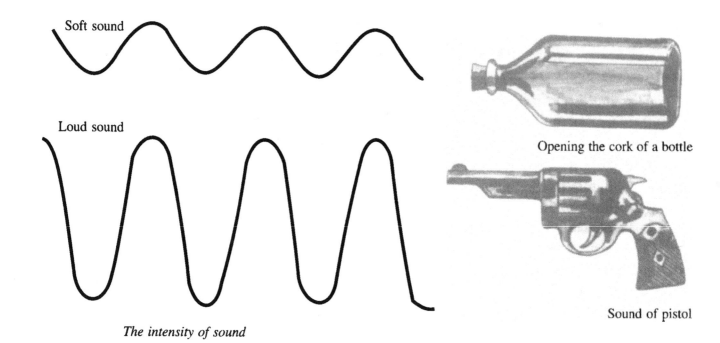

Soft sound

Loud sound

The intensity of sound

Opening the cork of a bottle

Sound of pistol

Quality or Timbre: The quality of the musical sound is that characteristic by which we can differentiate two notes of equal pitch and intensity emitted from two different musical instruments. We can recognize the sounds of musical instruments without looking at the instrument. It is because the quality of the note emitted by one instrument is different from that of the other.

This characteristic of sound depends upon the number of harmonics, their order and relative intensity of the note. If the number of harmonics in the note is more, the sound will be more melodious. This is the reason that sounds produced by the musical instruments with stretched strings are more sweet as compared to the musical instruments based on vibrating air columns. In stretched wire instruments all harmonics, first, second, third, fourth, fifth etc. along with the fundamental are present while in air column instruments only third, fifth, seventh harmonics are present.

Violin

Double bass

Viola

Cello

Some stringed instruments

RESONANCE

Resonance is the process of making objects vibrate at a certain frequency. When the period of the impressed force is equal to the natural period of the body, it immediately catches the vibration and begins to vibrate with a gradually increasing amplitude. The forced vibrations thus produced are called sympathetic vibrations or resonance.

When the frequency of the impressed force is equal to the natural frequency of the body, the impressed force acts in the direction of the vibration. In other words, the direction of the impressed force and the movement of the body are in the same phase. In such a case, the successive impulses of the impressed force continuously increase the kinetic energy of the vibrating body resulting in the increase of amplitude of vibration. The increase in amplitude is resisted by frictional forces and

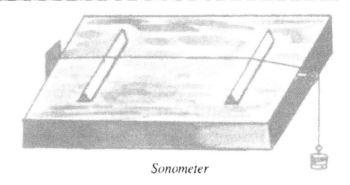

Sonometer

ultimately the body comes to an equilibrium position. At this position we say that the amplitude of vibration has reached its maximum value.

If the frequency of the impressed force is slightly different than the resonance frequency and reduces the amplitude of vibration drastically, the resonance is said to be sharp and on the other hand if the amplitude is reduced only a little, the resonance is said to be flat. The

Whispering sound can be heard when a conical shape object is held near one's ear

sharpness of the resonance depends upon the damping. Lesser is the damping, sharper will be the resonance and vice-versa. For example, the resonance for a sonometer wire is sharp because its damping is low while the resonance in a resonance tube is flat because the damping of air is high. This is the only reason that when we resonate the wire of a sonometer by a tuning fork, the amplitude falls drastically by slightly changing the length of the wire while in case of a resonance tube the amplitude does not change sharply over a long length and it becomes quite difficult to know the correct position of resonance.

Stringed Musical Instruments

In stringed instruments like sitar, violin etc. several wires are mounted by the side of the main wires. These wires remain tuned to different frequencies. When main wires are played, the side wires resonate and increase the intensity of the tone.

Sonometer: When a sounded tuning fork is placed above the box of a sonometer, the wire of the sonometer starts vibrating. When the length of the wire between the wooden bridges is adjusted to a particular length, the amplitude of vibration becomes maximum. Under this condition the natural frequency of the wire becomes equal to the frequency of the tuning fork. In other words, both are in resonance.

Vibrations of Medium: The medium surrounding us always vibrate with different frequencies and intensities. We hear only some vibrations and not others because their intensities are low. If we place a tumbler or some hollow thing near our ear and the air in this start resonating, we can even hear these low intensity vibrations due to the phenomenon of resonance. This is the only reason that by placing a hollow thing on the ear in the open we listen to certain sounds. For this reason the people having the defect of hard hearing put their palm at the back of the ear.

Vibrations: Some times you might have noticed that due to the sounds of radio, television, tape recorder etc. the tumblers or other utensils placed on tables start producing vibrating sounds. This happens because of the fact that their natural frequency coincides with the frequency of the music and due to resonance their amplitude of vibration becomes large and they start producing sounds.

Similarly you might have observed that at times while travelling in a bus, the bus starts producing vibrating sounds due to resonance when the frequency produced by the wheels of the bus becomes equal to the frequency of the bus. These are some examples of resonance which we observe in our daily life.

Due to certain special sound waves emitting from the television, glass, plates and other objects kept on the table start vibrating

10 Development of Chemistry

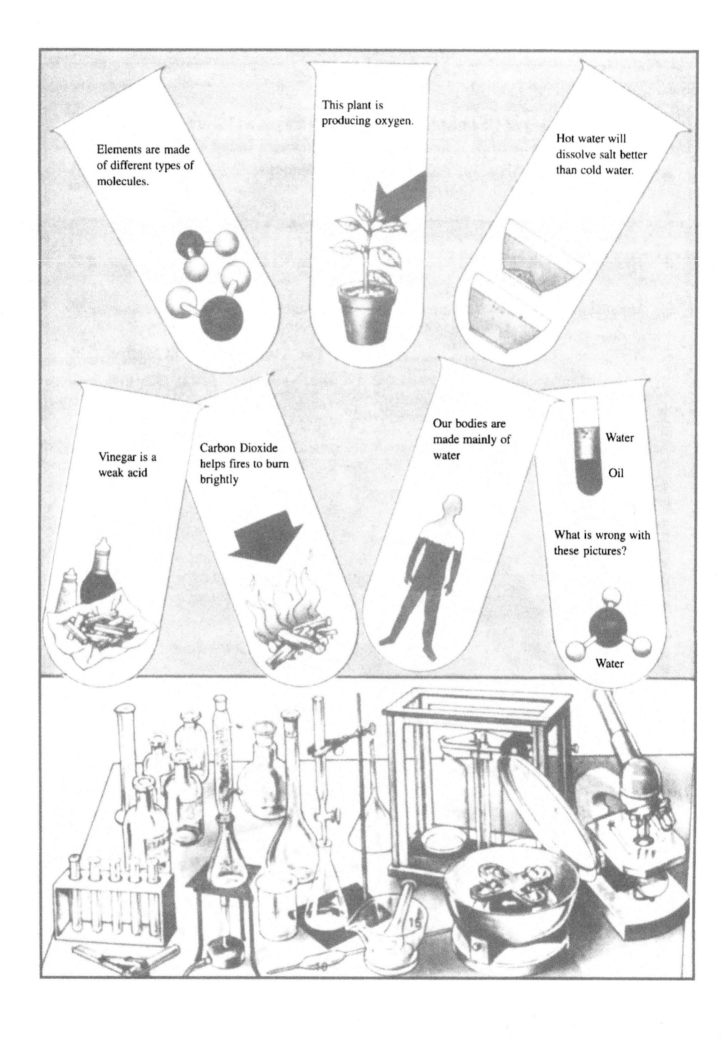

Elements are made of different types of molecules.

This plant is producing oxygen.

Hot water will dissolve salt better than cold water.

Vinegar is a weak acid

Carbon Dioxide helps fires to burn brightly

Our bodies are made mainly of water

Water

Oil

What is wrong with these pictures?

Water

THE FOREFATHERS OF CHEMISTRY

The prehistoric man lived in the wild hostile surroundings. He invented various objects to fulfil his needs. He also made weapons for self-defence as well as for hunting wild animals for food. The first weapons he made were of stone. Soon, he started making use of bones and horns. But it took him a long time to use metals in the manufacture of weapons and implements. The study of ancient civilizations shows that gold was probably the first metal which man put to use. The remnants of gold ornaments of the neolithic age bear testimony to the competence and skill of man of that age in fashioning gold. Likewise, priceless ornaments and utensils of pure gold were found in the grave of Tutankhamen, the 14th century B.C. Pharaoh of Egypt. His coffin was also made of gold, which weighed 110 kg. Interestingly, the Egyptians obtained gold by shifting the sands of river Nile.

Some 3,500-year-old copper utensils and other items were found during excavations in Egypt and Mesopotamia (present-day Iraq). These findings support the theory that copper was the next element invented after gold.

After copper, tin was invented at a later stage in the history. Tin as an alloy was in use around 300 B.C. in Egypt. Archaeological excavations in Egypt have yielded a tin bouquet of 1200 B.C. A tin finger-ring and a bottle belonging to

The ancient Egyptians had developed various methods of metal-work

281

Priceless items found in Tutankhamen's grave

roughly 1580-1350 B.C. were also obtained from a grave in Egypt. These items are believed to be the oldest among articles made of tin.

Bronze, a combination of tin and copper, was used by our ancestors to cast idols, utensils, ornaments, implements and weapons. The Bronze Age began in Egypt around 2500 B.C. During excavations in Egypt and in Ur (Mesopotamia), bronze utensils and weapons of 3000 B.C and 3300 B.C. respectively were discovered. Ur was the centre of the ancient Sumerian civilization.

The Sumerians had also perfected the art of separating silver from mercury. Nearer home, bronze and silver ornaments and utensils were retrieved from the Harappa and Mohenjodaro excavations. Among these, the famous bronze idol is *Nartaki*.

Interestingly, man could find iron at a much later stage than the metals we have talked about above. Initially, iron was extracted from the meteorites fallen from the sky. Since iron was scarce in Egypt, the Egyptians got it from the Hittites inhabiting the Black Sea area in the Asia Minor. The furnaces to melt iron were used even in 3,000 B.C. in Egypt. The iron weapons of 2,900 B.C. were discovered in the Cheops pyramid.

By 1400 B.C., Assyrians had started using iron on a large scale. In India, iron is believed to have come into use between 900-500 B.C., whereas in China its use began in 500 B.C. The iron pillar near Qutab Minar, which was erected in 415 A.D., contains 99.72% iron. Weighing 7 tonnes and standing 8 metres high, this massive pillar has a unique quality to it, viz. it is rust-proof.

The Egyptians were perhaps the first people to use lead. Six-thousand-year-old lead idols found in Egypt bear testimony to this fact. Lead coins of 300-200 B.C. were found in India, which show that Indians too had the know-how of using lead. Greek and Roman civilizations, however, seemed to have started using lead quite later, because lead items obtained from the places of these civilizations are of a much later stage.

282

Tutankhamen's countenance made up of pure gold

Although man had started making pots in the early years of his march towards civilization, he had not yet perfected this art. Initially, he made crude mud-wares, which were far from being artistic. Only after the invention of various metals did he develop the art of chiselling smooth and shiny utensils. While making weapons and utensils, man carried out many experiments with metals, which can be termed as his first step towards the science of Chemistry. Egyptians and Sumerians were the pioneers in trying out various compounds of copper. And they succeeded in manufacturing blue-and green-coloured utensils by 4,000-3,000 B.C.

Around 2500 B.C., the Egyptians started making glass by melting alkali and quartz. It did not take long for this know-how to reach the Sumerians, and from them to Rome, Spain, Germany and other parts of Europe. The coloured glasswares and ornaments of ancient Egypt and Mesopotamia show that a special kind of blue colour was obtained by heating silica and malachite with lime, which was spread over the glass to make these tinted wares. This art reached China in 600 B.C. The Chinese further perfected it and became the first people to manufacture glittering China claywares.

Colours and dyes for cloth and for other uses were obtained by different methods in different countries. Between 1700-1500 B.C., the Egyptians learnt to obtain the blue colour from the indigo plant. This led to the development of

283

An iron pillar near Qutab Minar in Delhi, which is more than 1600 years old, yet it has not rusted

of years ago, during the period of Aryans the Egyptians brewed wine from barley and the Indian version of wine, *somrus* – the juice of the moon plant, was consumed as a drink in 1000 B.C. The ancient man also knew about many chemical compounds.

Salt was made by evaporating sea-water and lime was obtained by burning limestone. Potash, nitre, alum, sulphur, carbon, perfume, turpentine oil, etc. were also known to the ancient man. Our forefathers were also experts in exploring natural resources like herbs, flowers, seeds and plants for medicines. In the Berlin Museum, a box containing medicines of about 2000 B.C. is preserved, which bears proof of ancient man's medical expertise. The Romans knew the process of manufacturing cement, which is also a chemical process.

The shine and gloss over the coloured designs of glass and tiles made by people of ancient Egypt is still intact even after centuries, which bears testimony to the Egyptian's knowledge of chemistry. In ancient India, chemistry was also taught to supplement the main subjects of metallurgy and *ayurveda*. The metals and chemicals were known to Indians even during the Vedic period.

We can say that the foundation of chemistry was laid by the ancient man. Though initially his knowledge was elementary, it served as a good starting point for the coming generations. The progress that we have made till today has its roots in the hoary past.

indigo plantation. In Crete, the violet colour was extracted from molluscs of the sea. For dyeing cloth, a combination of metals was used as a mordant. Antimonite came handy as a face-powder for the Egyptian beauties.

Another chemical process, that of brewing drinks, was also perfected by man. Thousands

■ ■

ALCHEMY

Alchemy was the earliest version of Chemistry.

In this kind of science, Alchemists were in search of a method to convert every metal into gold. Gold was valued, because it was a stable, precious and shining metal. There were two other things, besides gold, which they wanted to produce in their laboratories. One, a drink which once taken would make man not only young but immortal too; and the next, a liquid in which every material could dissolve.

In ancient Egypt, science was mainly concentrating on making gold from metals, extracting dyes and developing the art of colouring. 'Alexandria' was founded in 371 B.C. and from the study of Greco-Egyptian chemistry, it soon led to the first efforts made in alchemy.

The founder of Alchemy – Jabir ibn Hayyan

A glimpse of a page from an alchemy book written in the 18th century

285

GEBERI PHILOSOPHI AC ALCHIMISTAE
MAXIMI, DE ALCHIMIA.
LIBRI TRES.

Jabir was spell as 'Geber' in books written in Latin

The Egyptians used to call it *Chemia*, and those who practised it were called *alchemists*. Gradually, these words came into use in Greek. The English words *Chemistry* and *Chemist* are also derived from these. The German equivalent *Chemie*, the French *Chimie* and the Italian *Chimice* are all its derivates.

The Arabs acquired the knowledge of alchemy after their conquest of Egypt. Countries situated around the Mediterranean Sea came under the Arab influence when the Arabs started preaching and practicing their art and culture in these countries. Gradually, the knowledge of mathematics, geography, medicine and alchemy began to spread. A place called *Cordoba* in Spain became a centre for Arab culture in the world. From Cordoba, this culture spread to Italy and from thence, it travelled far and wide with Arab conquerors and emissaries.

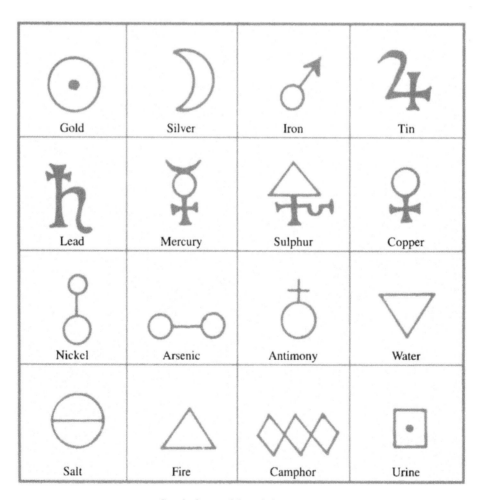

Gold	Silver	Iron	Tin
Lead	Mercury	Sulphur	Copper
Nickel	Arsenic	Antimony	Water
Salt	Fire	Camphor	Urine

Symbols used by alchemists

286

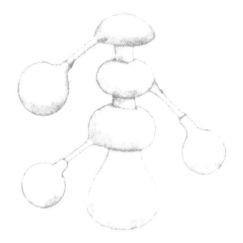

An apparatus used by alchemists

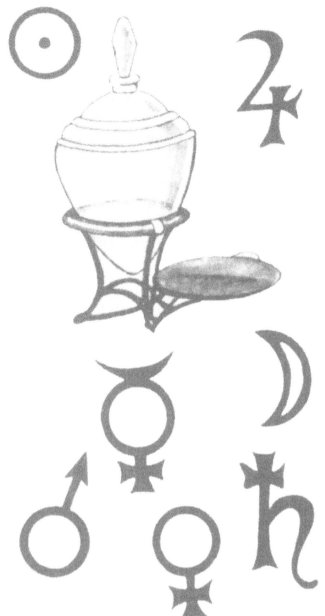

Jabir ibn Hayyan is known as **the founder of Arab alchemy**. Jabir (720-813 A.D.) came to Baghdad from Tus. His books, most of them in Arabic – though some are available in Latin too – contain a mine of information about the process of making artificial gold, manufacturing metals by chemical reactions, ways of cleaning metals, various kinds of furnaces and implements, sublimation of sal ammoniac, various methods of fermentation, manufacturing of vegetable oils, and the processes of making soda and soap. Besides, he discusses some salts and minerals like *Al-kaya* or *alkali*, *vitriols*, *Kuhl* or *grey antimonyore* and green and bluevitriols.

Some of the signs and symbols of alchemists are still used by druggists and chemists

The Baluka apparatus

The Deki apparatus

In India, knowledge of alchemy is described fully in Sushrit's books

Al-Razi (826-925 A.D.) was another famous alchemist of that period. He wrote many books, *Secret of Secrets* and *Cartinens* being the most important among them. His books first reached Europe via Spain and possibly it was from these that alchemy developed in Europe.

Another important name in alchemy is that of Ibn-Sina or Avicenna. His famous book is *De-Anima*. Ibn-Sina had expressed doubts about transmutation. According to him, what the alchemists made only resembled gold or silver.

The alchemists worked hard for centuries but they could not get what they were searching for. However, their efforts and hard work did not go unrewarded. Though they could not convert metals into gold, their apparatuses and the symbols they had developed came in handy for the *brewers* and *pharmacists*. While alchemists were searching for gold, *chemists* were busy in acquiring more knowledge about the processes of fermentation and chemical reactions. By 1150 A.D., alcohol was fermented at a place called Salermo in Europe. The word *alcohol* was used for the first time by Paracelsus in the 16th century, from the Arabic word *Al-kuhl*. The period of alchemy came to an end by the 16th century but *druggists*, chemists and pharmacists still use the symbols and apparatuses developed by them. After this began the era of Modern Chemistry, and Robert Boyle and Lyvasiye are regarded as the founders of Modern Chemistry.

■■

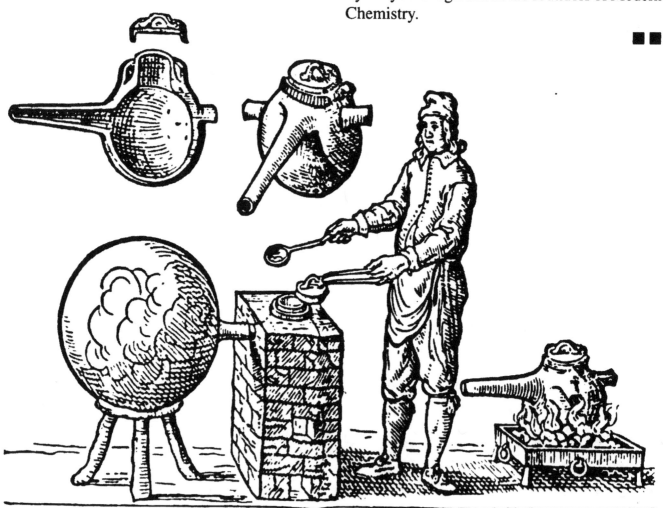

The 17th-century distillation furnace and instruments developed by J.R. Glauber

288

USES OF CHEMISTRY IN EVERYDAY LIFE

Chemistry has undoubtedly brought a revolution in the life of mankind. Chemistry has brought about innumerable developments in the industrial sector. Without the knowledge of Chemistry, perhaps, we would not be aware of so many chemicals, medicines, fertilizers, insecticides, pesticides, cosmetics, rubber, plastics and various other metals which have today become an indespensable part of human existence. In this chapter, we shall discuss in detail about the contributions of chemistry or the role of chemistry in our everyday life.

1. **Plastic:** Plastics are one of the greatest contributions of Modern Chemistry. The discovery of plastic has changed the life of man completely. Buckets, glasses, plates, water tanks, brush, mugs, plastic utensils, plastic or polythene bags that we use in our daily life, all are the gifts of Chemistry. Even rubber, teflon, nylon, etc are also the derivatives of plastics which are helping us to meet many of our daily needs.

2. **Fertilizers:** Man-made fertilizers are being used in agriculture for a long time. Now chemical fertilizers have increased the agricultural output to a great extent. There are many types of fertilizers such as: Urea, Phosphate, Nitrate, etc. Ammonium Sulphate, Nitro Phosphate, etc are being produced in great quantities in our country.

3. **Soaps and Detergents:** Today there are uncountable varieties of toilet soaps and washing soaps, washing powders, detergents, etc produced from these chemical industries which are meeting the requirements of about a billion people.

4. **Bacteria- free Disinfectants:** Now a days, dettol, savlon, boric powder, etc have been developed which have chemicals that can kill the bacteria that is prevalent in the atmosphere and these are of great use in hospitals, dispensaries, houses, etc.

5. **Pesticides and Insecticides:** These insecticides and pesticides are of great use in cleaning the floors of our houses, in drains, man-holes, etc to prevent the poisonous mosquitoes and insects from causing any infection or disease that can affect us adversely.

6. **Medicines and Antibiotics:** Today, there are more than 100 different kinds of antibiotics for various infectious and chronic diseases such as Penicillin, Tetracyclin, Chloromycetin, Erythromycin, Streptomycin, etc.which are also the gifts of Chemistry.

7. **Metals:** There are more than 80 metals that have been discovered so far in Chemistry. Motorcars, two-wheelers, three-wheekers, railway engines, aeroplanes, machines and machinery parts, household equipments, etc., all are the very many contributions of this wonderful field of science, called Chemistry.

8. **Rubber:** This is considered to be one of the greatest gifts of Chemistry. Tyres and tubes of different shapes and sizes are made up of rubber only. Almost all the vehicles running on the roads have rubber tyres fitted to them.

9. **Threads and Fibres:** Threads and fibres have solved the problem of clothes. Today there are innumerable varieties of clothes made up from Nylon, Terylene, Silk, etc., and these are even exported to various countries of the world.

10. **Coal:** Coal is just not used as a powerful source of energy and power but is also used for a number of activities such as cooking, running locomotives, etc.

11. **Petroleum:** Petroleum is the source of a number of products such as Petrol, Diesel, Kerosene, Grease, etc. There are a number of petroleum refineries across the world, especially in the Saudi Arabian countries. India too has a number of petroleum rifineries in Barauni, Guwahati, Mathura, Cochin, etc. Petrol and Diesel are the main food of the automobiles and also the huge aircrafts and helicopters flying high in the sky. So, you can well imagine that without petroleum products, life will certainly come to a standstill.

12. **Alcohol:** Due to the advanced techniques of Chemistry, several liquor plants have been established in our country, which are producing high quality alcohols and beer.

13. **Cosmetics and other Beauty Products:** Different kinds of creams, powders, lotions, oils (both body-oils and hair-oils), perfumes, deodorants, shampoos, lipsticks, foundations, etc., all are the gifts of Chemical Science or Chemistry to the human race.

In addittition to all these, Chemistry has provided us with innumerable other valuable products of everyday life and we can rightly conclude by saying that the human civilization is indebted to the countless contributions of Chemistry.

■■

BRANCHES OF CHEMISTRY

Although Lavoisier is known as *the father of modern chemistry*, it was Van Helmont who first collected the knowledge of alchemists and tried to give it a scientific form. Robert Boyle, Priestley, Shelley, Lavoisier were all later-day scientists to give a modern scientific touch to the ideas of alchemists.

Chemistry mainly developed in the 19th and 20th centuries. Carbonic compounds were discovered by Kolbe, Wohler, Kekule, Berthelot and Louis Pasteur. John Dalton, Rutherford, Niels Bohr, Thomson, Soddy, Henry Becqueral, Madam Curie, Einstein, Chadwick and Fermi discovered radioactivity, isotopes, atomic structure, nuclear fuel, etc., thereby enabling chemistry to make great strides. Now, with many new inventions, the field of chemistry has become very vast, extensive and deep. It has, therefore, been divided into various branches so that its study could become simpler and easier.

The Main Branches of Chemistry

- *Inorganic Chemistry* : This includes the study of elements and methods of preparation of their compounds (for carbon, only its oxides and sulphides), their compositions, properties, and uses.

- *Organic Chemistry* : It includes the study of various carbon compounds (except carbonates, oxides and sulphides), the methods of their preparation, their compositions, properties and uses. Benzene (C_6H_6) and its compounds are also studied under Organic Chemistry.

- *Physical Chemistry* : This includes the study of various laws, rules and principles on which the chemical reactions are based. It also includes the study of various instruments and measures used in Chemistry.

- *Analytical Chemistry* : This deals with the study of identification and analysis of various compounds.

- *Biochemistry* : This deals with the study of materials and their chemical reactions, which are related to living creatures such as plants and animals.

- *Industrial Chemistry* : It deals with the study of industrial manufactures and their uses.

- *Agricultural Chemistry* : It includes the study of agriculture and related substances, such as fertilizers, minerals, pesticides, insecticides, etc.

- *Nuclear Chemistry* : This includes the study of the nucleus of an atom, and related matters.

INVENTION OF GASES

In the beginning of the 16th century, a German chemist, Paracelsus observed that a highly inflammable gas was released when iron was added to sulphuric acid. The word 'gas' was, however, used for the first time in 1630 by Van Helmont. He referred to a number of gases, two of which he described elaborately, but he was not able to collect them. Robert Boyle was the first scientist who was able to collect a gas for the first time. He also measured the combustibility of the gas which was released by the reaction between iron and sulphuric acid.

Since it was not possible to learn the real form of gases scientifically without collecting them first, their study could not make any progress for a long time to come. It was in 1727 that an English priest, Stephen Hales invented a method of collecting gases over water. After the invention of the *pneumatic* bath, it was easier to study and experiment with them.

In 1766, Henry Cavendish (1731-1810) obtained a gas by reacting iron or zinc with dilute sulphuric acid. He studied its properties and termed it as an *inflammable gas*. He proved that this gas was a component of water. In 1783, Lavoisier (1743-1794) prepared water by combining this gas with oxygen and on the basis of this, it was named *Hydrogen gas*.

Cavendish obtained a gas by reacting a copper wire with a solution of salt. Priestley also collected *Acid air* over mercury. This acid air is today known as *Hydrochloric acid gas*. Cavendish mentioned about another gas known as *Flagistic* gas but did not publish his findings. In 1772, Daniel Rutherford invented a gas and after studying its properties, named it as *Mepheticair*. The properties of this gas were also studied by Lavoisier and he named it *Azote*, i.e. lifeless. In 1823, another Chemist, Chaptel named it as *Nitrogen*, a name that is still in use today.

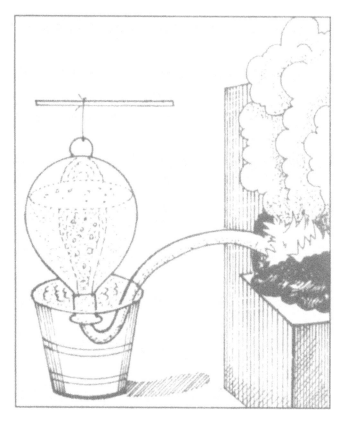

The process of collecting gases over water was invented by Stephen Hales in 1727

Henry Cavendish (1731-1810)

In 1772, a Swedish chemist, Karl Wilhem obtained a gas for the first time by heating red mercuric oxide with infrared rays from the sun. He named it as *Fire Air*. An English chemist, Joseph Priestley prepared another gas by the same method and named it *Oxygen* gas. (Oxus – acid, Gennao – to deliver). However, this gas is not an essential component of acids.

Shelley, in 1774, obtained *Chlorine* gas by reacting manganese dioxide with concentrated hydrochloric acid. This was called *Oxymuratic* gas. In 1810, Davy proved it to be an element and because of its greenish yellow colour (Greek *khloros* – greenish yellow), it is known as *Chlorine*.

In 1770, Priestley invented *Soda water*. He obtained *ammonia* gas in 1774 by reacting ammonium chloride and slaked lime. As it is obtained from ammonium chloride, it is called ammonia. In 1785, Berthollet proved that by

The father of Modern Chemistry :
A.L. Lavoisier (1743-1794)

decomposition of ammonia, nitrogen and hydrogen gases were produced.

In 1774, Priestley prepared a gas by reacting pure mercury with concentrated sulphuric acid. In 1777, Lavoisier named it as *Sulphur dioxide*.

Prior to this, Glauber had obtained a gas by reacting sodium chloride with concentrated sulphuric acid. In 1810, Davy proved that it was a compound of hydrogen and chlorine and named it as *hydrogen chloride* gas.

Some Other Gases Discovered

- *Argon* – Lord Rayleigh and Ramsay, 1894
- *Helium* – Ramsay, 1894
- *Neon, Krypton* and *Xenon* – Ramsay and Travers, 1898
- *Xenon* – Ramsay and Travers, 1898
- *Fluorine* – Moissant, 1896
- *Bromine* – A.J. Ballard, 1826

Since the molecules of gases are far apart from one another, as compared to that of solids and liquids, these are always in motion. The gases have no definite shape and form and can be compressed to reduce their volumes. By compressing, the gases are converted into liquids.

There are eleven elements that exist as gases at ordinary temperature. The gases function according to certain rules or laws which are commonly called as the Gas Laws or the Laws of Gases which have been explained briefly as under:

1. The Boyle's Law: The experiments carried out by Sir Robert Boyle have shown that, for nearly all gases, the volume of a sample gas at contant temperature is inversely proportional to the pressure; that is the product of pressure and volume under these conditions is contant.

PV= a constant (temperature constant, moles of gas constant.)

Mathematically, PV = constant value if the gas is behaving as an Ideal Gas. A

Practical math expression of Boyle's findings is as follows:

$$P_1 V_1 = P_2 V_2$$

Where the variables with the 1 subscript mean initial values before the manipulation

And the variables with the 2 subscript mean final values after the manipulation.

This equation expresses the Boyle's Law. The law was inferred from the experimental data by the English natural scientist, Robert Boyle. Whereas all of the common gases, such as Oxygen, Nitrogen, Carbon dioxide, Carbon monoxide and the rest behave in a way described by the Boyle's Law, there are a few gases that do not such as Nitrogen dioxide.

2. Charles's Law: Jacques Charles investigated the relationship between the Volume of a gas and how it changes with temperature. He noted that the volume of a gas increased with the temperature. Charles's Law states that the volume of a given amount of dry ideal gas is directly proportional to the Kelvin Temperature provided the amount of gas and the pressure remain fixed. When we plot the Volume of a gas against the Kelvin temperature it forms a straight line. The mathematical statement is that the **V / T = a constant**. For two sets of conditions the following is a math statement of Charles's Law:

$$V_1 / T_1 = V_2 / T_2$$

An example of Charles's Law would be what happens when a hot air balloon has air heated. The air expands and fills the balloon. Of course, other physical principles cause the balloon to rise against the gravitational force. As the air inside the balloon expands the balloon gets bigger and displaces more air. The displaced air produces a buoyant force that counters the gravitational force and causes the balloon to rise. **The Charles Law is sometimes called the Gay Lussac Law.**

3. Avogadro's Law: Avogadro's law is one of the gas laws. The law is named after Amedeo

Carl Wilhelm Scheele (1742-1786) preparing oxygen

Avogadro, who in 1811 hypothesized that equal volumes of gases, at the same temperature and pressure, contain the same number of particles, or molecules. Thus, the number of molecules in a specific volume of gas is independent of the size or mass of the gas molecules.

The minor aspect of the law can be stated mathematically as:

$$V/n = a$$

Where:

- *V* is the volume of the gas.
- *n* is the number of moles in the gas.
- *a* is a constant.

However, this above equation is just a trivial one, which is valid for all homogeneous substances, including homogeneous liquids and solids. This relation is easy to deduce and its validity was assumed before Avogadro's work.

Graham's law: It is also known as Graham's law of effusion, was formulated by

Joseph Priestley (1733-1804)

Thomas Graham. Graham found experimentally that the rate of effusion of a gas is inversely proportional to the square root of the mass of its particles. This formula can be written as:

$$\frac{\text{Rate}_1}{\text{Rate}_2} = \sqrt{\frac{M_2}{M_1}} \quad \text{where:}$$

- *Rate₁* is the rate of effusion of the first gas.
- *Rate₂* is the rate of effusion for the second gas.

Sir William Ramsay (1852-1916)

- M_1 is the molar mass of gas 1
- M_2 is the molar mass of gas 2.

Graham's law works for both effusion and diffusion.

All these above laws and many more have been quite helpful in understanding the behaviour of the gases.

MEDICINES

In the twentieth century, many chemical compounds have been used to make medicines. In the earlier attempt, in 1903, Fischer and Moring prepared *Veronal* or *diethyl barbituric acid*, which is a sleep inducer. Then in 1912, Heinrich Hoerlien invented *Luminol or ethyl phenyl barbituric acid*, which is now used to manufacture sleeping pills.

From time immemorial, arsenic acid has been used for the treatment of *syphilis*. In 1905, *Atoxyl* was prepared, which proved quite effective in curbing this dangerous disease. Then Paul Ehlrich and his associate, Alfred Bertheim invented an effective medicine named Salvarson to cure syphilis.

To cure malaria, Schulemann invented *Plasmochin* in 1924. Later on, Fritz Mietsch made another medicine *Atebrin* for malaria. In 1932, Gerhard Domagk invented *Sulphonamides* to cure this contagious disease.

The first antibiotic medicine was *Penicillin*, which was obtained by Sir Alexander Flemming (1881-1955) in 1928 from a fungus, the mould named *Penicillium notatum*. This medicine

Sir Alexander Fleming in his laboratory

proved very useful in treating various wounds and contagious diseases.

With the invention of Penicillin, a new method of treating diseases through antibiotics was opened up. In 1943, Selman Abraham Wakesman invented *Streptomycin*, which is used to cure T.B. In 1952, *Auromycin* and in 1953, *Actinomycin* were obtained by Benjamin Dugger and H. Bockman respectively. In 1952, Domagk was the first to invent synthesized *Neobeten* to cure T.B.

Some Other Inventions in Medicines

- *Morphine* – by Fredrick Surterner of Germany, 1805
- *Aspirin* – by Dresser of Germany, 1883
- *LSD* – by Hoffman of Switzerland, 1938
- *Chloromycetin* – by Birkholder of USA, 1947
- *Terramycin* – by Finley & others of USA, 1950

Till now, numerous medicines have been invented to cure various diseases throughout the world. Significant among them are—sulpha drugs, antibiotics, pain-killers, tranquilizers, and anti-allergic tablets.

Tuberculosis which was considered to be an incurable disease can be easily cured today. There are very many heart diseases that have found a cure due to the wonders of medicine and medical science. Today heart transplantations and impanting an artificial heart inside a human body if the original heart stops functioning properly is also quite common.

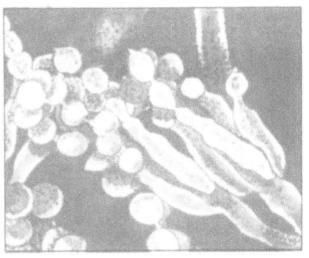

The first antibiotic medicine 'Penicillin', which was obtained by Fleming from a kind of fungus, the mould named Penicillium notatum

Hence, we can definitely conclude by saying that very soon the day will come when medical science will find a permanent cure for fatal diseases like **AIDS and Cancer**.

VITAMINS AND HORMONES

Vitamins and hormones are those chemical substances which carry on the various metabolic activities properly in a human body. According to Pole Casimir Funk, if an important ingredient were removed from the bran of rice, it resulted in a disease called *Beri-beri*. This important ingredient was a special organic compound and was named *Vitamin*. After this, Windaus invented Vitamin D in 1926. Vitamins D-1, D-2 and D-3 were obtained from Vitamin D. S.C.P. Jansen and W.F. Donath invented Vitamin B in 1926. Paul Karrer invented Vitamin A in 1931, Vitamin B-2 in 1935, and Vitamin 'K' in 1939. Vitamin C was synthesized by T. Reichstein in 1934. In 1937, R. Kuhn synthesized Vitamins A

and B-6. Most of the vitamins are obtained from food-grains. and food items such as **milk, meat, pulses (*dals*), fish, eggs and vegetables**. The important vitamins essential for our body are Vitamin **A, B1, B2, B6, B12, C, D, E and K**. These are essential for the good health and proper growth of the body. A deficiency of vitamins causes various diseases in the human body.

In 1906, Henry Starling established that our body has an active element which works as a chemical spokesman. It remains as an active material in the secretions obtained from specific glands of the body. These materials are called *hormones*. Actually, hormones are those materials which are produced in the specialized glands inside the body. In 1901, a hormone called adrenalin was discovered for the first time by Yokichi Tokamine, a Japanese and T.B. Aldrich. *Adrenalin* increases the blood pressure. Only two years later, Friedrich Stolz synthesized it and named it *Supra-Renine*.

Although E. Baumann had obtained *Thyriodine* from the thyroid gland in 1895, it was only in 1915 that Edward Kendall could obtain *Thyroxine* from the same gland, i.e. thyroid.

Kuhn, L. Ruzicka, ErnstLacquer, Adolf Butenandt together analysed the sex hormones – *testosterone* in males and *oestrogen* and *progesterone* in females. Their accurate synthesis was done in many laboratories in the middle of 1934-35. Today, man knows much about vitamins, hormones and their effect on the human body and on plants and animals, but we owe it all to those who worked in this field incessantly and opened up avenues for further research. Hormone deficiencies in the human body can result in a number of diseases such as **Goitre, Thyroid problems, Diabetes, etc.**

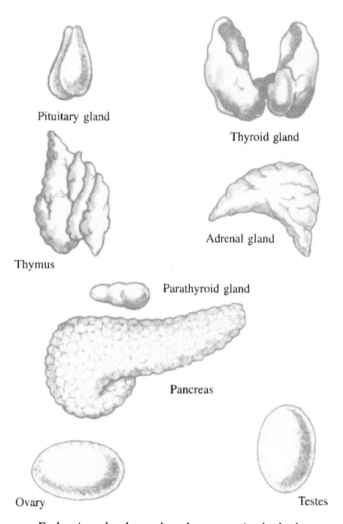

Pituitary gland

Thyroid gland

Adrenal gland

Thymus

Parathyroid gland

Pancreas

Ovary

Testes

Endocrine glands produce hormones in the body

ELEMENTS

According to the sages of ancient India, the world was made up of five elements – earth, water, fire, air and the sky. This is known as the principle of five elements. Aristotle and other Greek philosophers thought that the entire universe was composed of earth, air, water and fire. In the middle of the 17th century, Robert Boyle rejected the theories of four and five elements as baseless from the scientific point of view. He wrote a book Scoptic Chemistry in 1682 dealing with the subject. A century later, in 1789, a French scientist Lavoisier defined an element for the first time. According to him, an element is a substance which cannot be split by any known means into a simpler part than its own. During the same century, many new elements were discovered. Now, all the 92 elements existing in nature are known. Apart from these, many new elements, which do not exist in nature, have been manufactured by artificial transmutation. These are known as *transuranic elements*. These are manufactured either in a nuclear reactor, or obtained from the debris of explosions of the hydrogen bomb. Man-made elements are radioactive elements and these decay very quickly. The number of elements has now gone up to 109, and these are mentioned in the following table:

Sir Humphrey Davy (1778-1829)

Jöns Jacob Berzelius (1779-1848)

Atomic Element No.	Name	Discoverer/ Inventor	Year of Discovery/ Invention
1.	Hydrogen (H)	H. Cavendish	1766
2.	Helium (He)	Janssen and Lockyer	1868
3.	Lithium (Li)	Arfwedson	1817
4.	Beryllium (Be)	N. Vauquelin	1797
5.	Boron (B)	Guy – Lussac and Thenord	1808
6.	Carbon (C)		Prehistoric
7.	Nitrogen (N)	D. Rutherford	1772
8.	Oxygen (O)	Priestley	1771
9.	Flourine (F)	Moissan	1886
10.	Neon (Ne)	Ramsay & Travers	1898
11.	Sodium (Na)	H. Davy	1811
12.	Magnesium (Mg)	Guy Lussac	1808
13.	Aluminium (Al)	F.Wohler	1827
14.	Silicon (Si)	J.J. Berzelius	1824
15.	Phosphorous (P)	M. Brand	1869
16.	Sulphur (S)		Prehistoric
17.	Chlorine (Cl)	Scheele	1774
18.	Argon (Al)	Ramsay & Rayleigh	1894
19.	Potassium (K)	H. Davy	1807
20.	Calcium (Ca)	H. Davy	1808

Contd...

Madam Curie (1867-1934)

300

G.T. Seaborg (1912)

E.M. McMillian (1907)

21.	Scandium (Sc)	L. Nilson	1897	38.	Strontium (Si)	H. Davy	1838
22.	Titanium (Ti)	W. Gregor	1791	39.	Yttrium (Y)	J. Gadolin	1794
23.	Vanadium (V)	N. Del, Rio	1801	40.	Zirconium (Zr)	Klaproth	1789
24.	Chromium (Cr)	Vanquelin	1797	41.	Niobium (Nb)	C. Hatchet	1801
25.	Manganese (Mn)	Scheele	1774	42.	Molybdenum (Mo)	Scheele	1781
26.	Iron (Fe)		Prehistoric	43.	Technetium (Tc)	Segre & Perrier	1937
27.	Cobalt (Co)	G. Brandt	1737	44.	Ruthenium (Ru)	K. Klaus	1844
28.	Nickel (Ni)	Cronstedt	1751	45.	Rhodium (Rh)	Wollaston	1803
29.	Copper (Cu)		Prehistoric	46.	Palladium (Pd)	Wollaston	1803
30.	Zinc (Zn)		Prehistoric	47.	Silver (Ag)		Prehistoric
31.	Gallium (Ga)	Boisbaudran	1875	48.	Cadmium (Cd)	Stromeyer	1817
32.	Germanium (Ge)	Winkler	1886	49.	Indium (In)	Reich & Richter	1863
33.	Arsenic (As)	A. Magnus	Alchemy	50.	Tin (Sn)		Prehistoric
34.	Selenium (Se)	J.J. Berzelius	1818	51.	Antimony (Sb)	B. Vallentine	
35.	Bromine (Ba)	A. Balard	1825	52.	Tellurium (Te)	von Reichenstein	1783
36.	Krypton (Kr)	Ramsay & Travers	1898	53.	Iodine (I)	B. Courtois	1811
37.	Rubidium (Rb)	Bunsen & Kirchhoff	1861				

Contd...

D. Rutherford (1871-1937)

No.	Element	Discoverer	Year
68.	Erbium (Er)	C. Mosander	1843
69.	Thulium (Tm)	P. Clove	1878
70.	Ytterbium (Yb)	C. Marignac	1878
71.	Lutecium (Lu)	G. Urbain	1907
72.	Hafnium (Hf)	Coster & Hevesy	1923
73.	Tantalum (Ta)	A. Ekeberg	1802
74.	Tungsten (W)	J.J. & F.de Elhuyar	1783
75.	Rhenium (Re)	Noddack & Others	1925
76.	Osmium (Os)	S. Tennant	1804
77.	Iridium (Ir)	S. Tennant	1804
78.	Platinum (Pt)	D. de Ulloa	16th century
79.	Gold (Au)		Prehistoric
80.	Mercury (Hg)		Prehistoric
81.	Thallium (Ti)	Crooks	1861
82.	Lead (Pb)		Prehistoric
83.	Bismuth (Bi)	C. F. Geoffroy	1753
84.	Polonium (Pa)	P. & M. Curie	1898
85.	Astatine (At)	E. Segre' & Others	1940
86.	Radon (Rn)	Rutherford	1900
87.	Francium (Fr)	M. Perrier	1939
88.	Radium (Ra)	P. & M. Curie	1898
89.	Actinium (Ac)	A. Debierne	1899
90.	Thorium (Th)	Berzelius	1828
91.	Protactinium (Pa)	S.Soddy & Others	1917
92.	Uranium (U)	Peligot	1789
93.	Neptunium (Np)	Macmillan &Abelson	1940
94.	Plutonium (Pu)	G. Seaborg & Others	1940
95.	Americium (Am)	G. Seaborg & Others	1944
96.	Curium (Cum)	G. Seaborg & Others	1944
97.	Berkelium (Bk)	S. Thompson & Others	1949
98.	Californium (Cf)	S. Thompson & Others	1950
99.	Einsteinium (Es)	A. Ghiorso & Others	1952
100.	Fermium (Fm)	A. Ghiorso & Others	1953
101.	Mendelevium (Md)	A. Ghiorso & Others	1955
102.	Nobelium (No)	A. Ghiorso & Others	1957
103.	Lawrencium (Lr)	A. Ghiorso. & Others	1962
104.	Unnilquadium (Unq)	—	1969
105.	Unnilpentium (Unp)	—	1970
106.	Unnilhexium (Unh)	U.S.A.	Most recent
107.	Unnilseptium (Uns)	(Erstwhile) U.S.S.R.	Most recent
108.	Unniloctium (Uno)	G. Munzenberg & Others	1984
109.	Unnilennium (Une)	Germany	1982

No.	Element	Discoverer	Year
54.	Xenon (Xe)	Ramsay & Travers	1898
55.	Caesium (Cs)	Bunsen & Kirchoff	1860
56.	Barium (Ba)	H. Davy	1808
57.	Lanthanum (La)	C. Mosander	1839
58.	Cerium (Ce)	Berzelius & Hisinger	1803
59.	Praseodymium (Pr)	Welsbach	1885
60.	Neodymium (Nd)	Welsbach	1885
61.	Promethium (Pm)	J. Marinsky & others	1945
62.	Samarium (Sm)	Boisbaudran	1879
63.	Europium (Eu)	E. Demarcay	1901
64.	Gadolinium (Gd)	Marignac	1886
65.	Terbium (Tb)	C. Mosander	1843
66.	Dysprosium (Dy)	Boisbaudran	1886
67.	Holmium (Ho)	Soret & Delafontaine	1879

ORGANIC CHEMISTRY

Carbonic compounds have been in use since ancient times. Some of these are : the juice of the moon plant (somrus in Hindi), wine made from barley, indigo, etc. There is mention of methyl alcohol and acetone in Boyle's book *Sceptical Chymist*, written in 1661.

In 1675, Nicolas Lemery was the first to name the substances obtained from organisms as organic compounds. Earlier, it was thought that *organic compounds* are obtained only from living tissues by natural processes and cannot be obtained in the laboratory. On the basis of Lemery's theory, the study of products obtained from organisms (living beings) was called *organic chemistry* by the chemists.

In 1784, Lavoisier proved by analysing many organic compounds that every organic compound contains carbon; and that apart from carbon, it may also contain hydrogen, oxygen, nitrogen, sulphur, phosphorous and the halogens.

In 1815, Berzelius called this force as *vital force*. This theory of the vital force remained in existence for a long time. But in 1828, Friedrich Wohler(1800-1882) obtained urea by heating the mixture of ammonium sulphate and potassium cyanate. This was the first organic compound synthesized in the laboratory. This discovery by Wohler, who himself was a student of Berzilius, put an end to the vital force theory once and for all. After the synthesis of urea, Kolbe synthesized acetic acid in 1845. In 1856, Berthollet successfully synthesized methane. The formation of these compounds proved that organic compounds could be obtained from inorganic compounds by chemical reactions, similar to those by which inorganic compounds are obtained.

Organic compounds have played an important role in food products and other items of daily use like fuels, medicines and in agriculture. During the 19th century, researches in the organic compounds had been very fast and in a very short time, the number of organic compounds known went up to thousands.

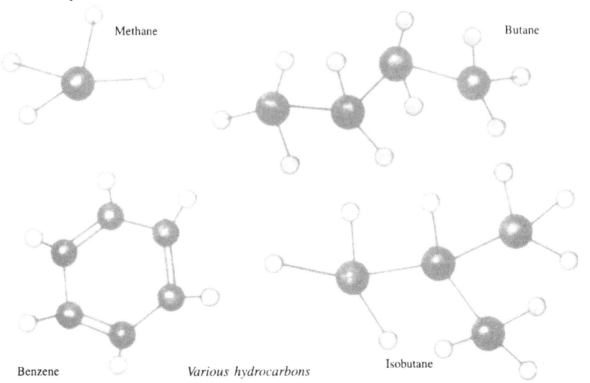

Methane

Butane

Benzene   *Various hydrocarbons*   Isobutane

303

PLASTICS

In the 20th century, the manufacture of plastic, among other synthetic carbonic compounds, was a revolutionary invention which changed the face of the modern world. Plastic is mainly made of unsaturated hydrocarbons of higher molecular weight. Plastic is manufactured by the polymerization of these unsaturated hydrocarbons. Most of the plastic and artificial fibres are *man-made* polymers. *Natural polymers* are: *rubber, proteins, cellulose, starch*, nylon etc.

Cellulose nitrate was manufactured in France, England and Germany for the first time in the middle of the 19th century.

An important manufacturing ingredient, plastic celluloid, was made in 1869 by John Wesley Hyatt. In 1903, A. Eichengrum and T. Becker patented the first plastic of the cellulose acetate class. This became the basis of thermoplastic plastics. In 1909, Dr. Leo H. Baekeland produced phenol-formaldehyde resin.

This was named Bakelite after Dr. Baekeland. In 1923, Fritz Pollock and Kurt Ripper invented urea formaldehyde plastic. Imperial Chemical Industries manufactured polythene, which is being widely used these days.

The use of plastic in the textile industry has not escaped attention. Artificial fibres are made from carbonic chemicals. In 1953, nylon was synthesized by Carothers of the USA and his associates. Terylene was invented by J.T. Dickson and J.R. Winfield.

Nowadays, clothes made from artificial resins like terylene, nylon, rayon, acrylic and stretchlon are more popular than cotton clothes. These are more durable and do not need ironing.

A number of household items, medicines and industrial products are also produced from plastics. Plastics are also used for manufacturing aircrafts.

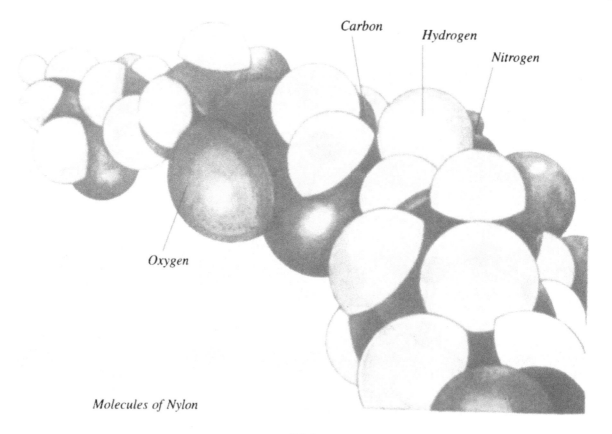

Carbon

Hydrogen

Nitrogen

Oxygen

Molecules of Nylon

CHEMICAL FERTILIZERS

The credit for meeting the food requirements of the ever-increasing population of the world by using chemical fertilizers goes to Justus von Liebig (1803-1873). He proved by systematically analysing the decayed vegetation that for the proper growth of vegetables, soil with minerals like nitrogen, phosphorous and potassium was essential. He showed, by experiments, the importance of artificial fertilizers in the cultivation of food grains.

In 1905, two Norwegians, Birkeland and S. Eyde, successfully used atmospheric nitrogen for the first time and laid the foundation of chemical fertilizers as an industry.

Fertilizers are added to the soil in order to increase its fertility. Fertilizers are of three types:

1. *Nitrogenous fertilizers :* These are: ammonium sulphate, calcium cynamide, urea, basic nitrate and calcium ammonium nitrate. All these predominantly contain nitrogen.

2. *Phosphatic fertilizers :* Super phosphate of lime. It is rich in phosphorous.

3. *Potash fertilizers :* These are obtained from natural deposits of potassium minerals. Potassium helps in the growth of plants.

Fertilizers in India

In India, fertilizers are mainly manufactured by the Fertilizer Corporation of India. In Sindri (Bihar), ammonium sulphate, urea and calcium super-phosphate are manufactured. Urea and nitro phosphate are manufactured in Tromby (Maharashtra). Urea and ammonium sulphate are also manu-factured in Gorakhpur (U.P.) and Namrup (Assam). Apart from these, a number of new fertilizer industries are being set up in other parts of the country, where chemical fertilizers are manufactured in order to meet the increasing demands of the agricultural sector in India.

Chemical fertilizers are added to the soil to increase its fertility

THE ATOMIC THEORY OF MATTER

The Greek philosopher Leucippus (5th century B.C.) and Democrates and the Roman philosopher Lucrates had assumed that matter was made up of small particles that were unchanging and indestructible and moved about in empty space. They called these particles atoma, meaning 'indivisible'. By the end of the 18th century, no scientist was able to give any definite opinion on the composition of matter. In 1803, John Dalton put forth a theory regarding the structure of matter for the first time. This is known as Dalton's atomic theory of matter.

According to Dalton, matter was made up of small indestructible particles called atoms. By the end of the 19th century and beginning of the 20th century, Thomson, Rutherford, and Chadwick proved by experiments that atom is not the smallest particle of a matter; rather atom itself is made up of three fundamental particles called electron, proton and neutron. Therefore, an atom is that smallest, fundamental, indivisible particle of the matter that can take part in a chemical reaction but which has no independent state of its own.

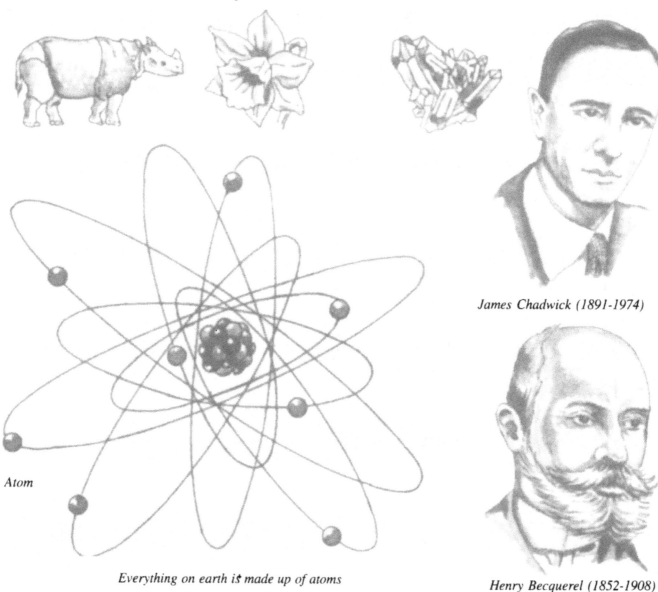

Atom

Everything on earth is made up of atoms

James Chadwick (1891-1974)

Henry Becquerel (1852-1908)

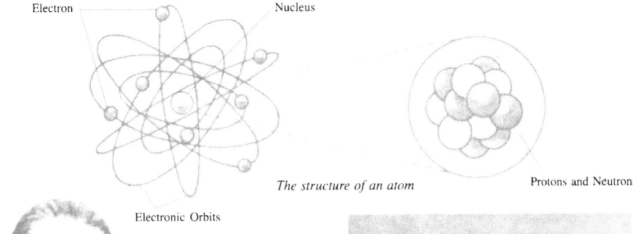

Electron
Nucleus

The structure of an atom

Protons and Neutron

Electronic Orbits

Niels Bohr (1885-1962)

*Fred Frederick Soddy
(1877-1956)*

Albert Einstein (1879-1955)

On February 24, 1896 Henry Becquerel observed that some invisible rays were emitted from uranium which affected the photoplates. These were called Becquerel rays. The substances which emit these rays are called radioactive substances and this natural process is called radioactivity. In 1898, Madam Curie and her husband attained two elements Radium and Pollonium from a mineral called pitchblende and proved the existence of radioactivity. The discovery of radioactivity destroyed many of the theories of elements the 19th century. The earlier theories of alchemists regarding transmutation, which were rejected by Boyle and other scientists, were proved right by the discovery of radioactivity. This transmutation always occurs in some elements in a natural way. Thus, in the 20th century, both natural and artificial processes of converting elements into other elements has

become possible. Scientists could not make gold out of other metals but a new category of artificial elements has been obtained.

The nucleus of uranium splits itself into two smaller parts and the process is known as nuclear fission. A tremendous amount of energy is released during the process. Uranium has, thus, proved to be a great source of energy. This energy is being used for making atom bombs as well as for generating electricity. In nuclear reactors, the neutrons obtained from uranium, start a

307

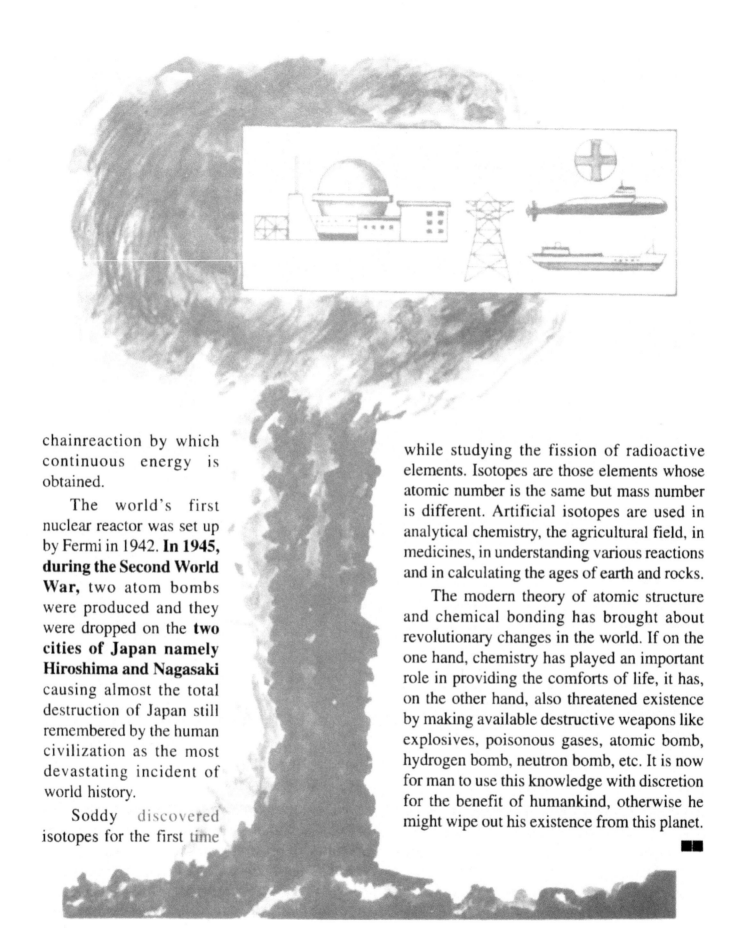

chainreaction by which continuous energy is obtained.

The world's first nuclear reactor was set up by Fermi in 1942. **In 1945, during the Second World War,** two atom bombs were produced and they were dropped on the **two cities of Japan namely Hiroshima and Nagasaki** causing almost the total destruction of Japan still remembered by the human civilization as the most devastating incident of world history.

Soddy discovered isotopes for the first time while studying the fission of radioactive elements. Isotopes are those elements whose atomic number is the same but mass number is different. Artificial isotopes are used in analytical chemistry, the agricultural field, in medicines, in understanding various reactions and in calculating the ages of earth and rocks.

The modern theory of atomic structure and chemical bonding has brought about revolutionary changes in the world. If on the one hand, chemistry has played an important role in providing the comforts of life, it has, on the other hand, also threatened existence by making available destructive weapons like explosives, poisonous gases, atomic bomb, hydrogen bomb, neutron bomb, etc. It is now for man to use this knowledge with discretion for the benefit of humankind, otherwise he might wipe out his existence from this planet.

■■

Along with the means of comforts and progress, terribly destructive war weapons have also been developed through chemistry

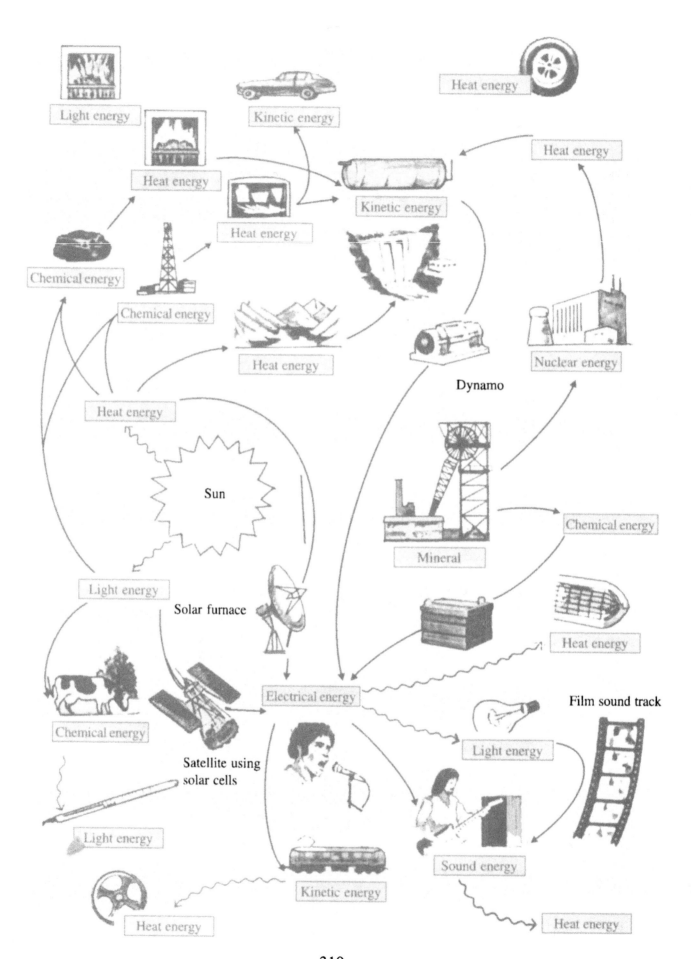

Light energy

Kinetic energy

Heat energy

Heat energy

Heat energy

Kinetic energy

Heat energy

Chemical energy

Chemical energy

Heat energy

Nuclear energy

Dynamo

Heat energy

Light energy

Sun

Chemical energy

Mineral

Solar furnace

Chemical energy

Chemical energy

Heat energy

Satellite using
solar cells

Electrical energy

Light energy

Film sound track

Light energy

Sound energy

Kinetic energy

Heat energy

Heat energy

ENERGY

The physical labour which we do in our daily life is called *work*. In scientific language, *the capacity of doing work is called energy*. All types of work in this world need energy. The water falling from great heights runs the generators to produce electricity, which is used

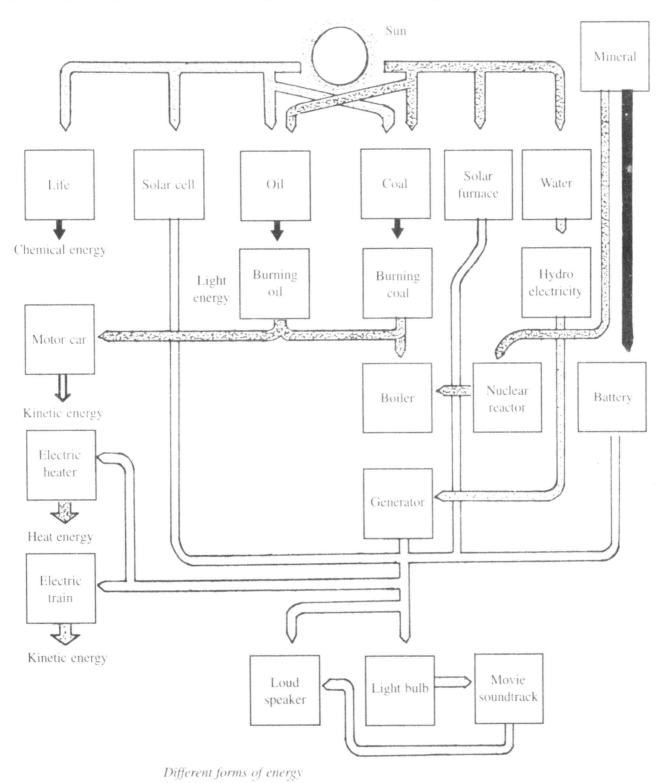

Different forms of energy

Joker in a box. On opening the box, the joker comes out, due to the force in the spring. This is an example of the inter charge ability of the potential and kinetic forms of energy

to perform various tasks. Water can be converted into steam by burning coal and the steam is used to run turbines. Electrical energy is used to run motors and the energy of a twisted spring is used to run the clock. So both in motion as well as in the state of rest, there is a capacity to do work, or we can say that energy is required to do work.

Energy can be of several kinds, for example, *mechanical energy, electrical energy, heat energy, sound energy, chemical energy, magnetic energy and atomic energy.*

Energy can neither be created nor can it be destroyed, it can only be converted from one form to another. So the total amount of energy retrains unchanged. Only its form keeps on changing. This is known as *the law of conservation of energy.*

All types of energies have two forms:

1. Kinetic energy; and 2. Potential energy.

1. Kinetic energy: Everybody has a capacity to do work due to its *motion*. This is called as *kinetic energy*. A bullet fired from a gun, a moving train or a motorcar, etc. have kinetic energy.

2. Potential Energy: The capacity of doing work in a body due to its particular *state* or *position* is called *potential energy* of that body.

The potential energy of a body can be of many types, like *gravitational, elastic, electrical, chemical* or *magnetic energy*. For example, if we lift a hammer lying on the floor, we have to do work against the gravitational force. This work gets stored in it, in the form of potential energy or it acquires the capacity to do work. If the hammer is now dropped, it can break an object.

In various kinds of fuels like coal, petrol, gas, etc., the energy is stored in the form of *chemical potential energy*. The nuclei of uranium or plutonium contain tremendous amounts of energy, which can be released by nuclear fission. This energy is called *nuclear energy*. Atom Bombs, Hydrogen Bombs and Neutron Bombs have nuclear energy which is highly destructive in nature. A Hydrogen Bomb contains 700 times more nuclear energy than an Atom Bomb.

SOLAR ENERGY

The life of all living beings on the earth exists only due to the heat and light of the sun. The sun maintains the water cycle between land and sky, it produces winds, and it has also produced oil and coal fields for us. It has created mineral resources for us from primitive vegetation. Except nuclear energy, all kinds of energy which man has used up till now have been obtained from the sun only. In fact, the sun is the biggest source of energy. We also get energy from coal, oil and natural gas, etc. If all our known resources of these fuels burn at the same rate at which solar energy is being supplied to us, the stocks of fuels will last only for less than three

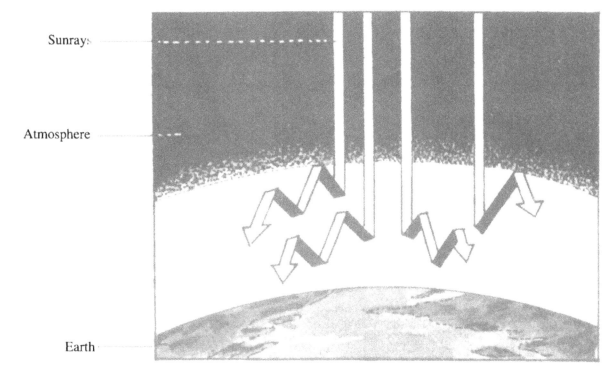

Sunrays

Atmosphere

Earth

The greenhouse effect

In a greenhouse, the temperature of the air inside is higher than that of outside

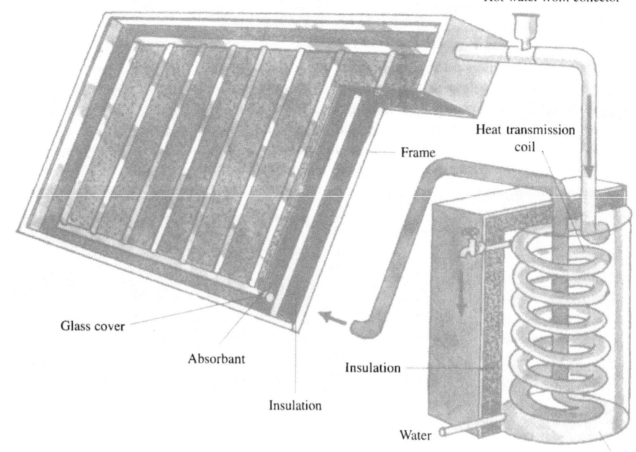

Heat collector

Hot water from collector

Frame

Heat transmission coil

Glass cover

Absorbant

Insulation

Insulation

Water

Water tank

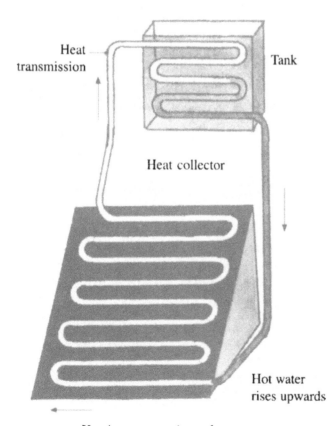

Heat transmission

Tank

Heat collector

Hot water rises upwards

Heating water using solar energy

days. The sun being such a big source of energy, man could learn its importance only very recently. The energy requirements are increasing day by day for industries and automobiles and these needs can only be met by accumulating and using *solar energy*. This can resolve the worldwide energy crisis to some extent.

The greenhouse effect

About 51 % of the sun's heat reaches the earth. First of all, the earth gets heated up and from the earth, this heat is transferred back to the atmosphere by *radiation*, *convection* and *conduction*. Heat from the earth radiates out in the form of infrared waves, of which 90% is absorbed by the *atmosphere*. This type of radiation is called *terrestrial radiation*, which is mainly responsible for heating the earth. The inability of the atmosphere in stopping the incoming solar radiations and the absorption of terrestrial radiations is called the *greenhouse*

A solar house

effect. On the same principle, the glass walls of the green house allow solar heat to reach inside the room, but heat from within is allowed to come out only at a very slow rate. In cold countries, seedlings are developed in greenhouses. Inside a greenhouse, the air remains comparatively hot. The family car is a good example of the greenhouse effect. The glass walls of a car make it a greenhouse. Due to the greenhouse effect, the inside temperature of the car in summer remains lower than the outside, while in winter it is more than the outside temperature.

Water heating through solar energy

Water can be heated through solar energy by using a special type of box. This box can be made up of wood, metal or plastic. Inside this shallow box, a coil of copper tubes is fixed. The box is coated with black concrete and covered with glass. Black concrete is used because it is a good absorber of heat. Water circulates in the coil of tubes and gets heated up through solar heat. This hot water is sent to the tank through pipes. In this way, hot water for domestic purposes can be obtained.

Many cold countries are using this technique to heat water. In Australia, about 60 to 95%

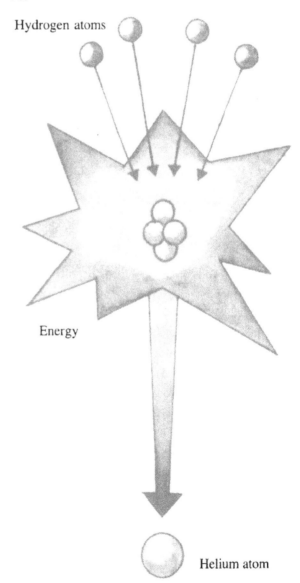

Hydrogen atoms

Energy

Helium atom

315

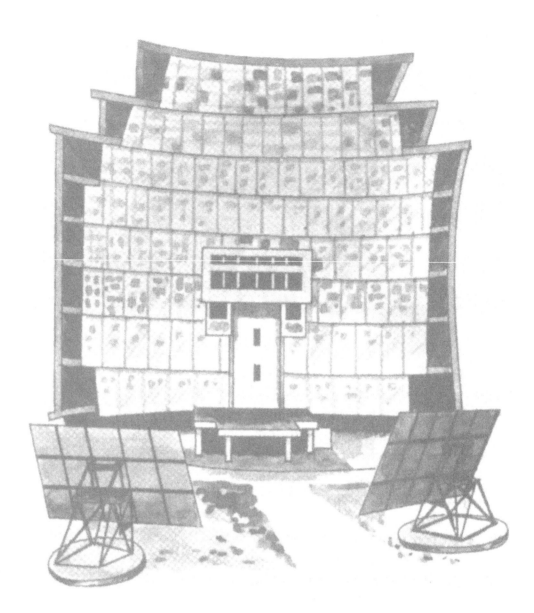

A solar furnace

families are using water heated by solar collectors. In the Florida city of U.S.A., buildings are getting hot water by using such devices. In our country too, some five-star hotels use this technique to get hot water.

The solar house

Builders have constructed different types of houses which are heated through solar energy. These types of houses are found in many countries. The U.S.A. has a large number of such houses. In these houses, solar heat is accumulated by solar collectors and used as and when required through some medium. Water, gravels or concrete can be used as a medium. These houses can be made warm in winter by using concrete in a special manner in the walls.

A heat collector is fitted at the top of the house from where heat reaches the concrete through a ventilator and gets absorbed. In winter, this heat can be used to keep the house warm.

Soft water from hard water

There is a great scarcity of drinking water in the Gulf countries. In these countries, sea water is converted into steam by solar energy and then condensed to obtain drinking water. A simple device is used for this purpose. Sea water is filled in a container and the container is converted with an inclined glass sheet. The water evaporates by solar energy and reaches the glass sheet, where it condenses into drops of drinking water.

These drops fall into another container from where drinking water is obtained. When the sea

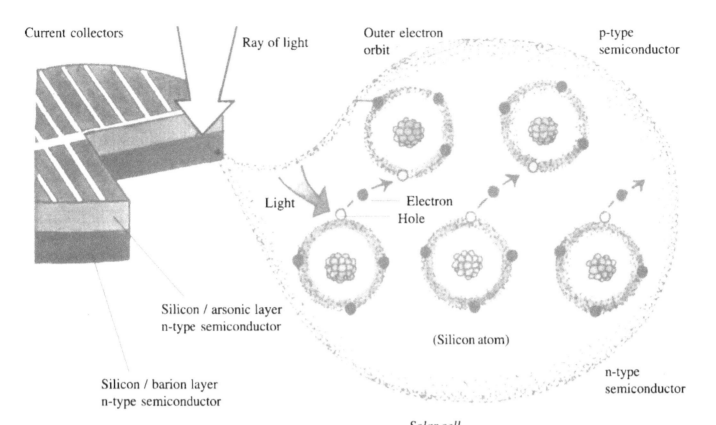

Current collectors

Ray of light

Outer electron orbit

p-type semiconductor

Light

Electron

Hole

(Silicon atom)

Silicon / arsonic layer n-type semiconductor

Silicon / barion layer n-type semiconductor

n-type semiconductor

Solar cell

water has been evaporated from the first container, salt is left at the bottom, which is also used for cooking purposes.

Solar furnaces

Scientists of various countries are developing solar furnaces, which are heated by solar energy. These furnaces are used to produce electricity. French scientists are using such furnaces for research work in the Pyrenees area. These furnaces are very big in size and make use of a large number of mirrors. One furnace needs about 9,000 parabolic reflectors and 11,000 plane reflectors. The heat is generated in between the reflectors. The energy reflected by these reflectors can produce a temperature of about 3,800°C. This temperature is sufficient to melt about 70 kilograms of iron per hour. The Russians, have developed a big solar boiler which can produce steam capable of running a turbogenerator of 1,000 kilowatt. It is hoped that solar furnaces will become very common in the near future.

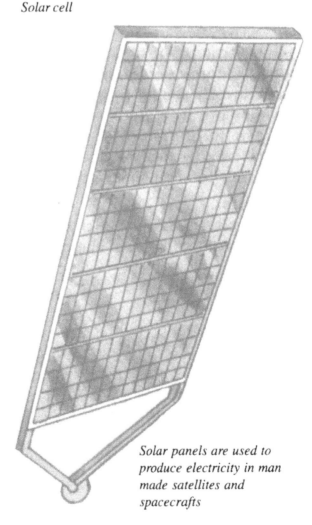

Solar panels are used to produce electricity in man made satellites and spacecrafts

A car powered by solar energy. It uses solar cells

Solar cells

The solar cell is a device which converts solar energy into electrical energy. Thin wafers of silicon or germanium crystals are used in these cells. When the sun's rays fall on these wafers, an electric current is produced. The first solar battery was developed in the U.S.A. in 1954. This battery contained 400 silicon cells and it could generate an electric voltage of 12 volts. After its first display, many solar batteries have been made. These batteries are being used on a large scale for producing electricity in artificial satellites and spacecrafts.

Solar cells are also used in photographic light meters. They are also used in telephone lines as current boosters and in calculators. Solar cells are very costly, but their price is likely to be reduced in the near future. Nowadays, solar panels are also in use to produce electricity.

Actually in calculators, there are 3 to 4 solar cells that produce electric waves and help to run the calculator smoothly. The solar panels are used to produce electric energy.

WIND GENERATORS

During the twelfth to eighteenth century, windmills were in common use in Western countries, for grinding wheat and for drawing water from the wells. *Holland* was known as *the country of windmills* at that time. In England also, there were about ten thousand windmills in the early nineteenth century. With the development of steam engines and electricity, the use of windmills was reduced. In the present age, the increasing demand of energy has again made scientists and technicians consider the utilization of wind power. Wind is a clean source of energy and does not produce environmental pollution.

In many countries, generators have been developed which run by the wind power. To drive the wind generators, vanes are not required. Its propeller is fitted with two or more aluminium blades and their rotations are set according to the direction of wind. These blades are hollow from inside and holes are made at their ends. When wind rotates the blades, these holes draw the air inside. The propeller is mounted on a steel pole and wind turbine is fixed at its base. Due to the pressure difference between the outer air and the inside air, the turbine starts rotating. The difference in atmospheric pressure is caused due to exit of air from holes. The turbine is connected to a generator, which produces electricity. There are big wind mills as well which have blades as long as 60 metres and their neck is as tall as 100 metres. These wind mills are mainly used to generate electricity for the small and big cities.

Today, several types of wind generators are available for domestic use, specially for those farmers who do not have power supply. These generators can produce electricity required for bulbs, tubes, refrigerators, fans, televisions, etc.

Some small wind mills are also used to draw water.

Wind mills were employed till the 18th century for grinding wheat and pulling water

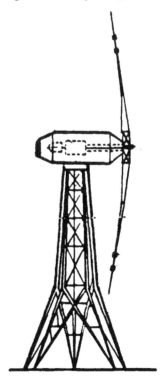

Wind generator

HYDROELECTRIC POWER

Efforts were made to control water power for human use in the Roman empire in about 100B.C. Initially, flowing water and then falling water was used to generate useful energy by making a wheel to rotate. In the outer periphery of these wheels, there were many small, flat spokes. For many centuries, this rotating wheel was used as flour mills. In 200A.D., the Romans developed a device which we may call the ancient 'Power Plant'. In the Aarley city of southern France, they established a plant of sixteen water mills. These water mills were used to run thirty-two other mills, which in turn ground 30 tonnes of flour every day.

Transmission

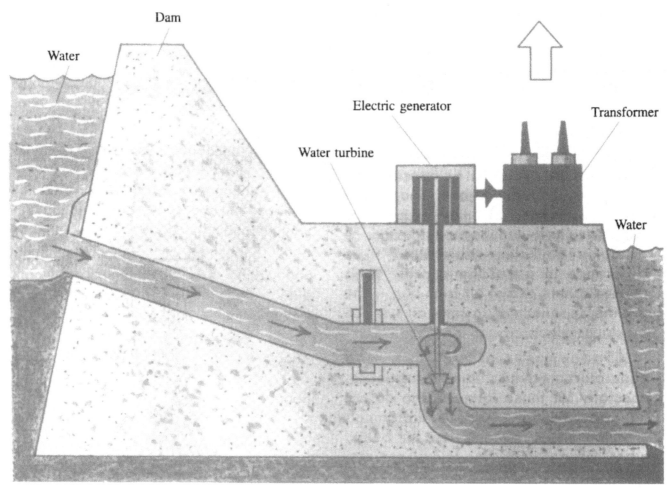

Hydroelectricity

Of the world's total energy production, 20% is by hydroelectric plants

The Arabs used water mills extensively for a long time, while the Europeans for got about them in the middle ages. This invention returned to the Western world around 800-1200A. D. and became the foremost energy source. It increased the industrial production by acting as the energy plant for flour mills, rice mills, hammer and mine pumps, etc.

Faraday discovered *electromagnetic induction* for the production of electricity, after which electricity became an important part of human life. In this process, turbines are used to run generators, which in turn produce electricity. Coal, mineral oil or nuclear energy is used to rotate turbines, but by using hydropower, we can produce electricity almost free of cost.

Due to the depleting resources of coal and mineral oil, the importance of hydroelectric power has increased tremendously. Electricity was first produced in the U.S.A. in 1858, using falling water. Flowing water was made to fall from a height to run a turbine and turn a generator for the production of electricity, and this was known as *hydroelectric power.*

The contribution of hydroelectric power in the total power production of the world is increasing day by day, because the burning of coal and mineral oil causes environmental pollution, while hydroelectric power production does not cause any pollution. Moreover, hydroelectric power is cheaper. However, there is a large initial expenditure in the construction of dam, but later on, the expenses are quite nominal. The overall average expenditure of hydroelectric power is cheaper than other methods of electricity production. By multipurpose projects, the expenditure of dam construction is distributed among flood control, irrigation and water transport plans, so that the electricity production is made further cheaper.

The geographical conditions are most favourable for hydroelectric power production in Africa. In the hilly areas of Asia, large water resources are found at the origin of big rivers. In Asia, Japan has been the most advanced country in the development of hydroelectric power. China and India are second and third respectively. Europe and North America have also made significant progress in the generation of hydroelectric power. One-third of the total hydroelectric power produced in the world is contributed by the U.S.A. and Canada.

Our country has a number of hydroelectric power projects among which the Bhakra Nangal and the Sutlej Vyas are the most important ones.

GEOTHERMAL POWER

The heat energy present inside the earth's crust is called *geothermal energy*. It occurs in three forms: *natural regions*, *hot water* and *hot rocks*. Today, man is using all the three forms of geothermal energy.

Iceland, Yellow stone National Park of U.S.A., Italy and North Island of New Zealand, are famous for hot-water springs and geysers. *Pohutu* is the biggest geyser of New Zealand, from which a 30 metrehigh fountain of steam erupts out. People of the Meori village of this area utilize this natural source of energy for cooking, washing and bathing. This geyser has reduced their fuel needs to a great extent. The hot water from this geyser is supplied through pipelines to a nearby city named Rotorua for heating homes.

In a small village called Larderello in Italy, geothermal energy is being used to produce electricity. The hot steam from the geyser of Larderello village rises as high as 50 metres. The temperature of this steam is even more than 190°C. To generate electricity, a hole is drilled in the rock about 150 to 450 metres deep from which pressurized steam is obtained. This steam is used to run steam turbines. Steam is a pollution-free source for generating electricity. Dams, reservoirs, coal, oil, etc. are not required to produce steam in this area.

In NewZealand, Japan, U.S.A., Russia, Italy, etc. geothermal energy is being used to produce electricity. In Rotorua and Wairakei also, geothermal energy is being used for many purposes. Scientists are also extracting energy from hot, dry rocks. In Russia and New Zealand, underground hot water is being used for air-conditioning. It is hoped that in the near future, geothermal energy will be used on a larger scale. ■■

Geyser

High pressure steam for the turbine

Heat from the earth

Geothermal energy can be used to produce electricity

COAL

Certainly coal may be considered as the base of modern industrialization. About half of the total energy consumed in the world comes only from coal. Coal is not only a source of energy and heat, but it also gives many valuable products such as coal gas, coal tar, pitch, ammonia, fertilizers, artificial dyes, waterproof paper, naphthalene, nylon threads, chemicals and several types of medicines. Coal is also used in thermal power plants to produce electricity. It is also used in steel plants.

Coal mines have been formed from the trees of jungles, which were buried under the earth's crust millions of years ago. At that time, the earth was ruled only by big reptiles and tall trees existed in marshy areas. All these trees were flowerless, and there were no birds and mammals then. In the course of time, these thick forests grew more and every year, countless number of leaves and branches fell down and got buried under the marsh. New trees continued to grow.

Gradually, these forests got buried under the earth and layers of sand and soil were deposited on them. These buried branches of trees remained as such for thousands of years and under high pressure and temperature, get converted into what we call coal today.

Whenever the buried ground came up again, new jungles grew on it. These new jungles were again buried due to the upheavals of the earth's crust and by the action of temperature and pressure were again converted into the second layer of coal. In this manner, several layers of coal were formed. These layers are called *Seem*. The period in which coal was formed is called the *Carboniferous period*.

The texture and different varieties of coal establish the ancient history of its formation. On the upper part of the layer of the coal, we can see the fossils of plant branches and in the lower part of the layer, the fossils of roots.

Millions of years ago, forests getting buried underground led to the formation of coal

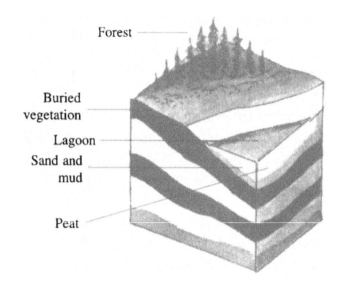

Forest
Buried vegetation
Lagoon
Sand and mud
Peat

Types of coal

Coal has been divided into four types, depending upon the amount of carbon present in it. These are - anthracite, bituminous, lignite and peat.

Anthracite

This variety contains about 95% carbon. It is considered to be the best quality of coal and gives maximum heat on burning. Anthracite burns with a bright blue flame.

Bituminous coal

In calorific value, it occupies the second place. Bituminous coal is of four types—steam coal,

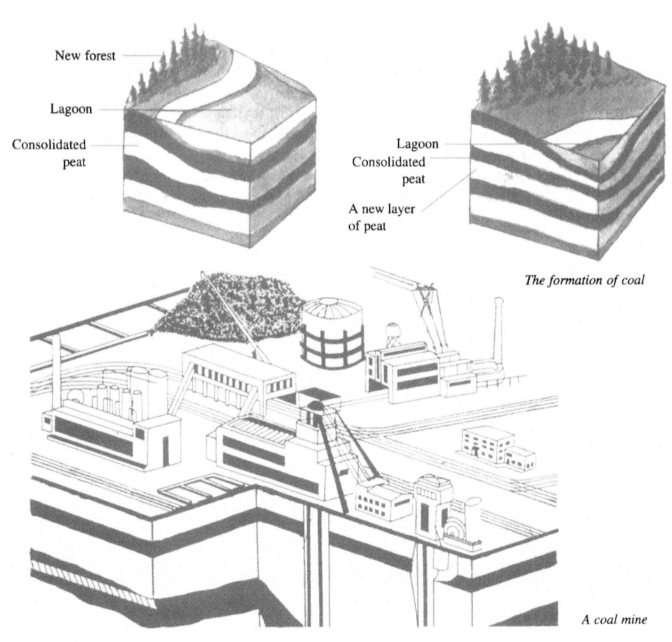

New forest
Lagoon
Consolidated peat

Lagoon
Consolidated peat
A new layer of peat

The formation of coal

A coal mine

324

domestic coal, cokerari coal and gas coal. Steam coal contains about 80% carbon. Coke is used to melt steel, which is prepared from high quality bituminous.

Lignite or brown coal

It produces more smoke and lesser heat on burning. Lignite contains about 40% carbon and 40% moisture.

Peat coal

This coal is fibrous and brown. It burns like wood and produces much smoke. It contains only one-third part of carbon. It is mostly used for domestic purpose.

Coal mines are generally very big and coal extraction is done by big machines and a large number of workers are involved in the mining work. Coal is used to produce coal gas. This coal gas is burnt to produce steam which is used in turbines to generate electricity. Petroleum products, natural gas, coke, coal gas, ammonia coaltar, etc. are all derived from coal. Our country has a number of coal mines such as in Chhattisgarh, Bihar, etc.

Lignite

Anthracite

Bituminous coal

Different kinds of coal

MINERAL OIL AND PETROLEUM

Millions of years ago, a large number of marine plants and animals were buried in the mud. Due to the high pressure and temperature and the action of bacteria, the hydrocarbons of these plants and animals were converted into *mineral* oil or *petroleum*. The word, petroleum is derived from the Latin word, *Petra* which means rock and *oleum* means oil. In this way, *petroleum* means the oil that is obtained from the rocks. The crude petroleum from the rocks is a very thick fluid like substance.

Man knew about petroleum hundred of years ago, but at that time he used it only for medicinal purposes or sometimes burnt it as fuel. The first

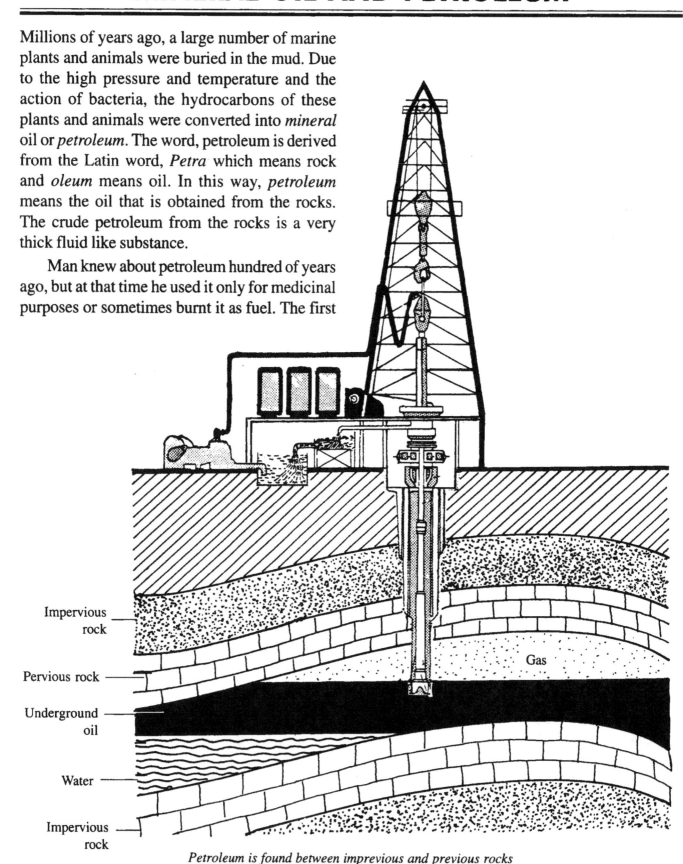

Impervious rock

Pervious rock

Underground oil

Water

Impervious rock

Gas

Petroleum is found between imprevious and previous rocks

326

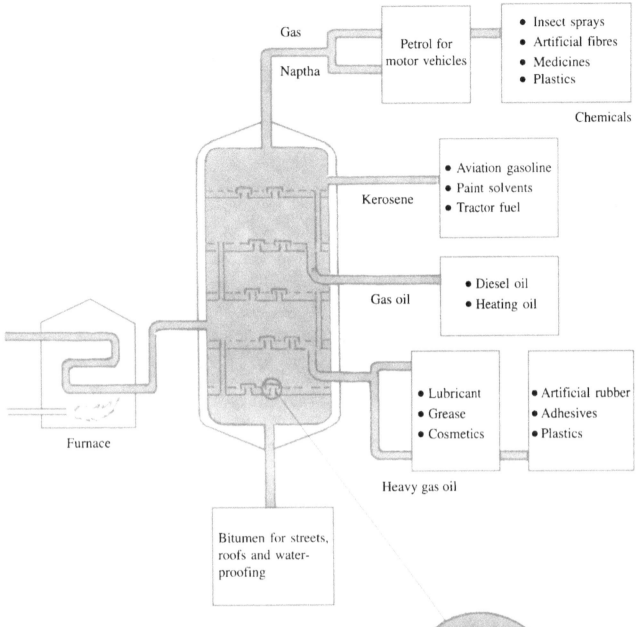

Gas

Naptha

Petrol for
motor vehicles

- Insect sprays
- Artificial fibres
- Medicines
- Plastics

Chemicals

Kerosene

- Aviation gasoline
- Paint solvents
- Tractor fuel

Gas oil

- Diesel oil
- Heating oil

Heavy gas oil

- Lubricant
- Grease
- Cosmetics

- Artificial rubber
- Adhesives
- Plastics

Furnace

Bitumen for streets,
roofs and water-
proofing

Fractional distillation of crude oil

petroleum well was drilled in 1859 in the U.S.A., which was about 20-21 metres deep. About 25 barrels of oil were extracted from this well every day. This marked the beginning of the petroleum industry.

Petroleum is an important fuel which produces heat and light on burning. It mainly contains hydrocarbons. It also contains oxygen, nitrogen and some quantity of sulphur. Today, a little less than half the energy of the world is obtained from petroleum. In addition to the energy, we get thousands of chemicals like

Waste gases are burnt in oil refineries

rubber, artificial fibers, fertilisers, plastics, grease, wax, etc. from petroleum.

Petroleum is found in the lower strata of permeable rocks. Usually, it is mixed with natural gas. Sometimes, it is squeezed out due to gas pressure. Wells are drilled at most of the places to get oil. It is taken by pipelines to a refinery. In the refineries, *fractional distillation* of crude oil is done to get different products. The boiling point of different hydrocarbons found in the mineral oil is between 0° and 400°C. By the fractional distillation of petroleum, several fractions are liquefied at different temperatures and can be obtained separately, e.g. noncondensed gases (below 18°C), crude naptha (18°C to 150°C), kerosene (150°C to 300°C), heavy gas oil (300°C to 400°C) and pitch (above 400°C.) These fractions further yield other products.

The products obtained from mineral oil are —gas, gasoline or petrol, aviation gasoline, diesel, LPG gas, white spirit, kerosene, wax, petroleum jelly, bitumen, artificial rubber, artificial fibres, plastics, olefins, benzene, toluene, gas oil, diesel oil, fuel oil, lubricating oil and grease.

The chief petroleum-producing countries are—the U.S.A., Venezuela, Saudi Arabia, Kuwait, Iran, Iraq, Abu Dhabi, Libya, Mexico and Russia. In India, Bombay High is a very popular source of petroleum. Besides this, petroleum and petroleum products have also been derived from Gujarat and Assam.

328

FUEL GASES

The gases which are highly inflammable and produce much heat on burning are called *fuel gases*. Natural gas, coal gas, water gas, producer gas and oil gas come under this category. They are used on a large scale in homes and industries.

Natural gas

Natural gas is produced by different petroleum-producing countries. Natural gas is generally obtained in water-oil-gas sequence from petroleum wells. The main constituents of this gas are methane, ethane and propane. These gases are highly inflammable, therefore natural gas is used as a fuel.

The chief producers of natural gas are the U.S.A. and Russia. The length of gas pipelines is more than 1 24 thousand kilometers in Russia. These lines supply gas to big industrial cities and metropolitan cities of Europe. Netherland is third in the production of natural gas in the world. In India, natural gas is found in Assam and Gujarat. Bangladesh has large reserves of this gas, which will last for the next 200 years.

The hydrogen gas that is obtained from this gas is used in fertilizer industries.

Coal gas

This gas is obtained by the destructive distillation of coal. It is a mixture of several inflammable gases like hydrogen, methane, carbon monoxide and unsaturated hydrocarbons. To produce this gas, coal is filled in a fire clay retort and heated up to 1 100°C to 1200°C. The gases produced by heat reach the hydraulic main through the ascension pipe. Here, some part of the gas is cooled and converted into tar and liquefied ammonia. The rest of the gas passes through iron pipes, which act as wind condensers. Here, some more part is liquefied and accumulated in a tar well. In tar-free coal gas also, some impurities are left. It is purified by passing it over cold water, slaked lime, moistferric oxide and heated nickel metal. Purified gas is collected in inverted tanks by the downward displacement of water.

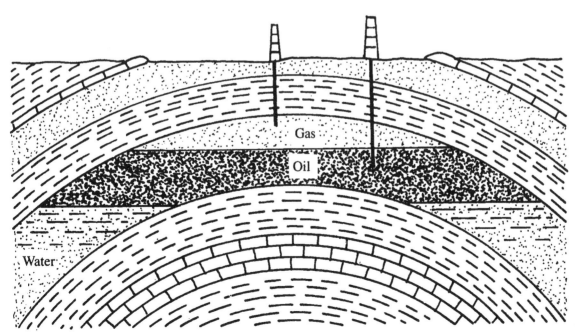

Natural gas

329

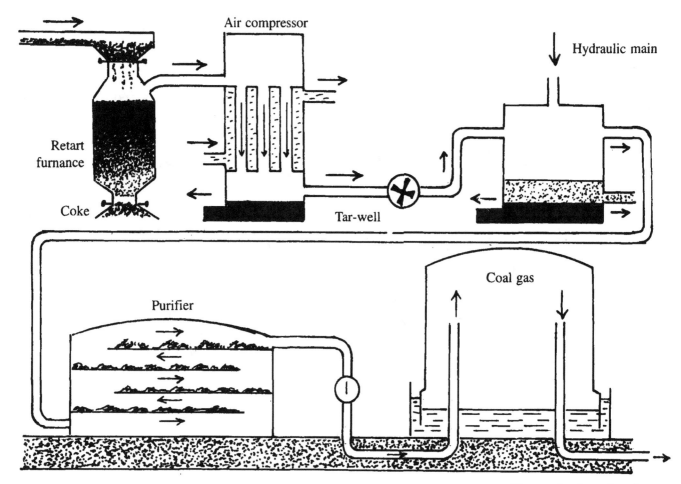

The production of coal gas

Biogas plant

This gas is used in Bunsen burners, for producing light and in metallurgy. This gas has more calories than other gases.

Water gas

This gas is a mixture of carbon monoxide, carbon disulphide and hydrogen. It is obtained by passing steam over red-hot coke. It is used alone or along with coal gas as a fuel. It is also used in producing hydrogen and methyl alcohol. It produces twice the amount of heat as given by producer gas.

Producer gas

It is a 1:2 mixture of carbon monoxide and nitrogen. It is obtained by passing air below red-hot coke. In this process, carbon dioxide is formed first, which after going up is reduced to carbon monoxide.

This gas is used as cheap fuel. On burning, it produces a high temperature. Producer gas is used in glass manufacturing and metal extraction. It produces less heat as compared to water gas, so a mixture of the two is used, which is known as *semi-water gas*. The mixture gives an average of the heat produced by the two gases.

Oil gas

This gas is a mixture of both saturated and unsaturated hydrocarbons (methane, ethylene, acetylene). This gas is obtained by the destructive distillation of petroleum or kerosene.

By spraying oil on a red-hot iron retort, the mixture of inflammable hydrocarbons is converted into gas. It is collected in tanks by the downward displacement of water. This gas is used in chemical laboratories.

Bio-fuels

Methane gas is a bio-fuel. This gas is highly combustible. It is formed by the decomposition of plants and animals, in marshy areas and oozes out in the form of bubbles, so it is also known as *marsh gas*. It can be easily produced in the digester. The digester is filled with all kinds of rubbish like grass, cauliflower leaves, banana-peel, cow-dung and sewage. This gas is supplied to houses through pipes from the digester. Nowadays, bio-gas is used as fuel in villages. Even electric generators are run by the heat produced by this gas. In fact, the bio-gas plant is a gift of modern biotechnology. Liquid fuels and gaseous fuels are produced by the fermentation of cow-dung, plants, fertilizers, sewage, sugar, potato, etc. Biomass, garbage, fodder, etc. have become well-known sources of energy today. This kind of fuel does not produce smoke but gives a lot of heat. Hence, it doesn't cause any pollution too.

In marshy areas, the decomposition of dead vegetation and animals causes bubbles of methane gas to emerge

NUCLEAR ENERGY

The energy obtained by the process of nuclear fission and nuclear fusion is called *nuclear energy*.

Nuclear fission

In 1905, *Albert Einstein*, proved that matter can be converted into energy. He established a relation between mass and energy which is known as *the mass-energy equation*, given by:

$$E = mc^2$$

(E = Energy, m = mass of matter, c = velocity of light)

In 1939, German scientists *OttoHahn* and *Strassman*, showed that when uranium is bombarded by slow-moving neutrons, barium and krypton are formed, whose atomic mass is less than that of uranium and a large amount of energy is released.

If even a milligram of matter were fully converted into energy, it could send a rocket to the space

Einstein gave the equation for the interconversion of matter and energy

This proved that the uranium atom (atomic number=92) was broken into two smaller fragments—barium (atomic number=56) and krytpon (atomic number=36). Later on, it was shown that in the fission process, uranium isotope having an atomic mass of 235

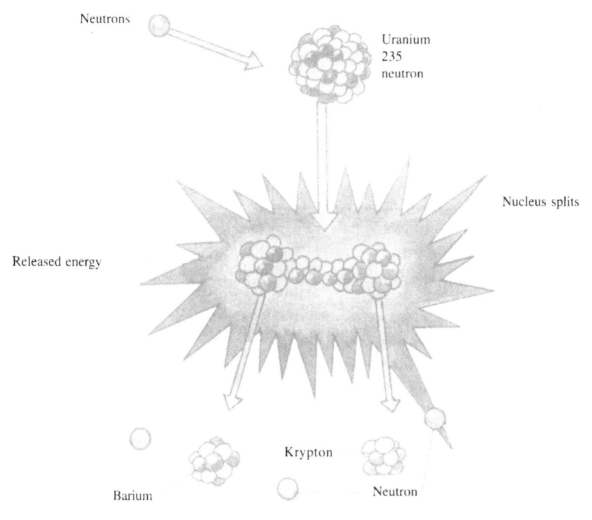

Neutrons

Uranium
235
neutron

Nucleus splits

Released energy

Barium

Krypton

Neutron

Nuclear fission

participates. A tremendous amount of energy (200MeV) along with some neutrons from the nucleus of U235 are released during fission. So, the breaking up of a bigger nucleus into two smaller nuclei by the bombardment of neutrons is known as *nuclear fission* and the energy released during this process is known as nuclear energy.

In nuclear fission, some mass of the fuel is converted into energy, according to Einstein's mass-energy relation.

Some of the neutrons produced in nuclear fission may further bombard other uranium nuclei, resulting in an *uncontrolled chain reaction*. This principle is used in the construction of the *atom bomb*, in which a lot of energy is produced.

Chain reaction

Three neutrons are liberated when a uranium nucleus undergoes fission by the bombardment of neutrons. Out of these three neutrons, two take part in further fission. If these two neutrons bombard two other uranium nuclei, then four fissionable neutrons are released. These four neutrons bombard four other uranium nuclei to produce eight neutrons and so on. The energy produced in a chain reaction is tremendous. Chain reactions are of two types—*uncontrolled* and *controlled*.

In an uncontrolled chain reaction, the number of neutrons released during fission is continuously increasing. The reaction becomes uncontrolled as the number of neutrons increases from one to two, two to four, four to eight, eight to sixteen and so on. The energy produced in an uncontrolled reaction, is so large that it can be highly destructive. This is nothing, but the atom bomb which was first dropped on Hiroshima.

In a controlled chain reaction, only one neutron takes part in each fission, the other two are absorbed. So the energy produced after each stage of fission remains constant. The energy produced is sufficient enough to be used constructively. A *nuclear reactor* is based on this principle only. These reactors are used for producing electricity and isotopes.

Energy liberated from fission

During fission, some mass of uranium is converted into energy. The energy produced by the fission of each nucleus of uranium-235 is 200 Mew, which is very large. If we calculate the energy produced by the fission of 1 gm of uranium-235, it becomes equivalent to the energy produced by 20 tonnes of TNT (tri nitro toluenel. The energy liberated from Nuclear fission is too high.

Nuclear fusion

The atomic mass of the isotope of any element is always less than the sum of the mass of free protons, neutrons and electrons present in it. The difference in the two masses is called *mass defect*. A large amount of energy is released, when two smaller nuclei combine to form a bigger nucleus due to this mass defect. The process of joining smaller nuclei is known as *nuclear fusion*.

Nuclear fusion results when the nuclei of elements are bombarded by fast moving particles like protons, deuterons, etc. Nuclear fusion is a reverse process of nuclear fission. Fusion reactions take place at very high temperatures of about 1,000,000°C. They are also called *thermonuclear reactions*. This temperature is maintained through the energy released during the fission reaction which is about 17.6 MeV.

The hydrogen bomb is based on nuclear fusion. Deuterium or tritium or a mixture of both

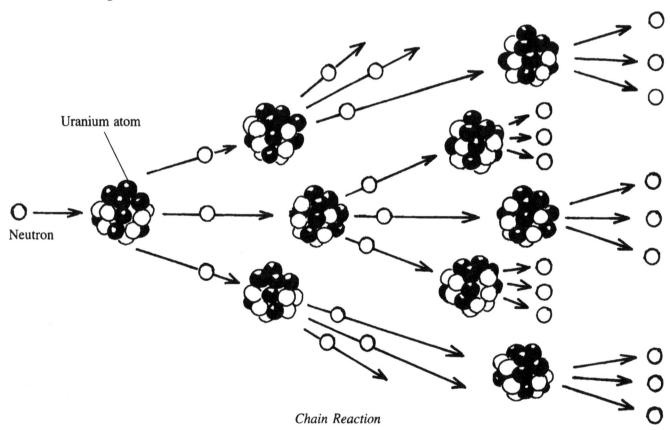

Chain Reaction

is combined into helium nuclei by the explosion of the atom bomb. A high temperature is produced in the explosion of an atom bomb, which initiates nuclear fusion. As a helium nucleus is formed by the union of deuterium and tritium nuclei, a large amount of energy is released. The neutrons produced again take part in the fission of plutonium or uranium. So, the destructive power of the hydrogen bomb becomes around 700 times higher than that of an atom bomb. The energy produced by the sun and stars is also a result of nuclear fusion.

Scientists are trying to control nuclear fusion so that it can be used to produce electricity. Scientists are using fusion reactions for constructive work through powerful *magnetic machines* known as *tokamaks* and *lasers*. It is hoped that scientists would be able to construct fusion reactors in the near future. Which can solve the shortage of electric power or electricity to a great extent.

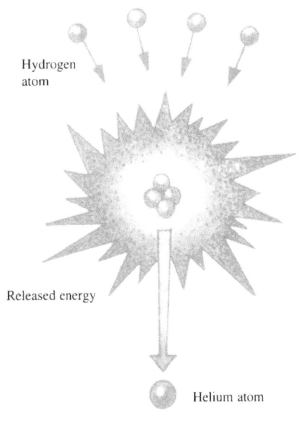

Nuclear fusion

In the sun and stars, energy is produced by nuclear fusion

335

NUCLEAR REACTORS

The first nuclear reactor based on controlled chain reaction was successfully tested by *Enrico Fermi* in 1942 on the squash court of Chicago University. In India, the first experimental nuclear reactor, Apsara, was installed by *Dr H.J. Bhaba* in 1956 in Trombay (Mumbai). Besides Mumbai, nuclear reactors have been established in Rajasthan, Narora (U.P.) and Kalpakkam (Tamil Nadu) for producing electricity.

Nowadays, three type of reactors are in use- *Pressurized Water Reactor (P W R), Advanced Gas Cooled Reactor (AGR)* and *Breeder Reactor.*

Reactors are used in many applications. The fast-moving neutrons produced during nuclear fission are used for studying their bombardment effects on other materials. They are used for producing artificial radio isotopes, which are used in the field of agriculture, industry and medicine. The heat produced during fission is used to convert water into steam. This steam runs turbines, which in turn produce electricity. They are known as *Nuclear Power Reactors.*

In an ordinary nuclear reactor, there is a block of graphite bricks in which natural uranium rods acting as nuclear fuel are inserted at fixed distances. Natural ranium contains 99.8% of U298 and 0.2% of U235 Cadmium rods are embedded in the holes of the block. These rods are known as *control rods.* Cadmium is a good absorber of neutrons. For controlled nuclear fission, more neutrons are absorbed by pushing these rods down. The nuclear reactor has some other rods known as *safety rods.* These rods automatically stop the reaction at the time of some accident or an emergency.

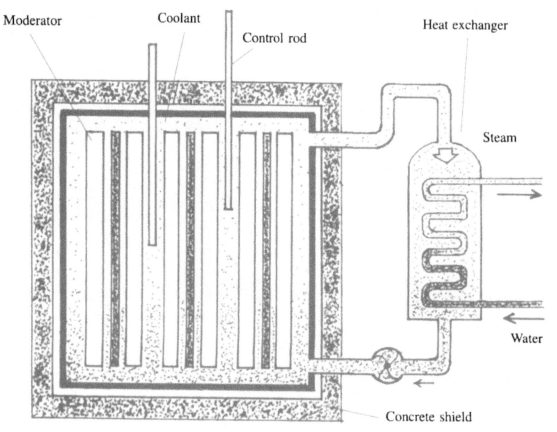

A typical nuclear reactor

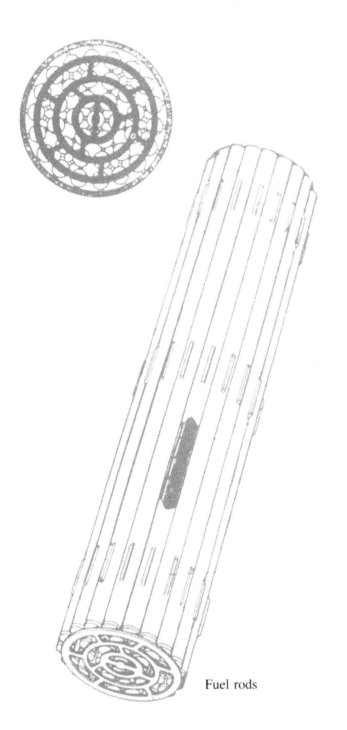

Fuel rods

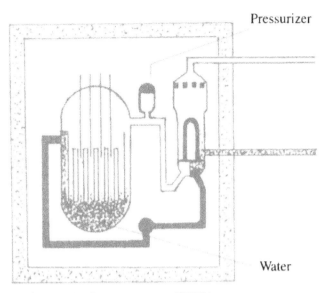

Pressurized water reactor (PWR)

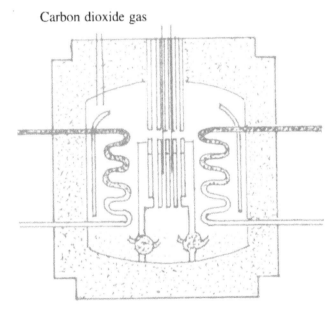

Advanced gas-cooled reactor

No neutron source is required to start the nuclear reactor. Some neutrons are always present inside it. To start the reaction, cadmium rods are pulled outwards so that the neutron absorption is reduced. These slow-moving neutrons bombard the nucleus of U235. The fast moving neutrons produced as a result of fission are slowed down by graphite. Then, they again bombard the other nuclei of U235. So it results in a chain reaction. To control this chain reaction, neutrons are absorbed by cadmium rods.

The construction of a nuclear reactor is a very complex undertaking. The main parts of the reactor-fissionable material like U235 and plutonium (PU239) are known as *nuclear fuels*. To slow down the speed of the fast-moving neutrons produced as a result of fission, graphite, heavy water (D_2O) or beryllium oxide are used.

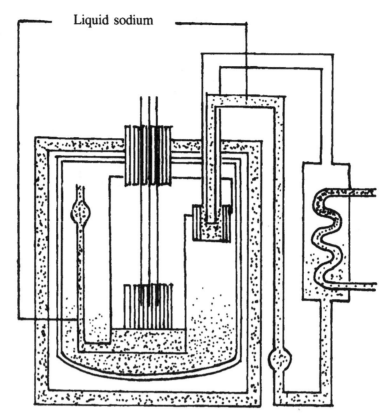

Liquid sodium

Breeder reactor

Heavy water is the best moderator, but the use of graphite is simpler. Cadmium rods are used to control the rate of the chain reaction. A large amount of heat is produced during fission, so water or CO_2 is circulated all around the reactor for cooling. During fission, some radioactive rays are emitted, which may be fatal. So a thick wall of concrete or lead is constructed all around. This several metres thick protective wall is called the *radiation shield.*

A Breeder Reactor is a nuclear reactor that produces more fissile material than it burns.

In other words, a reactor that consumes fissionable fuel (e.g. uranium) to create electricity but also produces more fissionable fuel (e.g. plutonium) usually by neutron capture and radioactive decay. Actually, it is a nuclear reactor in which non-fissile material is converted into fissile material by exposure to neutron radiation and in this process fuel is produced.

About 10% of electricity in the world is produced from Atomic energy.

338

12

Force & Movement

DISPLACEMENT, DISTANCE, SPEED & VELOCITY

Rest and Motion

When a body does not change its position with the passage of time, the body is said to be at **rest** and when the body keeps changing its position with the passage of time, then it said to be in **motion**.

Displacement

The vector distance from a fixed point which depicts the position of a body due to its velocity is called displacement.

Distance

The total length of the path traversed by a body in a fixed interval of time is called *distance*. It is a *scalar* quantity having only magnitude but not direction. In contrast, displacement has both magnitude as well as direction. Thus, displacement is a *vector* quantity.

This fact can be explained by an example. Suppose the distance between my house and my friend's house is 100 metres. If I go from my home to my friend's home and come back, the distance travelled is 200 metres but as such there is no change in my position, so the displacement is zero. So the difference between the final and initial position of a body gives its displacement. Similarly, a body completing one revolution around a circular path has *zero displacement* but distance travelled by it is equal to the circumference of the circle.

The speed of a car moving in a circle remains the same but its velocity keeps on changing

Speed

Speed is a *physical* quantity which has no direction but only magnitude. The distance travelled by a body in unit time is called its *speed*.

It is seen that when a car starts from a point, initially its speed is less but later, it keeps on increasing. Its speed is less on crowded places and turns, while it is more on wide and clean roads, so its average speed can be calculated by dividing the total distance travelled by the total time taken. The unit of speed is metre/second.

Distance

Speed or velocity

Speed

The only difference between speed and velocity is that speed is a scalar unit while velocity is a vector unit. It means that there is only quantity in speed but velocity has both quantity and direction

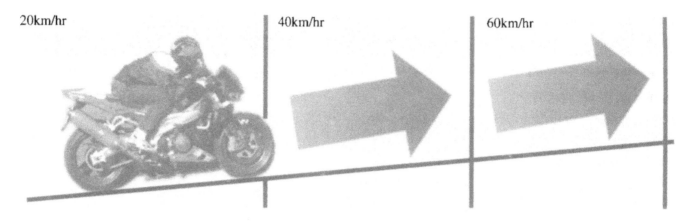

Soon after a motorcycle starts moving, its velocity increases every second. This is called accelerated motion or acceleration

Velocity

The unit of *velocity* is the same as that of speed. The only difference between velocity and speed is that direction also has to be specified in velocity. Velocity is a vector quantity so both magnitude and direction have to be specified. We can also define velocity as the rate of change of displacement in a particular direction.

A body may have two types of velocities, namely 1. *Uniform velocity*; and 2. *Variable velocity*.

When the body moving in a straight line traverses equal distances in equal intervals of time in a fixed direction, no matter how small these intervals may be, its velocity is called *uniform*. On the other hand, if the body traverses unequal distances in equal intervals of time its velocity is called *variable*.

When a moving body changes its velocity, it is called *accelerated motion*. The change in velocity may be either in magnitude or in direction or in both. If a body moves in a straight line, then its magnitude only changes. The rate of change of velocity of body is called its *acceleration* or it may be defined as the change in velocity in unit interval of time. If the velocity increases, acceleration is said to be *positive*, if it decreases, acceleration is negative. Negative acceleration is also called *retardation*. Acceleration is a vector quantity. The unit of acceleration is metre/second^2 or centimetre/second^2. These can be abbreviated as ms^{-2} or cm^{-2}.

If there is equal change in velocity in equal intervals of time, it is called uniform acceleration. If the change in velocity is unequal in equal intervals of time, it is called *variable acceleration*. In this case, the average acceleration is determined.

Acceleration due to gravity

The acceleration produced in the body on account of the force of gravity is called as *acceleration due to gravity*. It is denoted by 'g' at a given place. The value of acceleration is the same for all bodies irrespective of their masses. However, it differs from place to place on the surface of the earth. It also varies with altitude and depth.

CENTRIPETAL AND CENTRIFUGAL FORCE

Centripetal Force

A constant force always acts on a body moving around a circular path whose direction is always towards the centre of the circle. This is known as *Centripetal Force*. Circular motion is not possible without this force.

The nature of Centripetal Force is different for different systems. For moon to move around the earth in a circular path, the necessary Centripetal Force is provided by the earth in the form of the *gravitational pull*. For the circular motion of the electron around the nucleus, the necessary centripetal force is provided as the *electrostatic force* between positively charged nucleus and negatively charged electrons.

A bicycle rider negotiating a curve leans inwards to provide centripetal force required to push him sidewise towards the centre of the circle along which he moves or he takes a larger curved path. According to the *laws of friction*, the angle made by the scooter or bicycle with the vertical should not exceed the angle of friction. If it is more than the angle of friction,

If a bucket filled with water is moved fast in a circular path, water from it does not fall due to centrifugal force

the scooter will slip. So to avoid accident while turning, the rider either decreases the speed of the vehicle or negotiates a larger curved path.

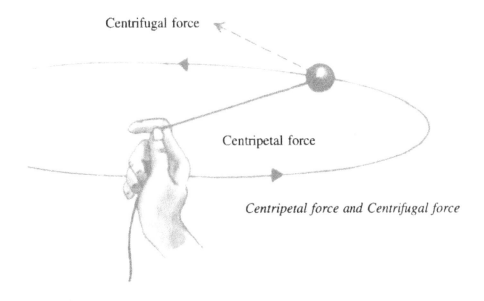

Centrifugal force

Centripetal force

Centripetal force and Centrifugal force

An example of centrifugal force

Centrifugal Force

If a person is sitting in a car against the door and the car takes a turn on circular road, while turning he feels that he may fall out if the door opens. In this case, the reactionary force acting outwards from the centre is called *Centrifugal Force*. The centrifugal force is equal to oppositely acting centripetal force.

If a bucket filled with water is swung in a vertical circle, the water does not fall. The reason is that a force acts on the swinging bucket which pushes the water inside.

Rollercoaster

344

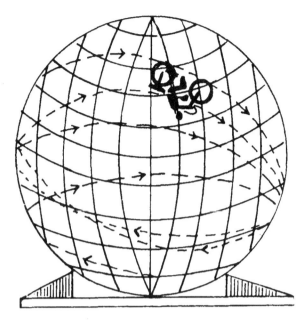

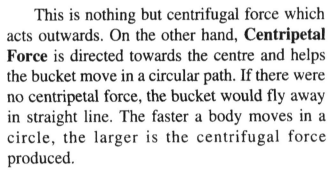

You must have seen in a circus the acrobatics of the motorcyclist. The rider moves fast in a spherical field and for a few moments is even vertically upside down but remains attached to the roof. This is a magic worked by centrifugal force

This is nothing but centrifugal force which acts outwards. On the other hand, **Centripetal Force** is directed towards the centre and helps the bucket move in a circular path. If there were no centripetal force, the bucket would fly away in straight line. The faster a body moves in a circle, the larger is the centrifugal force produced.

Many instruments like the centrifuge drier, centrifuge pump and cream separator have been devised on the basis of the centrifugal force.

You might have seen the roller coaster at Appu Ghar in which the car does not fall down even when the wheels are up. This is nothing but the magic of Centrifugal Force.

NEWTON'S LAWS OF MOTION

Motion

What is motion? What are the factors which produce motion? Why does earth revolve around the sun in an elliptical orbit? Why does a body fall with constant acceleration onto the surface of the earth? How do the factors responsible for motion affect it? These are some of the basic questions of physics which can be understood as follows.

The interaction between a body and its surrounding bodies leads to the production of motion. This interaction is called force.

Force

Force can be defined as a physical quantity which tends to produce or produces the motion of a body. In other words, *Force* can be defined as that physical entity or quantity that changes the state of an object which is at rest or in motion. *A Force* is actually a push or pull upon an object resulting in the change of state of that object.

Mathematically, Force is equivalent to the product of the mass and acceleration of the object or body.

$F = m \times a$ where 'F' stands for force.

'm' = mass and

'a' = acceleration of the body.

We have to apply force while weightlifting, playing football, hammering a nail etc. Force is that physical quantity that changes or tends to change the state of rest or of uniform motion.

Isaac Newton (1642-1726)

Force can alter the size or shape of an object

Force develops motion in objects

We can change velocity, direction of motion or both by force. This force produces acceleration or retardation in a body. Force can not only give motion to body at rest or change the state of motion, it can also change the shape and size of an object.

Newton's laws of motion

Newton is known as the *father of physics*. He propounded three laws of motion.

First law: If a body is in the state of rest or uniform motion, it will continue to be in that sate until or unless acted upon by some external force.

Second law: The rate of change of momentum is directly proportional to the impressed force and is always in the direction of the force.

Third law: To every action, there is always an equal and opposite reaction, or the mutual actions of two bodies upon each other are always equal but directed to contrary parts..

There are many examples of the third law of motion, e.g. the rifle recoils backward due to reaction when a bullet is fired. To balance the additional weight on our palm, we have to lift our hand upward with the same force. The rocket is propelled forward when the gases are shot out with very high speed. The higher the speed of the gases ejected, the higher will be the velocity of the rocket. The gravitational pull of the earth makes the moon revolve around it. The gravitational pull of the moon causes tides in the sea. The sun pulls the earth towards itself and so does the earth with the same force. This action and reaction results in the earth revolving round the sun. In the same fashion, the nine planets of the solar system are maintaining a balance with each other.

Momentum

It is a general experience that it is difficult to stop a moving loaded truck than an empty truck, both moving with the same velocity. The reason is that the *momentum* of the loaded truck is more than that of the empty one. If two bodies of different masses are moving with same velocity, we have to apply a greater force to stop the body with large mass. To change the state of motion of a body, the required force depends upon mass, velocity or product of both. The physical quantity which is the product of mass and velocity of a body is called its *momentum*. It is also a vector quantity. The momentum of a body remains constant in the absence of an external force.

Reaction

Action

When a person jumps from a boat towards the shore, he uses force to push the boat behind. As a reaction to his action, the boat exerts a force to send the person forward. Thus, the person manages to jump onto the land from the boat

347

JET PROPULSION

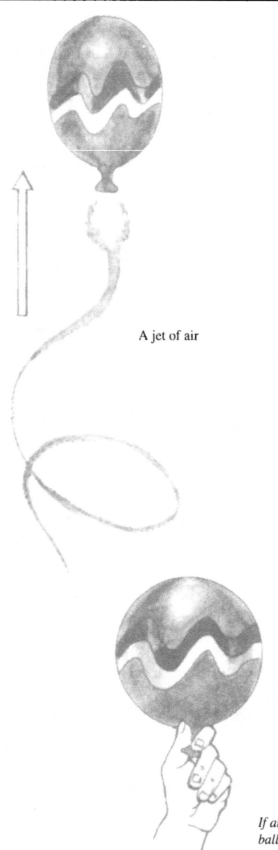

A jet of air

Jet propulsion is a consequence of Newton's third law of motion and *the law of conservation of momentum.*

Liquid or gas coming out of a narrow nozzle with great velocity in the form of a stream is called a jet. The vessel is propelled in a direction opposite to the direction of the jet if it is free to move. The propulsion of a body using a jet is called *jet propulsion* and the body propelled is said to be *jet-propelled.*

Fill a balloon with air. If the thread tied at its mouth is made a bit loose so that air comes out in the form of a jet, the balloon will at once move in a direction opposite to the direction of air jet. The rocket flies on the same principle.

Rocket propulsion

A mixture of fuel (oil gas) and oxygen is ignited at high pressure inside the combustion chamber of the rocket. The hot gases produced due to combustion shoot out from the exhaust nozzles. These gases exert an equal and opposite thrust (reaction) on the rocket and it starts moving upwards. Rockets can fly in vacuum also because they carry their own oxygen supply.

Jet propulsion plane

The jet plane works on the principle of jet propulsion. A bright white line which appears in the rear part of the jet plane and can be seen in the sky is called gas jet. Before the take off of the jet plane, the pilot starts a small motor which drives a compressor. It sucks the air inside. The compressed air, along with the required fuel (oil), is sent into the combustion chamber forming a *combustible mixture.* The mixture is burnt by producing a spark across spark plugs. By combustion, hot gases are produced. These

If air is released partially from a balloon blown with air, the balloon will fly in a direction opposite to that of the escaping air

gases strike the blades of the turbine with high pressure so that the shaft of turbine begins to rotate and the engine starts working. After passing through the turbine, these gases shoot out through the narrow nozzles with high speed. The jet plane gets a push in the opposite direction and starts moving forward. The jet planes can fly higher than aeroplanes but unlike rockets, they cannot fly in vacuum because they take oxygen from the atmosphere. A rocket is basically *a propulsion device* that produces thrust by ejecting stored material. The material that is ejected is called the **propellant**. Rockets can be classified by the type of propellant such as *Chemical, Electric, Nuclear* and *Solar*.

They can also be categorised by their application such as *Booster, Upper Stage* or *Sustainer, Altitude Control* and *Station keeping.*

They can be also classified by vehicle type: For instance, **Missile, Space Vehicle and Launch Vehicle**.

A Space Launch Vehicle is a kind of rocket which is used to launch spacecrafts into the space. Probably, the most recognised launch vehicle is the one that is used for the **Space Shuttle**.

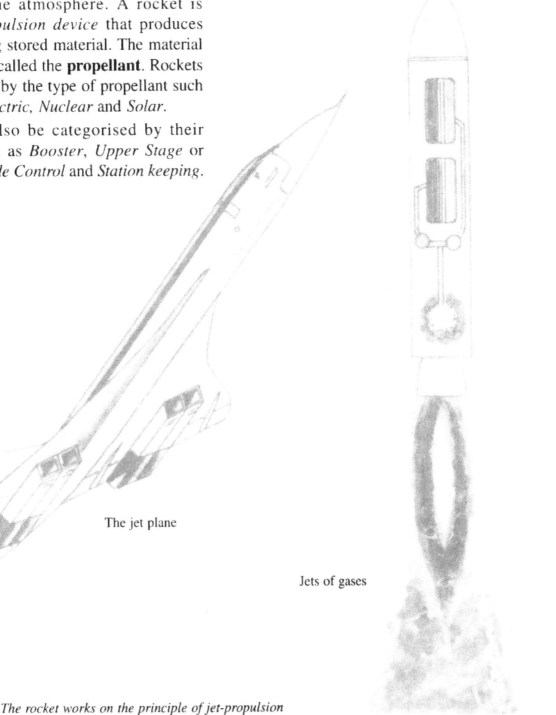

The jet plane

Jets of gases

The rocket works on the principle of jet-propulsion

349

CENTRE OF GRAVITY AND EQUILIBRIUM

Centre of gravity

It is a well-known fact that the earth attracts everything towards its centre. This is the reason why everything falls back to earth when thrown upwards.

Everything is composed of small particles and the earth attracts every particle towards its centre. The size of the earth is much larger than the size of the object. So the earth applies an attractive force on different particles of an object whose direction can be assumed to be parallel.

An object thrown upwards gradually slows down and returns to the earth

Owing to the force of earth's gravity, though we are inverted in the air, we do not fall off into the space

The resultant of the parallel forces is equal to the weight of the body and acts in a vertically downward direction from a fixed point inside the body. The algebraic sum of all moments of forces at this point is zero. The resultant force always acts at this point in whatever condition the body is placed. This point is called 'The centre of gravity'. So 'The centre of gravity' is a point where the total mass of the body is concentrated.

Equilibrium

If the forces are acting on a body in such a way that the body has neither linear nor rotational motion, then the body is said to be in *equilibrium*. For a body to be in equilibrium, the vertical line drawn from the centre of gravity should fall within the base of the body. If this vertical line fall outside the base of the body, it will topple down. So it is advisable to provide a wider base and low centre of gravity to keep a body in stable position.

The leaning tower of Pisa is still standing because the vertical line from the centre of gravity passes through its base. When the inclination of this tower would be such that the vertical line from its centre of gravity falls outside its base, it will fall.

When you walk with a bucket full of water in your right hand, you lean a little towards the left so that the vertical line from the centre of gravity would fall between your legs. In this way, your body maintains its equilibrium.

We lean forward while carrying a heavy weight on our back so that vertical line from the centre of gravity would pass through our legs. A person falls down when the equilibrium is disturbed.

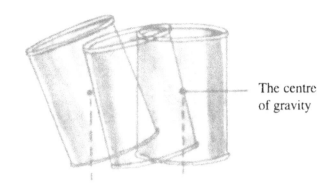

The centre of gravity

When an object, whose position has been slightly altered by a force, returns to the original position on removal of the force, it is said to be in stable equilibrium

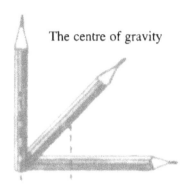

The centre of gravity

When an object fails to return to its original position, it is said to be in unstable equilibrium

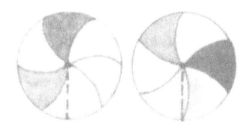

The centre of gravity

When an object, on force changing its position, makes no effort to return to its original position, it is said to be in neutral equilibrium

351

UNIVERSAL GRAVITATION

Universal Gravitation

According to *Newton's law of universal gravitation, every particle in this universe attracts every other particle*. The force of attraction between any two such particles is directly proportional to the product of their masses and inversely proportional to the square of the distance between them. This force is always directed towards the line joining the two points. This is known as Newton's law of universal gravitation.

In daily life, we do not feel the gravitational force because the force of attraction is very small but its value is very large between planets and satellites. Due to this force of attraction, the earth revolves round the sun and the moon around the earth. The stars and planets are orbiting due to this attractive force only.

Acceleration due to gravity

If a body is dropped from a height, it will fall towards the earth due to the gravity of the earth. This attraction brings about a downward acceleration (a constant increase in the rate of speed). The rate of change of velocity of the body per second due to force of gravity of the earth is called *acceleration due to gravity*. It is represented by '*g*' and its value is 9.80665 m/s².

Variation in the value of 'g'

The value of '*g*' is different at different places on the earth's surface. The value of '*g*' decreases as we move up or below the surface of the earth.

The value of '*g*' is greater at the Poles than the Equator. The reasons are the earth's shape and rotation around its axis.

The earth is not perfectly spherical. It is flat at both the Poles. Its radius at the Poles is about 21 km smaller than at the Equator. As '*g*' varies inversely to the square of the earth's radius, the value of '*g*' increases as we move from the Equator towards the Poles.

As we know, the earth is rotating with a velocity which is maximum at the Equator and minimum at the Poles. All the bodies on the earth are also rotating with it in a circular path. Due to rotation, each body experiences a centrifugal force which tends to take the body away from the earth and, therefore, effective gravity decreases. Centrifugal force is maximum at the Equator and minimum at the Poles. Hence the value of g is least at the Equator and greatest at the Poles.

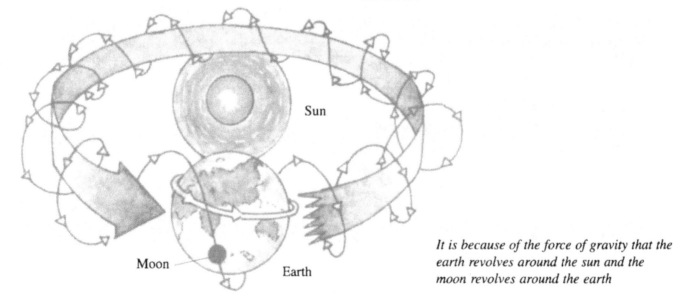

It is because of the force of gravity that the earth revolves around the sun and the moon revolves around the earth

The value of 'g' decreases as we move up from the surface of the earth. The value of 'g' decreases as the body is taken increasingly inside the earth or taken more and more above the earth's surface. The value of acceleration due to gravity at the centre of the earth is zero. The weight of the body—but not its mass—is zero at the centre of the earth. In fact, the vertical force exerted by a mass towards the earth as a result of the gravitational pull is called the **weight** of a body on earth or **weight** can also be defined as the force exerted upon an object by virtue of its position in a gravitational field. The weight of a body in *moon* is **1/6th** of its weight on our planet, *earth*.

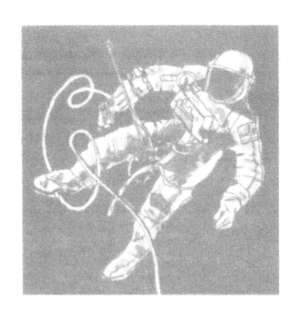

As an object moves away from the earth, the earth's gravitational force gets reduced. As a result the weight of the object also lessens

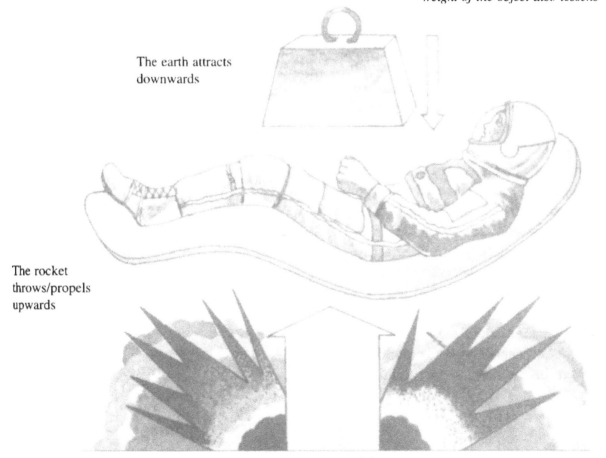

The earth attracts downwards

The rocket throws/propels upwards

The spaceman is caught between the earth's downward force of attraction and the rocket's propulsion. As a result, he experiences weightlessness

ESCAPE VELOCITY

Gravitation frontier

A body falls back to earth when thrown upwards due to the earth's gravitational attraction. A greater velocity is required to throw a body high up in the air so that it may not return. The least value of that velocity is called the *escape velocity*. The boundary above which objects do not return to earth, instead they orbit in space is called the *gravitation frontier*.

Why there is no atmosphere on moon?

It is a well-known fact that the earth has atmosphere while the moon does not. The average velocity of the molecules of the lightest known element hydrogen on earth is less than the escape velocity even at the highest temperature. The average *thermal velocity* of hydrogen molecules at 500 K is 2.5 km/s, while the *escape velocity* from the surface of the earth is 11.2 km/s, The average thermal velocity of oxygen and nitrogen molecules, which are heavier than those of hydrogen, is even less than 2.5 km/s at this temperature. So the molecules of these gases present on earth can't escape. That is why the earth has an atmosphere. The value of radius and acceleration due to gravity on moon is less than that of the earth. So the escape velocity on moon is hydrogen and other gases is more than the escape velocity at the temperature 2.38 km/s. The average velocity of moon, e.g. the average thermal velocity of hydrogen molecules at 1000 K is 3.5 km/s, which is more than the escape velocity at the moon. So molecules of gases cannot stay on moon and hence, there is no atmosphere. Similar conditions are found on other planets, e.g. on Mars the escape velocity is less than the average thermal velocity of gas molecules. Hence, there is no atmosphere. But on planets like Saturn, Jupiter, etc., the value of escape velocity being very large, a dense atmosphere is found.

The different escape velocities in the various planets of our solar system are as follows:

Planet	Escape velocity
Mercury	4.30 km/s
Venus	10.30 km/s
Earth	11.20 km/s
Mars	5.10 km/s
Jupiter	60.00 km/s
Uranus	22.00 km/s
Saturn	35.00 km/s
Neptune	25.00 km/s
Pluto	(Gravity unknown)

On the surface of the earth, the escape velocity is about **11.2 kilometres per second**. So the escape velocity of the spacecrafts flying from the earth's surface to the space should be more than 11.2 kms/sec.

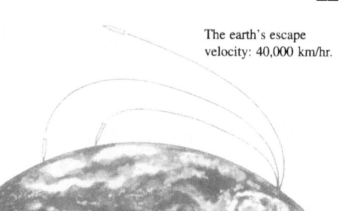

The earth's escape velocity: 40,000 km/hr.

If the velocity of a spacecraft is greater than the earth's velocity, it flies into space. If it is less, then it falls back to the earth

354

ARTIFICIAL SATELLITES

The process of launch

A body which revolves around the earth at a height of a few hundred kilometres is known as an *artificial satellite* of the earth.

An artificial satellites revolves round the earth under centripetal acceleration.

An artificial satellite is launched with the help of multistage rockets to establish it in the orbit of the earth. The satellite is put in the uppermost part of the rocket. Firstly, these rockets are launched from the earth in the vertical direction. When a rocket acquires a certain height (150-300 km), it starts bending in the horizontal direction and makes an orbit around the earth. When the fuel of the first stage of the multistage rocket is burnt out, it gets detached from the satellite. At this time, the velocity of the rocket becomes about 10,000 km/hr and now the second stage of the rocket takes over. Soon, the fuel of the second stage is also burnt out. This provides a velocity of 20,000 km/hr to the satellite. This stage also gets detached from the satellite and the third stage takes the satellite into the space. The third stage provides a velocity of 30,000 km/hr to the satellite and establishes it in its predetermined orbit.

If the horizontal component of the velocity of satellite is reduced due to some defect in it, the gravitational force acting on it becomes more than the centripetal force and the satellite loses

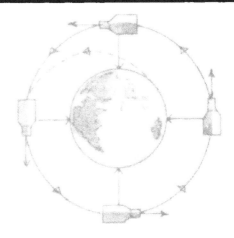

A rocket establishes a satellite into its orbit

its orbit and starts revolving round the earth in a spiral path. In this spiral path, the satellite enters into the earth's atmosphere and due to the atmospheric friction, it becomes red hot and burns out in the atmosphere itself. That is why, the orbits of the artificial satellites are always maintained at a considerable distance from the earth's atmosphere so that the satellite can revolve round the earth for a very long time.

Uses

Today, more than 1500 artificial satellites are revolving round the earth in different orbits. These satellites have been launched by several countries of the world, including India, and are being used for communications, naval purposes, weather forecasts and for finding the earth's resources.

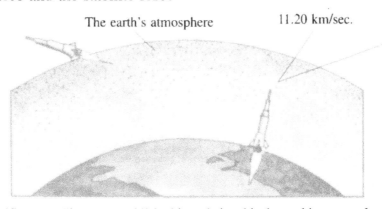

The earth's atmosphere   11.20 km/sec.

Beyond 1600 km of the atmosphere, the rocket enters the space

Artificial satellites are established into their orbits by multi-stage rockets

ROTATION

Linear and Rotatory motion

When a rigid body is subjected to external force, there can be two types of motion—*Linear* and *Rotatory*. In linear motion, a body moves from one place to another while in rotatory motion, it rotates around a fixed axis. In rotatory motion, the linear velocity of different parts of the body is different. The linear velocity of particles situated near the axis is less than those at a distance from the axis. But the angular velocity of every particle is the same. The spinning of the top is a good example of rotatory motion. All rotating bodies have an *Angular velocity*.

Moment of Inertia

Inertia is the tendency of a body to remain in state of rest or in motion until or unless some external force acts on it. This was propounded by Newton.

When we change the angular velocity of a body in a stage of rest or uniform rotation, it tries to oppose the change. The change in the state of the body can not be done unless some couple acts on it. So the *moment of inertia* is a property that a body possesses about the axis of its rotation.

The moment of inertia of a body not only depends upon mass but also the distance of various particles of the body from the axis of rotation.

The larger the mass of a body, the larger will be the force required to change its state of rest or uniform motion. So the mass of a body is its measure of inertia.

An example of rotatory motion

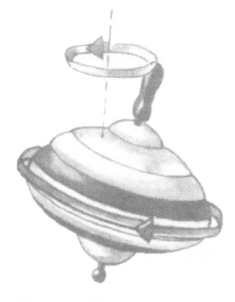

The rotation of a top is rotatory motion

While jumping from a moving bus, we are jerked forward, because of the sudden change in momentum

356

The earth is like a flywheel which, owing to its large amount of inertia, rotates on its axis at constant velocity

depends both on mass and the distance from the axis of rotation.

A body will continue to rotate with the same velocity and in the same direction due to its moment of inertia until or unless it is acted by a couple. Using this principle, a flywheel is attached with the axle of rail engine. The piston of the engine rotates this wheel, The flywheel takes out the piston from dead centres. Our earth is like a flywheel which due to its large moment of inertia is rotating around its axis with constant velocity. The wheels of the vehicles are so constructed that their moment of inertia is the maximum for the minimum possible weight to enable the vehicle to move uniformly.

■■

Boiled eggs can be distinguished from unboiled eggs by the principle of moment of inertia. A boiled egg will rotate on a table, the other will not

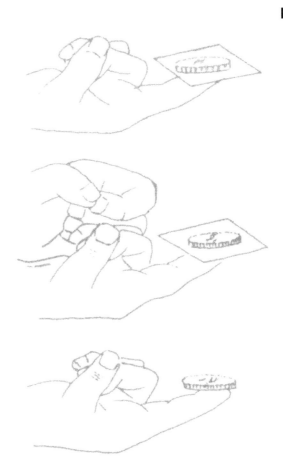

Place a coin on a card, at the tip of your finger. If you strike the card, it will fly off but the coin will remain at your finger due to its moment of inertia

A large moment of force is required to change the state of rotation of a body having a large moment of inertia, so it is clear that the moment of inertia plays the same role in rotatory motion as mass is rectilinear motion.

There is a difference between inertia and moment of inertia. Inertia depends only on the mass of the body while the moment of inertia

FRICTION

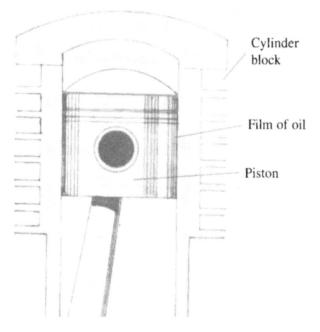

A matchstick can be lit by rubbing it hard against a rough surface

Cylinder block

Film of oil

Piston

In machines, to reduce friction, oily substances like grease are used

Friction and Heat

A car moves easily on a smooth road but not on a rough road. A force acting in opposite direction always tried to stop the motion. This force is known as the *force of friction*. The frictional force acts between two surfaces, of which one is moving, and opposes the motion of the moving surface. The chances of slipping on a smooth surface are more than on a rough surface because the force of friction is less for a smooth surface than for a rough surface.

Friction produces heat. In ancient days when there were no matchsticks, man produced fire using friction. Even now, a matchstick can be lighted by striking it against a rough surface. The functioning of a cigarette lighter is also based on friction. A spark produced due to friction between flint and a small metal piece, fires the gas.

The moving parts of a machine get heated and wear out due to friction. Lubricants are used to reduce friction. Ball bearings or roller bearings also reduce friction. Grease is also a lubricant. One can also reduce friction by polishing the surface of objects. The body of aeroplanes, motorcars, etc. are also made smooth to avoid friction.

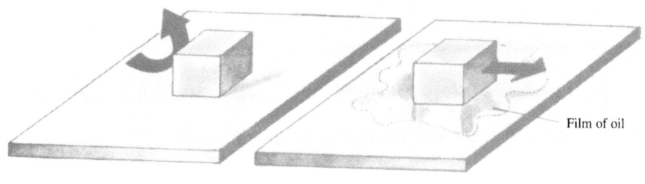

Film of oil

On oily surfaces, the force of friction is less

Friction is basically of two types:

1. Sliding Friction

When two solid objects are in contact and a force is applied to slide one object against the other, the sliding friction force resists the motion. If 'F' is the force pushing on an object and 'Fr' is the force of friction, the relationship between 'F' and 'Fr' will determine whether the object will slide or not.

On the other hand, when a ball or a wheel is in contact with a solid object and a force is applied to the wheel, it will start to roll due to the friction at the point of contact with the other surface. This is called the **Starting Friction** for a wheel. Once, the wheel starts rolling, there is a resistive force that slows the wheel's motion on the other surface. This is known as **Rolling Friction**.

2. Atmospheric Friction

To protect a satellite from atmospheric friction while entering the earth's atmosphere, a special *heat shield* is used. The shooting stars due to friction become red-hot and burn out when they enter the earth's atmosphere. They are broken into small pieces and get scattered in the atmosphere due to heat and friction.

Friction is also useful to us. The friction between shoes and floor prevents us from slipping. Brakes work on friction. Friction prevents car tyres from slipping on the road. We can neither walk nor run without friction. In fact, friction is a necessary evil.

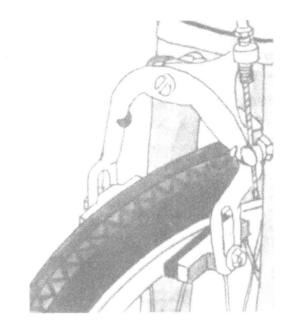

A cycle brake

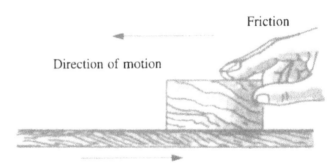

The force of friction exists between two substances in contact

ELASTICITY

Elasticity and its extent

There is a change either in the shape or size or both when a body is subjected to an external force. This external force is called the deforming force. There is a tendency in every body to oppose this force. Due to this tendency, a body tries to regain its original state when this deforming force is removed, e.g. rubber gets elongated when it is stretched but regains its original length when released. Similarly, a rubber ball gets squeezed when pressed but regains its original shape when released. The deformation produced on applying force and regaining the original state on removal of force is called *elasticity*. Actually, elasticity is the property of a body by virtue of which it opposes any change being produced in its shape or size by the external force, and tends to regain its original shape and size after the removal of external force.

If a body regains its shape and size fully after removal of the deforming force, it is called *perfectly elastic*. There is no perfectly elastic material. Steel etc. are elastic up to a certain limit. They get permanently deformed by a large force.

The body which does not possess the property to oppose the deforming force is called *plastic*. They do not regain their original shape or size when the external force is removed, e.g. wet sand and wet flour. There is no perfectly plastic material. *It has been observed that steel is more elastic than rubber.* Elasticity or Deformation is basically of three types: *Longitudinal or Tensile Deformation, Shear Deformation* and *Bulk Deformation.*

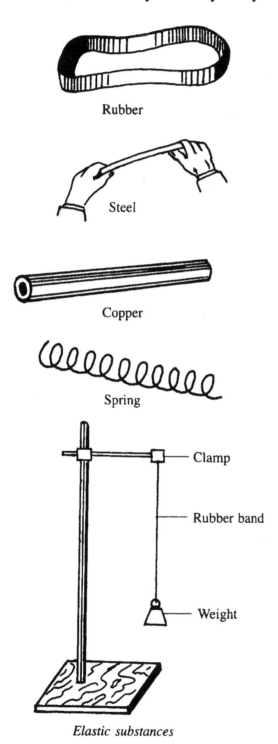

Rubber

Steel

Copper

Spring

Clamp

Rubber band

Weight

Elastic substances

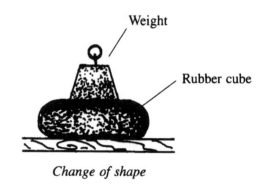

Weight

Rubber cube

Change of shape

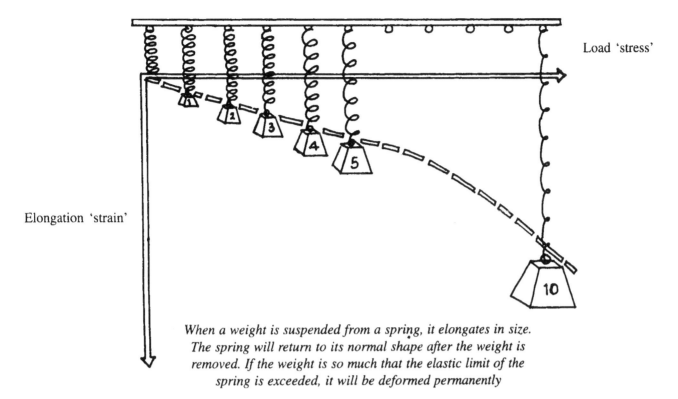

Load 'stress'

Elongation 'strain'

When a weight is suspended from a spring, it elongates in size. The spring will return to its normal shape after the weight is removed. If the weight is so much that the elastic limit of the spring is exceeded, it will be deformed permanently

Strain

Whenever a deforming force is applied on a body, there is a change in shape, size or both due to the displacement of molecules. It is called *strain* in the body. The change in the dimension of an object due to deforming force is called strain. Strain is of three types depending upon the force applied to deform it—longitudinal, volume and shearing. Strain is a ratio, therefore it has no units.

Stress

A body is said to be strained when it undergoes a change in its shape or size on the application of a deforming force. To oppose this change, according to Newton's third law, an internal reactional force is produced in each section of the body which tries to bring back the body to its original shape and size. This internal reactional force opposes the external force. The restoring force per unit area of the body is called *stress*. It is measured as the external force per unit area. The unit of stress is Newton per metre square.

Elastic limit

When a weight is hanged on a spring, its length increases. It comes back to its original state when weight is removed. If we keep on increasing the weight continuously, a stage will come when it will not regain its original state but will be deformed permanently. Spring will remain elastic up to a fixed value of external force. The limit beyond which the object is deformed permanently is called its *elastic limit*. Sometimes a phenomenon called **Elastic fatigue** or **Metal fatigue** has been discovered in ductile metals in which the material deformed fails to return to its original shape.

■■

PERIODIC MOTION

Planetary revolution

Any motion which repeats itself about an equilibrium position in equal intervals of time is called *periodic motion*. The fixed time interval is called *time period*.

The revolution of planets around the sun, the oscillation of the pendulum of a wall clock are examples of periodic motion. The earth revolves around the sun and takes 365 days to complete one revolution. This motion of earth is called periodic motion, Whose time period is 365 days. Similarly, the moon revolves around the earth and completes one revolution in 27.3 days. So, the time period of the moon's periodic motion is 27.3 days. The revolution of artificial satellites around earth is also an example of periodic motion.

Oscillation

When vibrations are produced in a violin string, it starts oscillating from its equilibrium position. The motion of swing is also oscillatory. When a body possessing periodic motion moves to-and-fro about a fixed point, its motion is called *oscillatory motion*. Oscillatory motion is also a periodic motion.

When a blade is struck, it will make to-and-fro oscillatory movements around its mean position of rest

When the second end of an iron strip clamped at one end is pressed and released, it starts oscillating about its equilibrium position. Similarly, when a string with weight at its end is stretched and released, the weight starts oscillating up and down.

Simple harmonic motion

Simple harmonic motion of a simple pendulum is a well-known example of periodic motion. In simple harmonic motion, the body moves to-and-fro along a straight line about the mean position. Simple harmonic motion can be easily represented as the projection on any diameter of a point moving in a circle with uniform speed.

The motion of pendulum can be understood as follows. When the bob of the pendulum is displaced from its mean position to a new position, its potential energy increases. When the pendulum is released, it tends to go to its mean position. During this process, the stored energy is converted into kinetic energy. At mean position, potential energy is minimum and kinetic energy is maximum. According to the law of inertia, the bob crosses its mean position and its centre of gravity moves up, resulting into increase in its potential energy. This increase in potential energy is brought from kinetic energy, i.e. its kinetic energy is reduced. When the displacement of the bob is maximum from the mean position on either side, whole of the kinetic energy is converted into potential energy. In this condition, the bob comes in a state of rest for a fraction of a second. To balance the centre of gravity, the bob again moves towards the mean position, resulting in increase in kinetic energy at the expense of potential energy. The interconversion of potential and kinetic energy will continue indefinitely under the condition that no force of friction of air acts on the bob.

SIMPLE MACHINES

Simple and complicated machines

Machine is a device in which by application of force at one mint, we can lift a heavy load at the other.

Today, man is using many types of machines. A complicated machine consists of many simple machines, e.g. wheel, gear, bearings, shafts, pistons, springs, rods, etc. There is a definite linkage between the different parts of a machine. They synchronize together to work as machine though their own work is very simple. They only work like a simple machine. Though the nail/extractor and knife not look like machines, they act as simple machines.

Pulley

It is a device by which we can lift a heavy load with little effort. An opposing force always acts when a body is displaced from its position. If this opposing force is absent, the body can be easily displaced. A *pulley* works against these opposing forces. To lift any weight by a pulley, we have to pull the rope more in the downward direction as compared to the weight in the upward direction. Therefore, effort travels more distance as compared to weight. The work done in lifting a weight by a pulley is not less than the work done by lifting it straight. In fact, the work done by the effort is equal to the work done on the weight.

Archimedes once wrote to the king that he had invented a machine by which a huge weight could be lifted with very little effort. He demonstrated this by single-handedly pulling a large ship full of royal forces to the shore. This work was a magic achieved by pulleys

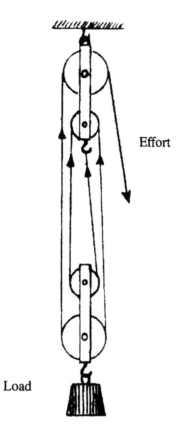

The lesser the effort you want to use,
the more should the number of pulleys be

A man can even lift a car engine using an ordinary pulley because his whole weight acts as effort. Pulley systems are of several types, in which one or more than one pulley are used.

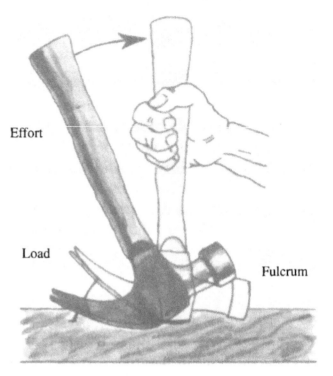

The claw hammer is a simple machine

The see-saw is in equilibrium when the moment of force on both the sides is equal

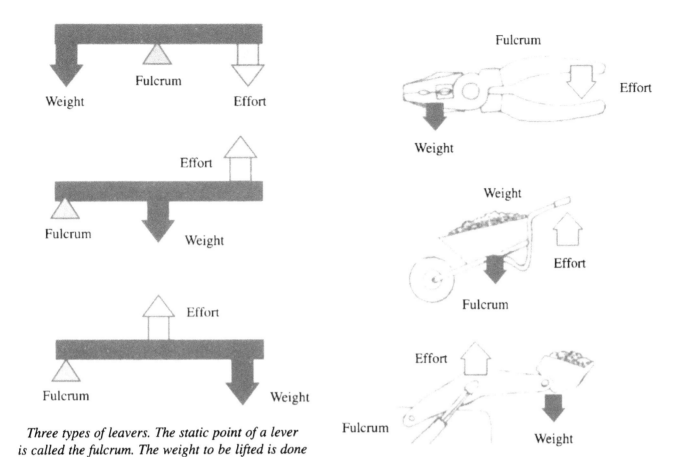

Three types of leavers. The static point of a lever is called the fulcrum. The weight to be lifted is done so by the effort. Various machines work on the principle of the lever

To balance a see-saw in this position, the boy on the left would have to move nearer to the fulcrum

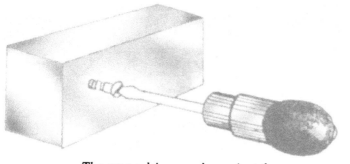

The screw driver works against the strong force of friction in the wood

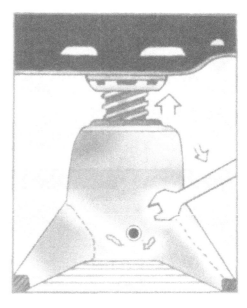

Screw jack

By using a screw jack, cars and houses are lifted up

Lever

It works on *the principle of moments*, e.g. the ordinary balance in which the weight placed in one pan is balanced by the item in the other. When the correct weight is placed on both pans, then the moments at both ends become equal because the distance from the fulcrum is the same.

Heavy loads can be pushed with less effort using levers. *Levers* are of three types—in the first type of levers, the fulcrum is in between the load and the effort, e.g. crowbar. In the second type, the load is placed between the fulcrum and the effort, e.g. the wheelbarrow. In the third type, effort is applied between the load and fulcrum, e.g. a pair of tongs.

Screw

The effort arm is much longer than the load arm in the *screw*, so we can apply larger force with very little effort. It is also a simple machine which works against the force of friction through the screwdriver in the wood. The screw jack also works on the principle of screw. By rotating its handle, the car can be raised up.

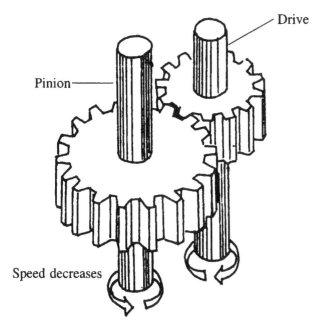

Speed gears

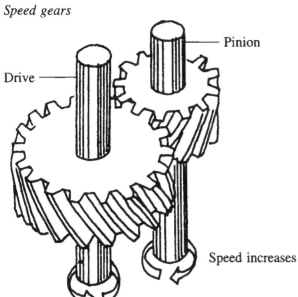

Helical gears

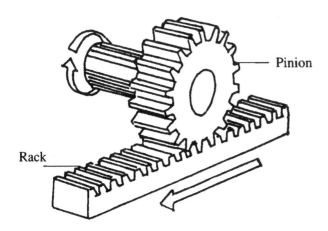

Rack and pinion gears

Gears

A wheel with teeth cut into its outer rim is called a *gear*. Gears are connected to shafts. They are also a kind of simple machines which not only convert smaller forces into larger forces but also larger forces into smaller ones. Gears can be used not only to increase or decrease the speed of a moving body but also to change the direction of motion. Gears are used in almost all the motor vehicles.

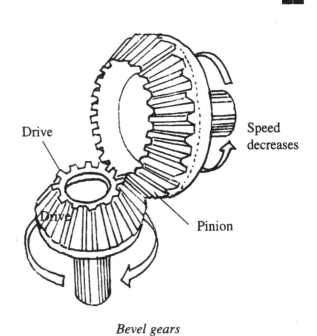

Bevel gears

367

SYSTEMS OF UNITS

Two systems are in use to measure the physical quantities nowadays. They are—the **Centimetre -Gram-Second system (C.G.S. system)** and the **Metre-Kilogram-Second system (M.K.S. system)**.

Units of Different Quantities in the C.G.S and M.K.S. Systems. The third system of units is the **Foot, Pound and Second unit or the (FPS)** system which was used in ancient times but now it has become obsolete and people rarely use this system, today.

Quantity	CGS system	MKS system
Length	centimetre	metre (m)
Mass	gram	kilogram (kg)
Time	second	second (s)
Velocity	centimetre per second	metre per second (ms^{-1})
Acceleration	centimetre per second^2	meter per second^2 (ms^{-2})
Force	Dyne	Newton $(1N=10^5$ Dyne$)$
Work	erg	Joule $(1J=10^7$ ergs$)$
Power	ergs per second	Watt=Joule/ second (JS^{-1})

M.K.S.A. system

A third new system of units is the M.K.S.A. *(Metre-Kilogram-Second-Ampere system)*.

The M.K.S. system, based on the three units of length, mass and time is called the M.K.S. system. It is very important in the field of mechanics. In 1958, International Committee of Units devised a Giorgi system for *mechanics, electricity* and *magnetism.* This system was based on the fundamental units of these four fundamental quantities—length, mass, time and electric current.

(a) Metre (m) (b) Kilogram (kg); (c) Second (s); and (d) Ampere (A).

In 1960, the International System of Units was recommended. This system has six fundamental and two supplementary units. These fundamental and supplementary units are as follows—Metre (m), Kilogram (kg), Second (s), Kelvin (K), Mole (mol) for amount of substance, Candela (cd), Ampere (A), Radian (rad) for plane angle and Steradian (sr) for solid angle.

In this system, we can write units of any physical quantity, e.g.

Megametre per hour	Mm/h	Mmh^{-1}
Metre per second	m/s	ma^{-1}
Kilometre per hour	km/h	Kmh^{-1}
Gram per cubic centimetre	g/cm^3	gcm^{-3}
Metre per second	m/s^2	ms^{-2}
Newton per square metre	N/m^2	Nm^{-2}
Kilogram force per square centimetre	Kgf/cm^2	$Kgfcm^{-2}$
Joule per degree celcius	$J/°C$	$J°C^{-1}$
Joule per gram per degree celcius	$J/g °C$	$J/g^{-1} °C^{-1}$

In chemistry, the quantity of chemicals is measured using mole. One mole=gram atomic mass or gram molecular mass

$= 6.023 \times 10^{23}$ atoms or molecules

$= 22.4$ litres

At standard temperature and pressure, 1 mole can be expressed in all the three units of mass—number of atoms or molecules, or volume.

13

COMMUNICATION

COMMUNICATION

The word 'communication' means exchange of information. Animals communicate with each other by special sounds. In olden times, man could communicate by shouting or blowing a horn or beating a drum or flashing a light. Man gradually developed the art of talking and writing and started expressing his complex thoughts and information through language and writing. In the Mughal period, pigeons were used as carriers of messages. In the early 16th century, the postal system had begun. But the riders were used till 1830 for carrying letters and parcels to distant places.

Communication took a new turn with the development of technology. In 1837, Cooke and Charles of England and Samuel Morse of America developed the *electric telegraph*. The information was sent in the form of electrical signals or codes through a cable by this instrument. The dots (short clicks) and dashes (long clicks) were used to represent the letters of the alphabets. In 1876, Alexander Graham Bell invented the *telephone* by which it became possible to communicate by voice over long distances through cables. In 1894, Guglielmo Marconi of Italy invented the *wireless telegraph* by which messages could be sent across a long distance without wire. The efforts of these great

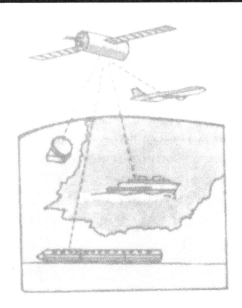

For distant communication, nowadays communication satellites are used

scientists made it possible to send or receive messages from any part of the globe within no time.

The electric telegraph led to the development of the *telex system* while the wireless telegraph gave birth to the *radio*. Telephone networks have advanced very much. Besides messages, we can also send pictures and documents by fax.

Radio waves have played an important role in the communication systems. Radio waves are

Around 5,000 years back, messages were conveyed by means of inscribing pictures on rocks

371

used in radio, telex, telephone, and television systems. Very short radio waves or *microwaves* are used to send messages across sea using *satellites*.

The development of *microelectronics* and *computers* have added new dimensions in the field of communications. Computers can control thousands of messages simultaneously without any mistake.

The fax machine can send and receive both words and pictures

In future, everyone will have a compact telephone

Books, newspapers, radio, television, cinema, etc. are other well-known means of communications. Through these means, we can receive information from any part of the world within no time.

■■

PRINTING

Printing is a process in which a large number of copies of text and pictures is produced on paper in a short time. This art was developed in the sixth century in China. After about 500 years, movable type was also invented in China. From China, this technique became popular in European countries. In the fifteenth century, Johannes Gutenberg of Germany developed the printing process in Europe. In 1476, William Caxton introduced the first printing press in London.

Platen press, which used to print sheet by sheet, was replaced by *rotary press*. A rotary press continuously prints a paper roll. The printing which uses movable type is called *letter press*. In *lithography*, smooth plate is used. Text or picture is taken on a greasy surface, while the remaining part of plate is covered with grease-repelling material. When greasy ink is applied onto the plate, it adheres on the greasy parts only.

Offset printing is a modified form of lithography. The main parts of an offset machine are: a plate cylinder, a blanket cylinder and an impression cylinder. The plate cylinder marks an impression on rubber and it prints on paper.

The invention of computer has brought many improvements in printing. Nowadays, edition of text, printing and designing of page is done with the help of computers. The publishers of this book, Pustak Mahal have made all the pages of this book by using *DTP (Desk Top Publishing)* based computers.

Word Processing

The type and designing of any text, newspaper or book can be done very fast and easily in required size in a *word processor* with the help of a computer. A word processor makes use of electronic system by which written matter can be edited on its own. We can alter the sequence

The German, Johannes Gutenberg established the typing process in Europe

A page from Gutenberg's Bible

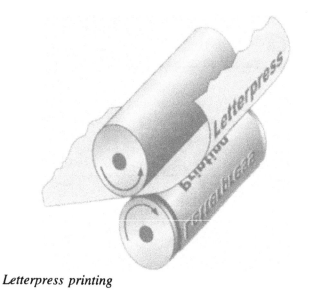

Letterpress printing

of words and sentences, even delete or add words without typing the total matter again.

Whatever you type on the keyboard gets displayed on the *visual display unit VDU* like a TV screen. Operators can do any change or correction in the text by using the *editing key* and *movable pointer*. Text gets stored in the memory of the word processor, which can be arranged accordingly by looking at the screen. Operators can store the text permanently on disk or tape, which can be used afterwards whenever needed. Word processor has a link with the electronic printer, by which a typed copy of text stored in memory comes out.

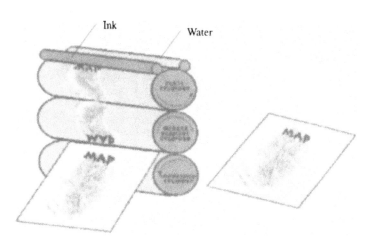

Lithography printing

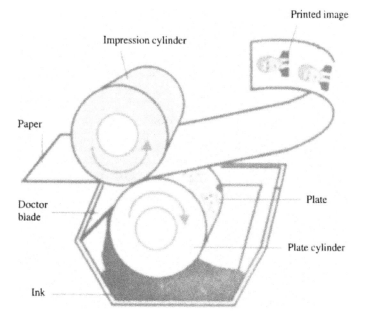

Gravure printing

Offset machines

374

VDU

Keyboard with
editing keys

Word processor

Word processor also helps to set the text into pages. Line length, spacing, indentation, margins, etc. can be set according to will. When it is printed, the machine can arrange the text on its own, if a word is longer and is coming out of the line, the word processor brings this word in the next line and adjusts the spacing of the first line. In the same way, the size of the letters etc. can also be changed.

PHOTOGRAPHY

The technique of recording the image of an object on a *photosensitive* film with the help of a camera is called *photography*. In the beginning, metal or glass plates were used to records the images. *Plastic roll films* were developed in 1889. In the middle of the nineteenth century, several other chemical methods were invented for developing images in addition to Daguerreutype and calotype negative-positive methods. The negative-positive method is in use even today.

L.J.M. Daguerre (1787-1851)

All the cameras have a lens by which an inverted image of the scene is formed on the *photofilm*. A photosensitive layer of silver bromide is coated on photofilm. This is called *exposing*. After exposing, the film is developed in chemicals and then fixed. In this way, a permanent negative of a scene is made on the film. From this negative, prints of required size can be made on bromide paper using an enlarger.

In coloured photography, two types of films are used. The first one is colour-reversal film, on which colour positive, slide or transparencies are made. The other one is colour-negative film, which is used for making colour prints. Both of these films are sensitive to the three primary colours i.e. blue, green and red. In one colour film, we have three layers of photosensitive emulsion. The first layer is sensitive to blue colour, second for green and third for red. Each emulsion layer absorbs or subtracts a definite amount of light. This is called *subtractive process*. The coloured films are developed in a special developer and prints of desired size are made from them.

RADIO

Radio is one of the most effective means of sending messages, news, music, etc. to distant places. In this process, radio waves are used, which are a type of *electromagnetic waves*.

The credit of inventing radio goes to Guglielmo Marconi of Italy. In 1901, Marconi succeeded in sending radio from England to Newfound land.

The frequency of the radio waves used in radio broadcast lies between *150 kilohertz* and *30,000 megahertz*. The messages are transmitted through a *transmitter* which is connected to the *radio station*. Transmitter consists of instruments like *microphone, modulator, amplifier, oscillator, antenna*, etc. The person in the radio station speaks in front of a *microphone*, which converts sound waves into electrical signals. These signals are mixed with carrier waves in a modulator, amplified and transmitted.

The modulation of radio waves is of two types: *amplitude modulation* and *frequency modulation*. The messages transmitted by transmitter are received by the aerial of our radio set. A radio set converts the electrical signals into sound and we hear the original programme.

Guglielmo Marconi (1874-1937)

Radio is used in telecommunications, navigation, etc. The radio services started in 1936 in our country. Today, there are more than 100 radio stations in our country.

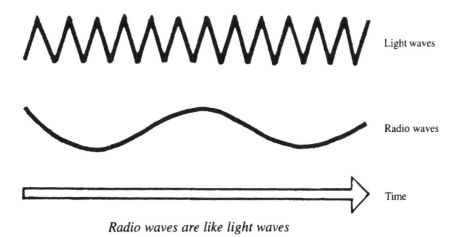

Light waves

Radio waves

Time

Radio waves are like light waves

RADAR

The instrument used to detect the position and distance of an object is called *radar*. Bad weather, darkness and smoke do not effect the performance of a radar. Radars are mostly used to determine the position, speed and distance of aircrafts and ships.

An echo is produced when sound waves are reflected after striking an object. Similarly, radio waves produce an echo when they get reflected after striking a surface. The invention of radar is based upon the same principle. This is called the *echo principle*.

The word RADAR stands for Radio Detection and Ranging. The first successful radar was developed by Robert Watson Watt and his colleagues in 1930 in Britain.

The *transmitter* of radar sends short pulses of high frequency from a rotating antenna. These pulses are reflected back after striking an object and are received by a *receiver*. The *screen* of the radar shows the position of the object. The time taken by the wave in going from the transmitter of radar to the object and then after reflection back to the receiver is determined. When this interval of time is multiplied by the velocity of light, we get double the distance of the object from the radar. The display unit fixed in the radar shows half of this distance. Besides aeroplane and ship control, radar has played an important role in missile control, transport control, army, weather forecast, spacecraft control, etc.

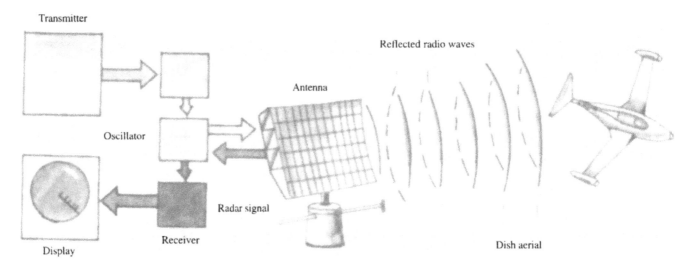

A radar can detect enemy planes

CINEMATOGRAPHY

The moving pictures of cinema are based on the *persistence of human vision*. In fact, pictures do not move but still pictures appear on the screen one after the other. If we are looking at an object, then even after removing the object, its image persists in our eyes for about 1/30 of a second and it gives an impression that the object is there before our eyes. When pictures are shown very fast, one after the other on the screen, our mind is under the misapprehension that the pictures are moving. The second picture comes into the brain before the mental picture of the first is vanished. Then after a fraction of a second, a new picture appears which is slightly different from the previous one. In this way, the process continues and we get an *illusion* of moving pictures. The cinema pictures are shown at a speed of 24 scenes per second.

A few scenes from films made by the Lumiere brothers

In a cinema film, we have a long strip of thousands of still pictures, which is wound around a *spool*. The spool is mounted on a *projector*. The projector shows every picture or frame on a screen. Every second, 24 frames or pictures are projected on the screen so that the scene looks alive. Images on the film are recorded by a *movie camera*. This camera works like a still camera, the only difference is that it takes 24 pictures per second on the film. On one side of the film, sound signals are recorded. In the positive print of the film, sound and picture are both synchronized. This is called *synchronization*.

The first cinema or movie was made by Thomas Alva Edison. This instrument was called *kinetoscope*, in which moving pictures were seen through a small hole.

The credit of showing cinema film goes to the Lumiere brothers of France. They made the first cinema projector by which film could be shown to a large number of audience at a time.

In March 1895, they showed their first short film with the help of a camera and projector in their factory. The world's first cinema hall was built in France. Posters were pasted on the walls of the city on which photographs of the Lumiere brothers were printed and 'Lumiere's Cinema' was written in bold letters.

Cinema was started in 1895 and for 34 years silent films were made. The first talkie was made by the Warner brothers of America in 1927. The name of the film was *The Jazz Singer*. In the year 1930, coloured films appeared. After that, cinema developed very fast. The wide screen cinemascope films were made in the middle of this century. Today, we can see real scenes due to the techniques like *cinerama* and *technirama*.

■■

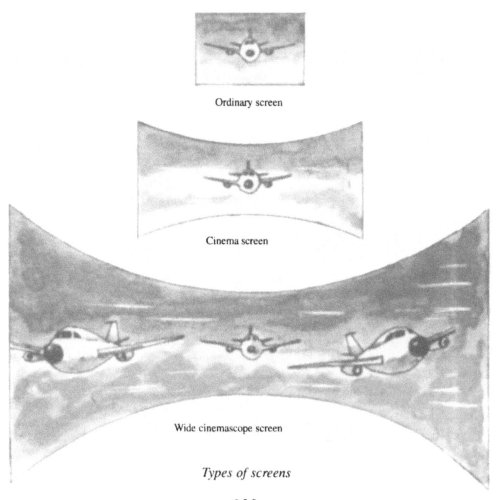

Ordinary screen

Cinema screen

Wide cinemascope screen

Types of screens

REPRODUCTION OF SOUND

Sound is a kind of energy which travels in the form of waves. It can be recorded on a *tape* or *disk* by converting it into electrical signals. It can also be reproduced and heard from the tape or disk. This whole process is called *sound recording* and *reproduction*. Sound waves are recorded on tapes in the form of *magnetic patterns*. On ordinary record disks, waves are

Thomas Alva Edison was the foremost person to record sound using a phonograph. The first sentence recorded was -"Mary had a little lamb."

recorded in the form of *grooves*. In a *compact disk*, sound is recorded as spiral patterns of microscopic pit. All these systems of recording sound are different from each other but all require a *microphone* for recording and a *loudspeaker* for reproducing the sound.

The microphone converts the sound waves into varying electric current. A simple microphone consists of a thin disk or diaphragm connected to a piezo-electric crystal. When this crystal is pressed by sound waves, a feeble electric current is generated. Sound waves cause vibrations in the diaphragm of the microphone. The vibrating diaphragm exerts a little pressure on crystal and, as a result, a feeble electric current is generated. The current produced varies as the amplitude of the vibrations and produces

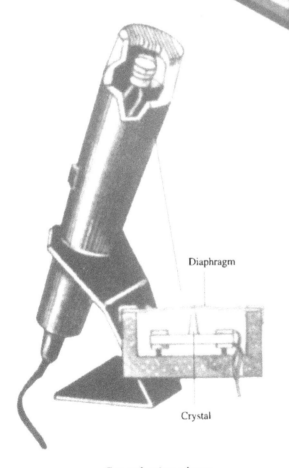

Diaphragm

Crystal

Crystal microphone

Cone

Magnet

Moving-coil loudspeaker

a sound wave pattern which is recorded on tape or disk.

To listen to this sound again, electric current is produced from the disk or tape and this electric current is sent to the loudspeaker. The loudspeaker converts these electrical signals into sound waves. An ordinary loudspeaker has a big cone of paper connected to a coil of wire. The cone is fitted in between the pole pieces of a permanent magnet. When electric current is passed through this coil, a magnetic field is produced which causes vibrations in this cone and sound is produced. In this way, the recorded sound is reproduced and heard.

Record Disc

The credit of recording sound on an ordinary disc goes to the American scientist, Thomas Alva Edison. He invented the *phonograph* in 1877.

In 1887, the production of modern type of disc gramophone started. Commercial disc recording started in 1895.

Record player

If you see the disc with a magnifying glass, you will observe wavy grooves on it. These wavy grooves are formed according to the intensity of vibrations of sound. Sound waves are recorded as an image in these grooves. After recording, the disc is rotated with a needle placed on it. Due to the up-and-down movement of the pin vibrations are produced in it and thus, the original sound gets produced. Today, many developments have been made in the field of recording and reproduction of sound. Nowadays, amplifiers are also used. Mechanical reproduction has been replaced by electrical pickup system. *Long play* and *stereo records* have also developed.

Tape Recorder

Tape recorders were started by Valdemar Poulsen in 1899. Sound is recorded on a plastic tape which remains wound like a reel in a cassette. Most of the cassette tapes are coated with iron oxide, which is a magnetic material.

In a tape recorder, first of all, electrical signals coming from a microphone are converted into magnetic signals. This is done by a small electromagnet which is called the *recording head*. Tape is passed through the head with the help of a motor. The iron oxide coated on it changes to a magnet by the electrical current generated due to sound waves. In this way, sound is recorded on the tape in the form of magnetic field.

To listen to the sound recorded on the tape, it is again passed through the head. The changing electric current due to the head resembles the electric current of the microphone at the time of recording. This electric current is amplified by

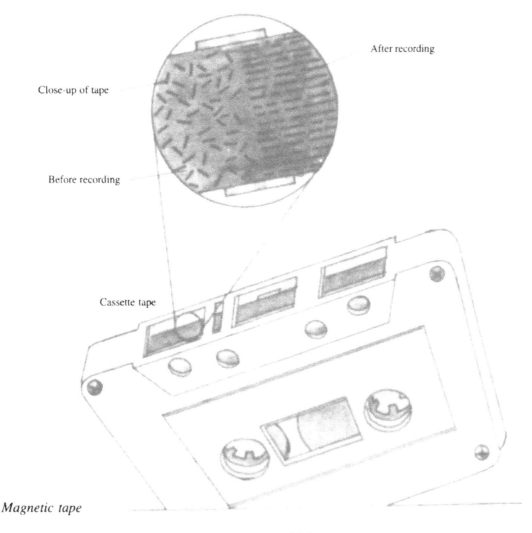

Close-up of tape

After recording

Before recording

Cassette tape

Magnetic tape

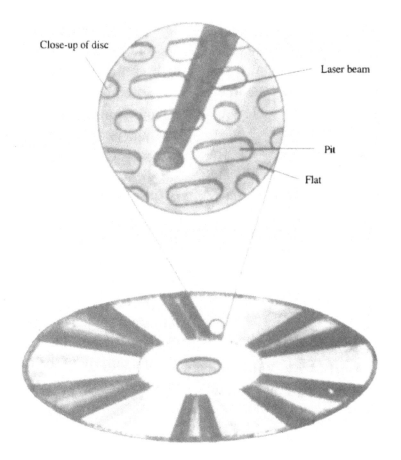

Compact disc

an amplifier and then the original sound is produced by the loudspeaker.

Compact Disc

The electric signals of a microphone are recorded on a compact disc by *laser beam*. Laser is also used for playback. The signals are recorded in the form of digits (in the series of 1 and 0). We have the best reproduction of sound by this method. A compact disc measures only about 12 cm but in comparison to 30 cm-long playing record, its playing time is much more.

384

TELEPHONE NETWORK

Telephone network is the most effective and useful two-way telecommunication system. With the help of a telephone, people can talk with each other. The use of computers in telephones has added new dimensions to telecommunications. *Viewdata, teleshopping, electronic banking, electronic offices,* etc. are all born out of computers. In fact, computers have added a new chapter to the phone network, which has made communication systems very easy and fast.

How Phone Works

When we speak, our sound creates vibrations in the air. The microphone fitted in the mouthpiece of the phone changes these sound waves into electric current. This current reaches the earpiece of the receiving phone through cable. In the earpiece of this phone, the electric current again changes to the original sound waves. In this way, your voice reaches the destination. This system is called analog broadcast because in this system,

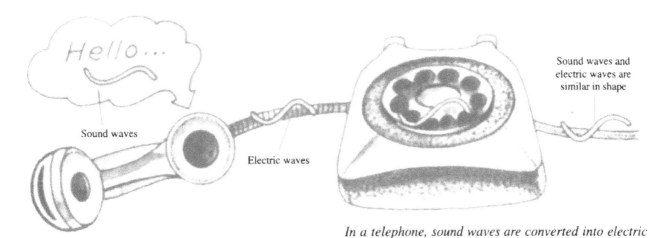

In a telephone, sound waves are converted into electric waves and then reconverted into sound waves

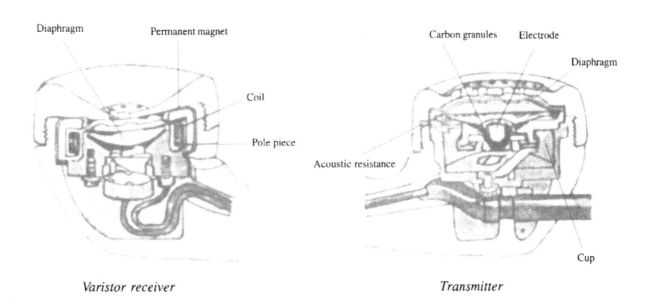

Varistor receiver

Transmitter

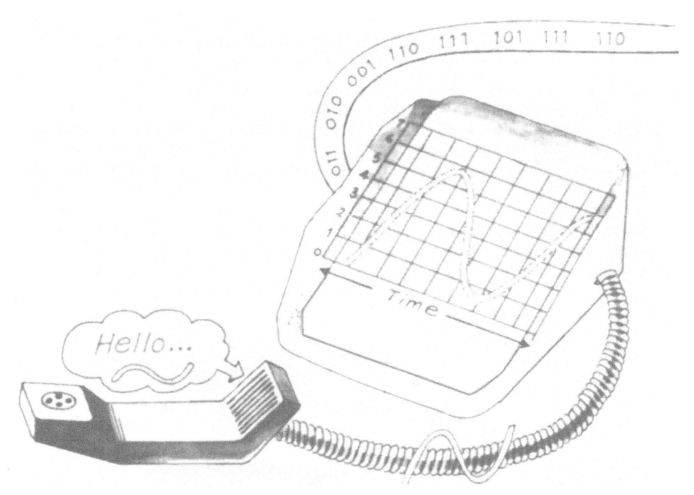

A digital phone firstly converts sound into electric waves. Then these are developed into a number in the binary code. For listening to the sound, the digital signals are converted into the analogue form.

electric current is 'analogous' or identical to sound waves.

Digital Phone System

The *binary digital system* was developed in 1960. In the digital phone system, sound signals are converted into binary digits. Analog signals are converted in the form of 0 and 1 which represent the on and off states. Binary digit is called a *bit*. Seven bits are generally used in telephone communications. The receiving phone converts these digits back into sound waves. Binary digits are fast and are of the same kind, so it is very easy to convert them into their original form. Nowadays, these systems are used on a large scale along with computers.

Fibre Optic Cables

Most of the phone calls travel through copper cables in the form of electric current, but nowadays these calls are also sent through optical fibre cables in the form of light. These cables have proved better than copper cables. Through this cable, about 10,000 telephone calls can be communicated at a time. In this system, laser beam and cable made of optical fibres are used.

Computer Exchange

The telephone system has not been totally digitalized in our country yet, only a few exchanges have been computerized. These exchanges control much more information in less

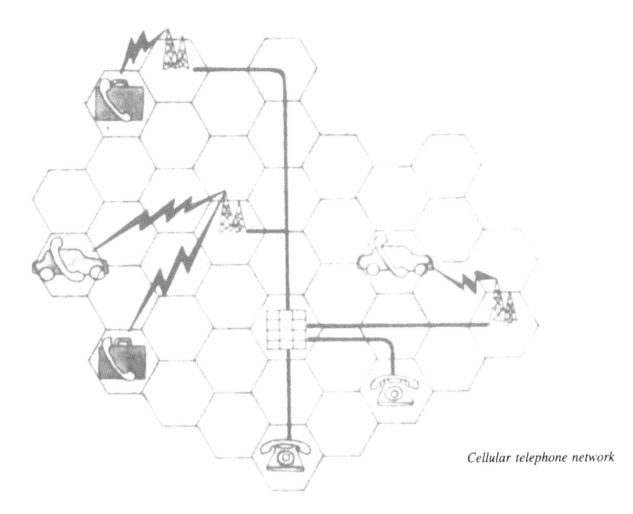

Cellular telephone network

Base station

Microwave or landline link

Electrical signals

Car phone

Portable telephone

Private telephone exchange

Public telephone exchange

time in comparison to the old mechanical exchanges. The chances of misconnections, crossed lines, interference or lost calls in these computerized exchanges are extremely low. Such exchanges also provide additional facilities like rerouting of a phone call for another phone, viewdata phone, number directories, automatic monitoring of call, etc.

Cellular Radio and Telephones

Cellular radios or telephones are used for mobile communications. Telephones remain linked with the cables and the person using the cordless phone can use it only within a cell of a few metres from the *base station*. In a few parts of big cities, special types of radio channels or transmitting frequencies are used. In cellular telephone network, an area is divided into many cells. Every cell is at a distance of five kilometres from each other. When a person using a cell

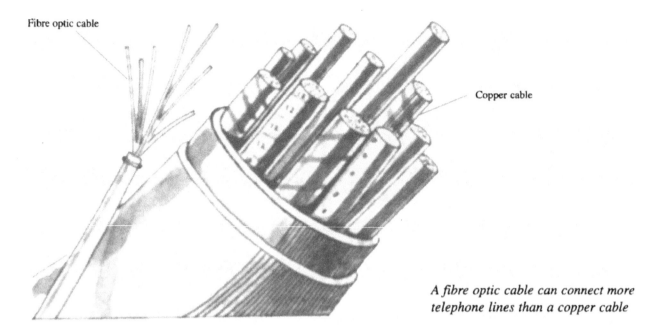

Fibre optic cable

Copper cable

A fibre optic cable can connect more telephone lines than a copper cable

phone makes a call, a radio signal is sent to the nearby base station. The base station sends it to the nearest *mobile telephone exchange*. After that, the call is automatically transferred from the transmitter of one cell to the other cell. In this system, many calls can be made simultaneously.

■■■

FAX

Fax or *facsimile transmission* is a new system of sending information on paper through telephone line. It is different from other communication systems like *telex* and *electronic mail*. All types of documents, either printed or handwritten, line diagrams or photographs can be sent or received using the fax machine.

A modern fax machine takes less than 30 seconds to send a written paper. In the beginning, fax machines were of the analog type but now digital machines are in common use.

A fax machine scans the document by light and the image is changed into electrical signals by *photocells*. The message travels through the telephone line and is received by the fax machine at the other end. After decoding, the machine produces a printed copy of the original document.

The coding and decoding system of all the fax machines in the world is the same. So you can send the document from your fax machine to any other fax machine. Every fax machine has a number like a telephone, which has to be dialled before sending a message.

Electric signals are transmitted by the phone

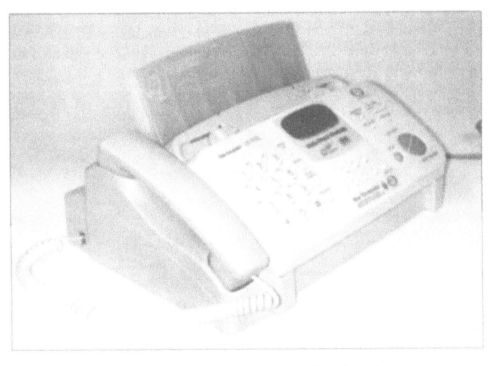

A fax machine can send or receive every sort of word or picture

SATELLITE COMMUNICATIONS

Telecommunication messages are not only broadcast through telephone cables but through satellites also. Everyday, thousands of phone calls, TV signals and computer data are transferred from one part of the world to another through satellites. A *communication satellite* can handle more information than cable and communicate them much faster.

Sending Signals

Messages are sent to the satellite from any earth station in the form of *microwave signals*. Microwaves are a type of short radio waves and can travel through space with the speed of light. Signals are sent and received through dish-shaped antennas. The messages sent from the earth station are received by the satellite and are retransmitted to the earth. The antennas located on earth receive the messages and send them to the destination. These messages can be for telephone, radio or television.

Using Satellites

We can send telephone, radio and television messages to any part of the world with the help of communication satellites very easily. Any event can be telecast live from any part of the globe. Now, it is possible to transmit a telephone call which can travel around the globe and can be received at the same point from where it was transmitted with the help of communication satellites.

Satellites have proved very useful for *multinational companies*. Different offices in the world can be linked to each other. Office work stations have communications between each other.

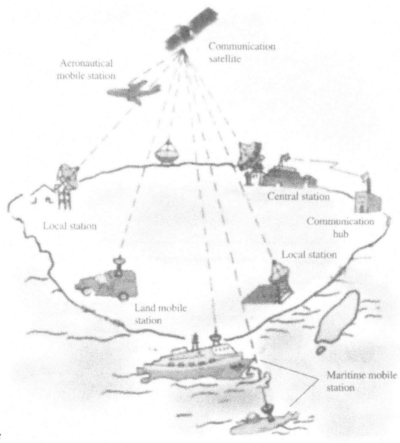

Communication satellite

390

TELEVISION

The first public demonstration of television was given by John Logie Baird of Britain in 1926. The television which displays the different objects and scenes on the screen is a result of the systematic and joint efforts of the scientists.

V. Zworykin of America made a significant contribution in this field. In 1928, he developed an electronic system, which superceded Baird's mechanical system. It was a revolutionary development for telecasting programmes quickly and correctly.

To telecast TV programmes, sound and scenes are first converted into *electromagnetic waves*. These waves are again converted into sound and scene by the TV set. A TV camera consists of an *orthicon tube*. The image of the scene formed through a lens falls on a photosensitive plate of this tube. Electrons are emitted from the plate according to the intensity of light. After that, it is *scanned by a cathode ray tube*. Scanning changes the image into electric current. It is called a *video signal*. It is telecast after amplitude modulation. In addition to this, sound is converted into electric current by a microphone and is transmitted after frequency modulation. It is called an *audio signal*. These electro-magnetic waves of scenes and sound hit our TV antenna. These are received by the TV set where they are converted back into the original scenes and sound.

The working of colour TV is almost the same as of a black-and-white TV. The light coming from the scene is divided into the three primary colours by three filters fitted in the

Red, green and blue dots in an image

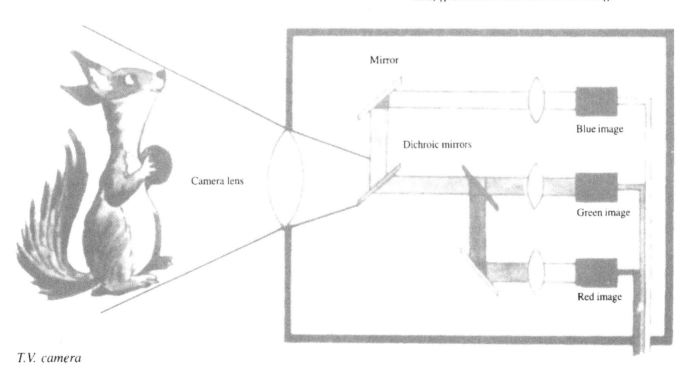

T.V. camera

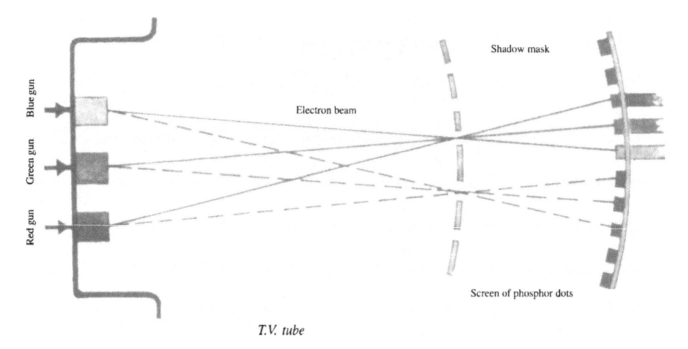

T.V. tube

camera. One filter allows only red colour to pass, the second only blue and the third only green. The light of each colour falls on different camera tubes. Each tube contains a separate glass plate and electron beam. The three signals from the tubes reach the transmitter. The TV transmitter mixes these three signals into one. A black-and-white signal is also mixed with this signal. After that, these signals are sent to the transmitting antenna and telecast. This signal reaches our TV sets. Three electron guns, one for red colour, the second for blue and the third for green are fitted in the TV set. A layer of about 125 lakh phosphor dots of the three colours is painted on the screen of the TV set. These dots are arranged in a pattern of three and emit light when electron beam falls on them. Out of these dots, one emits red light, the second blue and the third green. The colour emitted from each group of dots depends upon the intensity of electron beam. The coloured scene is created on the screen by mixing of these three primary colours in different ratios.

DBS TV

Artificial satellites brought about a revolution in the art of TV transmission. TV stations telecast programmes for different countries with the help of satellites. Nowadays, direct transmission is also done from the satellite. This is called *DBS (Direct Broadcast Satellites)*. In this system, your TV set receives signals directly from the satellite. This system requires a special *dish aerial*.

Cable TV

In Cable TV, the signals reach our TV set through a cable. Nowadays, a dish aerial is installed in a

Dish aerial for D.B.S. T.V.

colony and programmes can be viewed by the people of the colony on their TV sets through a cable. More channels can be transmitted through cable than ordinary transmission. TV signals can be sent to hills, valleys and even tall buildings with the help of the cable. Cable TV has become very popular these days.

VIDEO

Video recording is a modern technique in which sound and picture are simultaneously recorded on a tape or disc electronically. Television studio makes use of video to record programmes and feature films. Video films can be purchased from the market and can be taken to home to see on one's own TV set. Video tape costs less than ordinary films to make domestic movies. It is also very simple to use them. Disk is the latest development but it is different from the tape.

In 1956, US Ampex Corporation developed the first video recorder. In 1976, Philips Company of Europe manufactured the first video recorder for domestic use. The most popular home video system is *VHS (Video Home System)*.

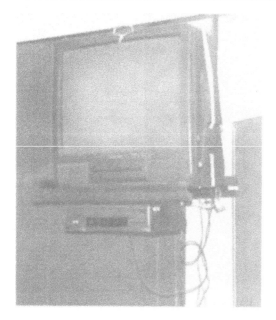

Video cassette player/recorder

Video camera

This was developed by JVC in Japan in 1970. In this, a 12.65 mm wide tape is used.

Camera and Recorder

Video camera converts light and sound into electrical signals. Recorder converts these signals into magnetic pulses and records them on magnetic tape.

Video Cassette Player/Recorder

Video cassette player again converts the magnetic pulses which are recorded on tape into the original electrical signals. These signals are again converted into original picture and sound in a TV set. Most of the domestic video cassette recorder machines can record as well as play the tapes. You can record TV programmes on these tapes and watch them afterwards.

Video Disc

Video disc resembles *long play record (Audio LPS)*. You cannot record your own programmes on these discs. They are pre-recorded. A special type of disc player is required to play them.

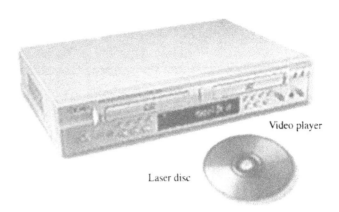

Video player

Video disks are of two types: one is used like audio players and for the other, a thin laser beam is used. Laser disc is like a silver mirror which shows rainbow colours. It is played by *laser beam*.

The programmes recorded on video disc are very clear. They can be preserved for a long time. They are also cheaper in comparison to video tape. These are used in the field of education, training and industries. A few video discs can be controlled by the domestic computer also.

COMPUTER

Computer is an automatic machine which can do calculations, write and solve complex problems within seconds without making any mistake. Not only this but it can also perform many functions at a time. A single computer can do millions of calculations within a second.

Computers are being used successfully in telecommunications, space research, engineering, science, industries, medicine, education, banks, transport, post offices, railways, sports etc.

The first electronic computer was ENIAC, which was invented in 1946 by Eckert and Mauchly of Pennysylvania University. There were 18,000 electron tubes, 70,000 resistors and 10,000 capacitors in it. Each tube was of the size of a small bottle. This computer was controlled by a team of trained operators. In 1950, the transistor replaced the vacuum tube. Nowadays integrated circuits or chips are used in computers which have reduced the size of the computer considerably. Computers are of two types – the first type of computer is called *analog computer* which measures one quantity in terms of the other and the other one is *digital* in which various problems are solved by using digits.

A computer has three main parts –*Input Unit, Central Processing Unit (CPU)*, and *Output*

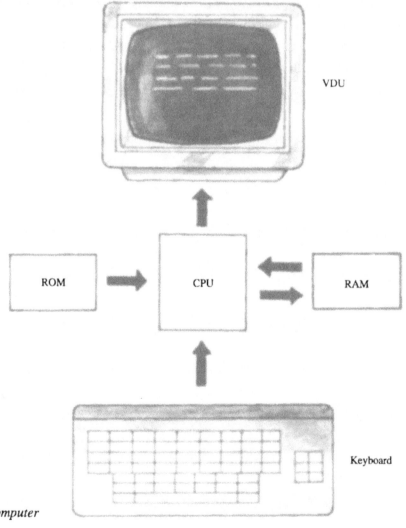

VDU

ROM

CPU

RAM

Keyboard

The main parts of a computer

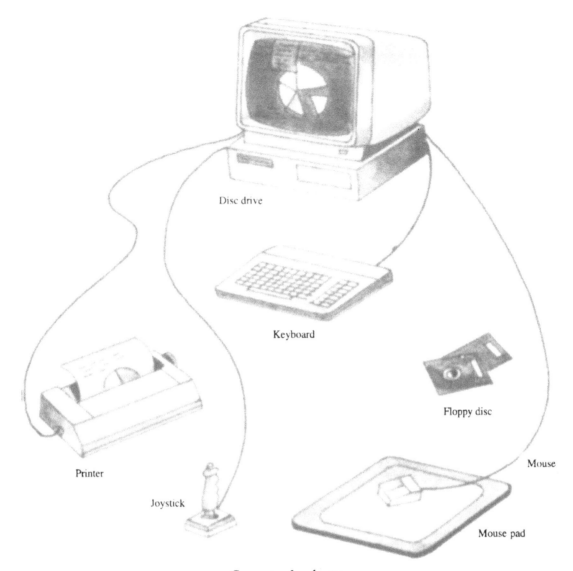

Computer hardware

Unit. In the input unit, magnetic tape, punch card paper or tapes are used. There is a memory in CPU which stores the information. Its controlling system gives directions and arithmetic system performs various operations. Printers, cathode ray tube or sound indicating units are used as output devices. In fact, the input of computer is just like our eyes and ears, CPU functions like our brain and output unit functions like our hand and mouth.

Digital Computers

In digital computers, all information is sent through programme and these bits of information guide the functioning of computer. The data of the programme are sent to the computer through an input machine. This data is processed in the CPU. The data of programme gets stored in the memory. Calculations are done in the arithmetic and logic units. Storage and calculations are controlled by control unit and the solution of the problem is given by the output unit.

Home Computer

The heart of a home computer is only one chip which is called a *microprocessor*. Different parts of a chip perform different functions of the computer. The chip of a home computer has two types of memory, one Central Processing Unit (CPU) and one clock.

Read Only Memory or *ROM* carries necessary messages for the computer. This cannot be changed. *Random Access Memory* or *RAM* is a temporary memory and it is used to give instructions and messages when computer operates. The instructions given to the computer are called a *programme* and messages kept in it are called *data*. Experts wiring a computer programme use special languages which the computer can understand. There are several computer languages like *COBOL, FORTRAN,* etc. but *BASIC (Beginner's All Purpose Symbolic Instruction Code)* is most commonly used.

Computer Software

The programme and data fed to the computer is called *computer software*. We can type words and decimal figures with keyboard which go to Random Access Memory (RAM) of the computer. These do not go into the memory in the form of words and decimal figures because memory cannot store them in that form. Computer changes them into binary code. In this code, all the characters are changed into 1 and 0.

Two binary digits are called *bits*. Most of the computer data are controlled and stored in eight-bit units, which is called a *byte*. The RAM of a single home computer can control about 1,28,000 bytes.

Computer Hardware

The physical parts of a computer which we can touch are called *hardware*. In the main processing unit of a home computer, there is a *Visual Display Unit* (VDU) resembling the TV screen, one *keyboard, mouse, joystick, printer* and *modem*.

Binary code

Magnetic disks are fitted inside one or two slots of the processing unit. Programmes and data are stored on disks. These are called *floppy disks*.

Mouse is a device which helps you to make a design directly on the VDU. Joysticks are used in video games.

The programme given to the computer can be received in the form of a copy printed by the printer or it can be stored on floppy disk or magnetic tapes. These can be transferred to the other computer through telephone cables. This you can do by modem, which converts computer output signals into audio signals. The modem on the other end reconverts them into original form.

Scientists have made various types of computers today but the fastest and the most powerful among them are *Super Computers*.

ELECTROSTATICS

The electric charge is present all around us. Still, man remained unaware of the electric power forages. He tried to explain the lightning in the sky through various imaginary interpretations which were not satisfactory. Then the Greek philosopher, Thales lifted the curtain from this mystery. He discovered, around 600 B.C., that the same effect which was visible during lightning in the sky also appeared in a piece of amber when it was rubbed against fur. The friction against fur produced in amber the property of attracting light substances like straw, feather and paper. The Greek word for amber is 'elecktron'. So, Thales called this mysterious power 'electric'.

About 2,000 years after Thales' discovery, William Gilbert, an English scientist, made another discovery. Experimenting with amber and loadstone, he discovered the basic difference between electric and the magnetic attraction. William used the term 'electrica' for the things like glass and sulphur which behaved like amber. And he used the word 'electricity' to describe the natural phenomena associated with these things.

Here is an easy experiment to make you understand the phenomenon of lightning. Brush your hair with a comb in a dark room. Then, bring the comb close to your thumb. You will

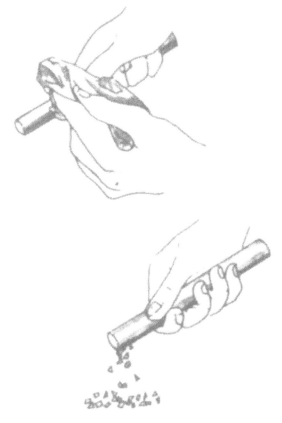

When a rod of glass is rubbed against a piece of fur, the glass rod acquires the electric charge to attract small pieces of paper towards it.

notice a small sparkle. This sparkle occurs due to the energy of the charged comb which appears as light from the air atoms between the comb and your thumb. Similar is the cause of lightning in the sky.

A plastic comb, when used to brush hair, also gets electrically charged. It can also attract small pieces of paper towards it, if placed near them.

Now, brush your dry hair with a plastic comb and try to listen. What did you hear? A crackling sound! Bring the comb close to small bits of paper. These are attracted towards the comb! The reason: when you brush your dry hair with the comb, it gets charged and can attract small bits of paper.

A glass rod and an ebonite rod can also attract small bits of paper when they are rubbed against a piece of silk and cat's hide, respectively. When an object receives the property of attracting other objects, it is called 'charging'. This may happen when the object is rubbed against another object, or comes close to it, or simply gets in contact with it. The charge in the glass rod is different from the charge in the ebonite rod. The glass rod has the positive charge, and the charge in the ebonite rod is called the negative charge. When two similar charges are brought together, they cause repulsion. But, two opposite charges attract each other, neutral rod gets charged when brought close to a charged rod. With the removal of the charged rod, the other rod again becomes neutral. This effect is called the Electrostatic Induction. What is the nature of the charge, and the difference between the positive and the negative charges? Man still does not know.

■■

A plastic or pen, rubbed against wool, can be used to separate pepper from o mixture of salt and pepper. Pepper particles, being lighter, are drown to the pen.

402

ATMOSPHERIC ELECTRICITY

You must have seen a flash of light in the sky. It is lightning. It may frighten you. It kills or injures thousands every year. It can also turn vast jungles into ashes and raze lofty buildings to the ground. But it has also played an important role in the origin of life. It has created certain chemicals which are essential for life.

What causes lightning? In 1708, William Wall of Britain discovered that the lightning phenomenon is similar to the discharging of an electrically charged object through a conductor. The dry air does not let electricity pass easily. But, in the presence of vapours, the air becomes an easy conductor. Still, electricity requires a lot of energy to pass through the air speedily. When the electricity is dispersed, the air surrounding its route becomes very hot. The temperature of the surrounding air rises up to about 10,000°C. This phenomenon of dispersal of electricity appears as 'lightning' in the sky. The raised temperature causes the air to expand. When the air cools down, it abruptly contracts. This sudden expansion and contraction of air produces thunder. Although lightning and thunder occur at the same time, it is the flash that we observe first. Why? Because light travels faster than sound.

According to Benjamin Franklin, an American scientist, lightning was simply an extraction between two objects with different electric potential—such as, cloud and earth. He observed that lightning generally struck high-rise buildings and trees. To prove this point, he experimented with a kite and a key during the rains in 1752. With the occurrence of lightning, he started flying a silk kite which had an iron end. He tied a long key at the end of the kite string. He flew the kite through a thundering cloud. As he brought his finger close to the key, there was a spark between the key and his finger.

In 1752, the American scientist, Benjamin Franklin (above) conducted his famous kite and key experiment.

403

Thus, he could prove his point through this dangerous experiment. Franklin and his son would have been killed, had the lightning struck his kite.

Franklin's experiment led to an invention which could save high-rise buildings from lightning. He fixed a metal rod at the highest point of a building, and attached a wire to it. The wire was taken down along the building and buried under the ground at the other end. When lightning struck the building, the electric charge was quickly absorbed by the earth through the wire. And, the building remained safe.

Franklin's invention, though very effective in saving man from nature's wrath, could not get religious sanction. People believed that lightning was God's weapon to punish man for his sins and man had no right to escape from God's punishment. But, despite religious boycott, all the public buildings in Philadelphia, the former capital of America, had Franklin's lightning conductor by 1782. The French embassy was the only exception. Incidentally, the French embassy was struck by lightning in the same year. One official was killed. Then people realized the importance of the lightning conductor. Today, all the high-rise buildings in the world have lightning conductors installed atop them.

A metal lightning conductor at the top of a building, which transmits electric current to the earth, without causing any damage to the building.

404

ELECTRIC SUPPLY

Can you imagine the modern world without electricity? No, it's not possible. Our modern lifestyle is heavily dependent on numerous technological inventions which just cannot work without electricity. Be they the toaster and refrigerator of your mother's kitchen, or your father's office computer, or the heavy machinery of big industries or the electric trains—electricity is the lifeline of all these things.

Electricity is generated by huge generators at the power station. Two parallel wires are used to supply electricity to farflung areas. There is a wastage of electric power during its transmission through wires. Transformers, which increase the voltage, are used to lessen this wastage. A transformer converts the low-voltage and high-current electricity into a high-voltage and low-current electric power. In this way, the transmission losses of electric power are reduced. Generally, an electric generator generates 100 kilowatt (Kw.) electric power at 6,000 volts. This voltage is increased up to 1,32,000 volts and transmitted through wires. These wires are fitted on pylons. A network of wires is called grid. The electric power is supplied at 33,000 volts through the grid to towns. At the town, electric substations, the voltage is again brought down to 6,600.

For domestic supply, the voltage of electric power is further reduced from 6,600 volts to 220 volts by a transformer. The domestic power is supplied through a pair of wires, called line wire. Bulbs, tubelights, fans, refrigerators, heaters and other appliances are fitted in parallel to the line wire. Each of these appliances has its own on/off switch. To use an appliance, you simply have to turn on its individual switch. All these appliances have the same voltage, but different electric currents. If the electricity is supplied at 220 volts, the potential difference between the

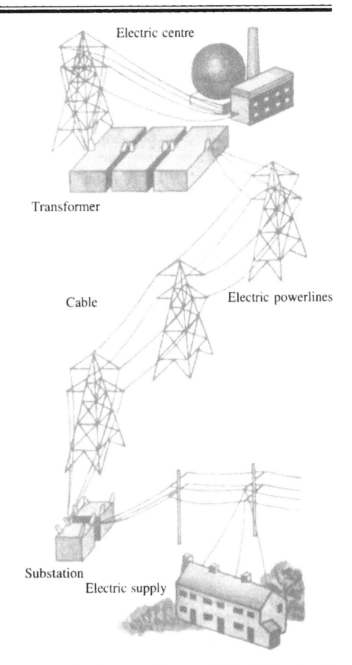

Electric centre

Transformer

Cable

Electric powerlines

Substation

Electric supply

two ends of any appliance will be 220 volts. A transformer can transmit only alternating current (A.C.). It cannot transfer direct current (D.C.). Today, only alternating current is generated all over the country.

The quantity of electricity consumed in a house or a factory is recorded by a meter. We pay our electricity bills according to the readings measured by the meter.

ELECTRIC SHOCK

Today, electricity has become a household necessity. As compared to our forefathers, we are living an easy and comfortable life, thanks to electric appliances, which are useful domestic aids. But a defective electric appliance could cause us a tremendous shock. During electric shock, we experience an extreme stimulation of the nerves, muscles, etc. This stimulation is the result of the passage of electric current through our body. It needs at least a 1000th part of an ampere of electric current to pass through our body before it could give us a shock. When our skin is dry, our body resistance is at about 50,000 ohm. But a wet skin brings the resistance down to 10,000 ohm. It will need at least 50 volts of electric supply to give shock to a man's dry skin. That is why the torch-current does not give us any shock. The domestic power is supplied at 220 volts. Therefore, a contact with it could cause a shock, even death. If we are wet, a mere 10 volts of electricity is enough to give us a shock. In wet condition, touching a 220-volt current is an invitation to death.

Why do the birds, perched on electric wires, not suffer electric shock? You must have wondered. This is so because touching a single wire by the birds does not allow the passage of electric current through their bodies. The circuit of electric current is not complete then. But, if a person touches both the wires at once, he will get a tremendous shock. Because in this case, the electric circuit will be complete and the current will flow through his body. Sometimes, we can also suffer electric shock while touching one wire only. It happens when a bare-footed person touches a wire. Reason: the current from the wire goes to the earth through his body, thus completing the circuit. Therefore, we must not touch any electric appliance when bare-footed or with wet hands. We must keep an electric tester at home so that we could check whether an electric appliance is having a short-circuit. If the appliance indicates a short-circuit, we should call for an electrician at once.

■■

220 volt

A bird perched on an electric wire does not get an electric shock

SOURCES OF ELECTROMOTIVE FORCE

The electromotive force represents the energy which is passed on to the free electrons through cells or any other electric source. A source, which can convert any other form of energy into the energy required for the flow of electric charge, is called the source of electromotive force. Some major sources of electricity or electromotive force are:

1. **Electric Cell:** Voltaic cell, dry cell, lead accumulator cell and mercury cell are some sources of electricity. In these cells, the chemical energy is converted into electric energy through chemical reactions.

2. **Generator or Dynamo:** Generator or dynamo is used to generate electricity on a large scale. In a generator, the electromotive force or electric voltage is generated by electromagnetic induction. To achieve this, coils are made to move in a magnetic field within the generator.

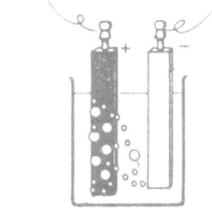

Electric cell

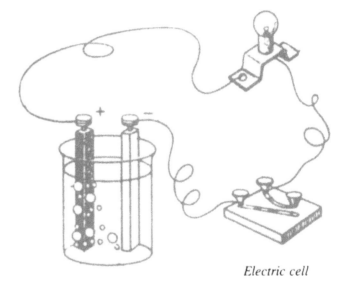

Electric cell

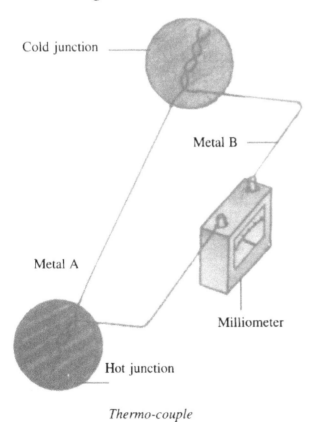

Thermo-couple

Cold junction

Metal B

Metal A

Milliometer

Hot junction

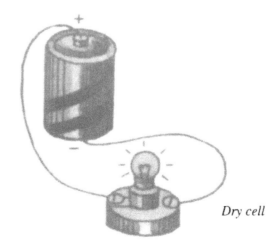

Dry cell

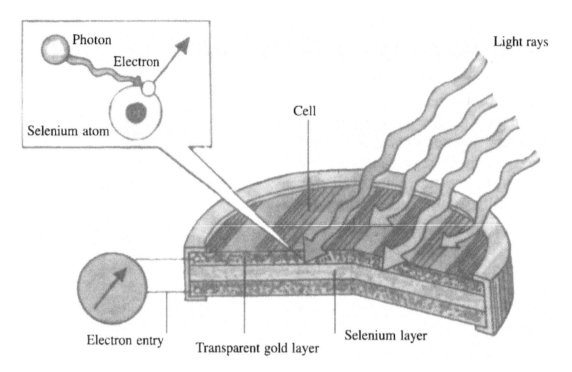

Photo-electric cell

3. **Thermo-couple:** A thermo-couple uses two conductors of two different metals to convert thermal energy into electric energy. The two conductors are connected at two ends. When one conductor is kept cool and the other is heated, the electromotive force is produced between these conductors. The thermo-couple is used to measure temperature.

4. **Photoelectric cell:** This cell converts light energy into electric energy. When light falls on the surface of selenium, potassium, silicon, zinc, etc., the electrons are freed. The movement of these electrons produces electric energy. Thus, the photoelectric cells transform the radiation energy into electric energy. These cells are used in satellites, space shuttles and cameras.

5. **Piezo-electric source:** This source transforms mechanical energy into electric energy. When pressure is applied on two sides of a quartz, electromotive force is generated between the other two sides. This energy-conversion technique is used in special very sensitive microphones, gramophone pick-ups, etc.

DRY CELL AND LEAD ACCUMULATOR

Some electric or electronic devices like transistor, calculator, torch, etc. require a low voltage. These devices function with the help of electric cells. There are two types of electric cells: (i) Primary cell; and (ii) Secondary cell. In a primary cell, the chemical energy is produced through the reactions between chemical substances. This chemical energy, in turn, generates electric energy. Voltaic cell, Daniel cell, Bunsen cell, Bichromate cell are all primary cells. A secondary cell uses electric current from some other electric source to transform electric energy into chemical energy. This process is called the 'charging of the cell'.

When a charged cell is used, it converts chemical energy back into electric energy. A secondary cell 'stores' the electric energy in the form of chemical energy. Therefore, it is also called a storage cell or accumulator. This type includes Alkalie cell or Nife cell and lead accumulator.

Dry Cell

A dry cell has a cylindrical zinc casing which functions as a negative electrode. There is a carbon rod in the middle of the casing which functions as a positive electrode. The carbon rod is sunk in manganese dioxide. The space between the zinc casing and manganese dioxide

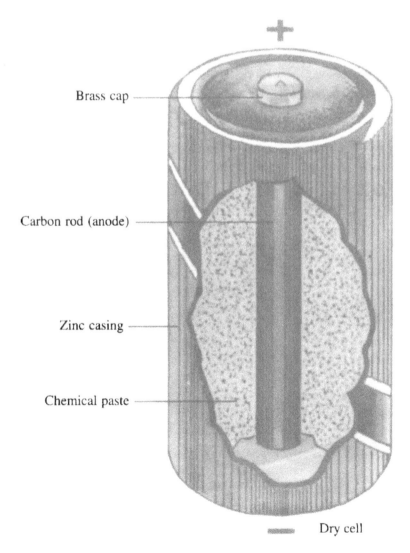

Brass cap

Carbon rod (anode)

Zinc casing

Chemical paste

Dry cell

409

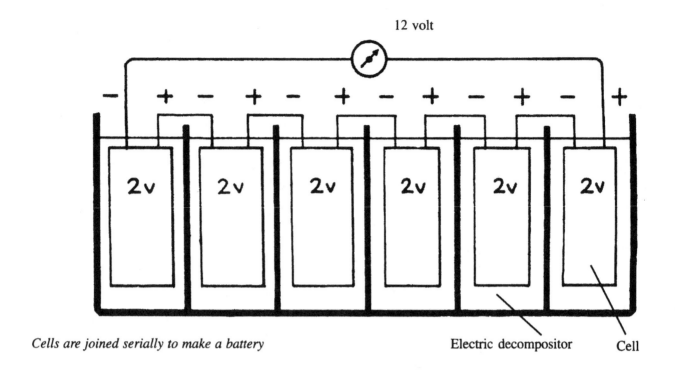

12 volt

Cells are joined serially to make a battery

Electric decompositor

Cell

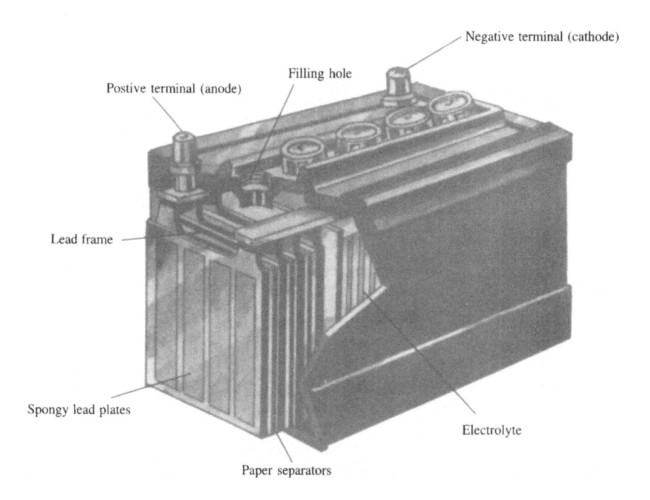

Negative terminal (cathode)

Filling hole

Postive terminal (anode)

Lead frame

Spongy lead plates

Paper separators

Electrolyte

Car battery

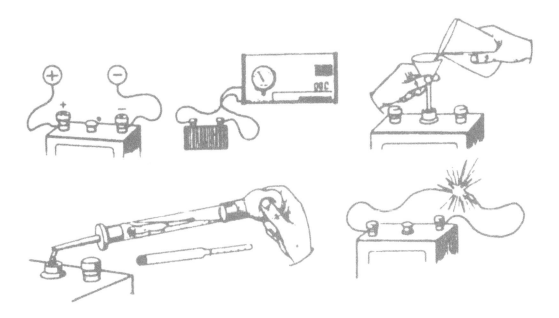

When the electromotive force of a lead storage cell remains 1.8 volt only, the cell should be recharged. Do not let the volume of the solution get less. Take care not to short-circuit the battery.

is filled with the ammonium chloride glue. The upper end of the carbon rod is capped with brass. The mouth of the cell is sealed with wax or pitch. This cell has electromotive force of about 1.5 volts. A dry cell produces electric current through reactions involving its various chemicals. When the chemicals have been used, the cell gets weak and needs replacement.

Lead Storage or Lead Accumulator

A lead storage can be recharged for further use. High potential difference batteries are made by connecting each cell in a series sequence. These batteries are generally used in motor cars. A lead storage has two sets of six or twelve plates each. These plates are made of lead grid. The plates of each set are joined together. One set works as anode while the other functions as cathode. The anode plate grid is filled with the porous powder of lead dioxide and the cathode plate grid is filled with spongy lead. The two sets of plates are put into a sulphuric acid solution of 1.25 relative density contained in a rectangular case of glass or hard rubber. The opening of the case is sealed tightly. Anode and cathode ends stick out of it. The cell has a small hole through which water or acid may be poured. A lead accumulator can also be made with just two plates.

A lead storage has 2.2 volts of electromotive force. When the cell's voltage is reduced to about 1.8 volts, it can be recharged for further use. For recharging, the cell is connected to another source of electromotive force in such a way that the electric current flows in the cell in a reverse direction—from anode to cathode. After the cell's electromotive force is restored to 2.2 volts, it is removed from the charger for further use. Repeated recharging of a lead storage results in vapourisation of the solution to some extent, which can be replaced with distilled water.

A car battery gets charged by the dynamo of its engine. Electric motor vehicles also use lead accumulators. These vehicles do not cause pollution.

411

ELECTRIC BULB AND ELECTRIC HEATER

When an electric current is passed through a resistant conductor, free electrons collide with the conductor's atoms. As a result, a major part of the electrons' kinetic energy goes to the atoms. Therefore, the internal kinetic energy of the conductor is increased. Consequently, its temperature rises and it becomes red hot. This rise in temperature is called 'the thermal effect of electric current'. The electric bulb, heater, press, electric arc, etc. were developed making use of this effect.

Electric Bulb

A bulb is a hollow ball of glass whose opening is sealed with a bad conductor. Two wires of a good conductor reach into the bulb through the opening. A fine tungsten filament is joined between these wires. The upper ends of these two wires are connected with metal points at the bulb's opening through which electric current flows into filament. Air is taken out of the ordinary and low power bulbs before sealing them. High power bulbs are filled with a mixture of nitrogen and argon gases.

When electric current is passed through the filament, its temperature rises up to 1500°C-2500°C. At such high temperature, the filament becomes white and exudes light. Generally, only 5% to 10% of electric energy flowing into a bulb is turned into light. The rest of it is transformed into heat, which the bulb discharges as heat radiations.

Electric Heater

The thermal effect of electricity is utilised in various ways. Electric stove, water heater, immersion rod, press and electric radiator are some examples of the usage of the internal effect of electricity.

All these heaters use a coil of the Nichrome alloy wire. Its ends are connected with two screws through which the current is passed into the coil. The current produces heat and the temperature of the coil wire reaches up to 800°C-1000°C. These devices are also fitted with heat controllers to prevent the excessive heating.

■■

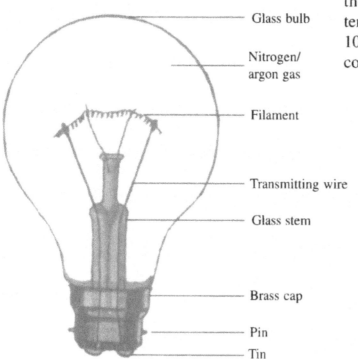

Glass bulb

Nitrogen/ argon gas

Filament

Transmitting wire

Glass stem

Brass cap

Pin

Tin

ELECTROLYSIS

Electrolysis is the decomposition of an electrolyte by the action of the electric current. It takes place when the electric current is passed through the solution of a salt, alkali or acid. In other words, electrolysis is the chemical effect of electricity which transforms the electric energy into chemical energy.

When electric current is passed through the solution of an electrolyte, it decomposes into its positive and negative ions. The positive ions move towards negative electrode (cathode) and the negative ions move towards positive electrode (anode). The ions collecting on cathode are called cations, and the ions depositing on anode are called anions. The container in which electrolysis takes place is called electrolytic cells or voltameter. Electrolysis has assumed even a greater importance in this industrial age. The chemical effect of electricity is utilised in electroplating, electrography, extraction and purification of metals, etc.

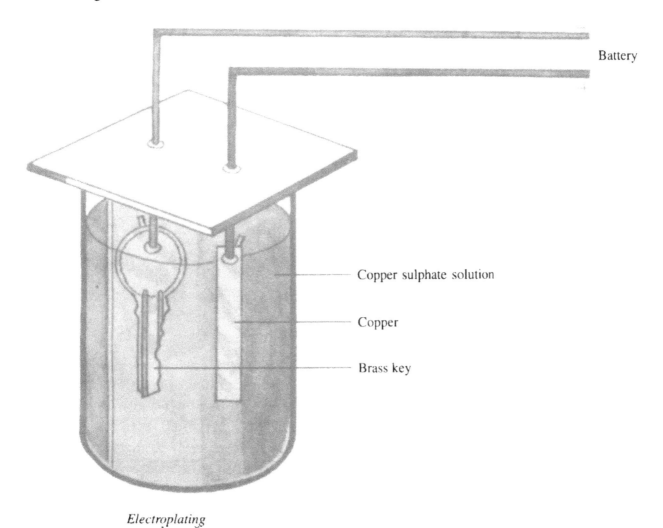

Electroplating

Battery

Copper sulphate solution

Copper

Brass key

SIMPLE CIRCUIT

The path through which the electric current flows is called electric circuit. When there is no current in the circuit, it is called open circuit. And, the closed circuit is one through which the current is flowing.

Electric circuit is a combination of various electric devices and instruments. We put electricity to various uses by making it flow through a circuit.

Electric Current

The movement of electrons or charged particles produces electricity. In metal conductors, electric current is produced only by the movement of electrons. But in gases or in a solution of a salt, alkali or acid, the flow of ions produces electricity.

There are two types of electric charge—(i) Positive charge; and (ii) Negative charge. The flow of either of the two electric charges is called electric current. But for both the charges, the direction of the current flow is opposite to each other. The current is supposed to flow in the direction of the movement of the positive charge, and against the direction of the movement of the negative charge. Thus, the electric current flows in the opposite direction to the flow of the free electrons in a conductor.

Series circuit

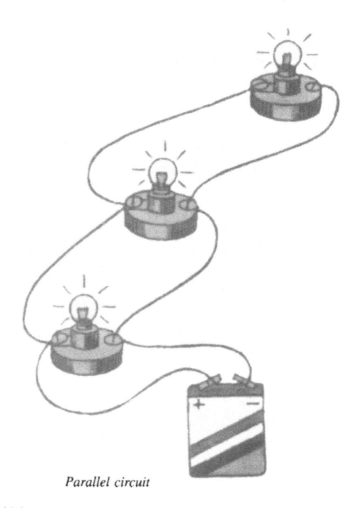

Parallel circuit

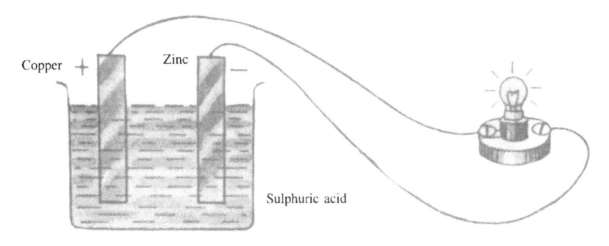

Copper + Zinc −

Sulphuric acid

In a unit volt cell, electricity is produced by the chemical process.

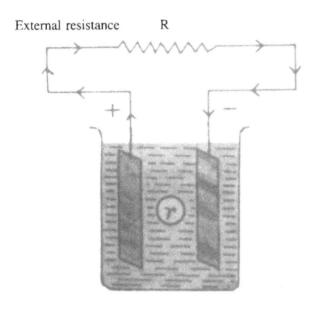

External resistance R

+ −

Internal & external resistances

Cell

Electric cell is a device which transforms chemical energy into electric energy through chemical reactions. When two different solid conductors are put into the solution of an electrolyte, a potential difference develops between them. This potential difference affects the charges or generates electric current. For example, when a copper plate and a zinc plate are put into a sulphuric acid solution, and the copper plate is connected to the zinc plate outside the solution, an electric current starts flowing.

Direct Current

When the direction of a current remains unchanged in a conductor, it is called Direct Current or D.C. In an electric cell, the positive and negative electrodes are fixed. And, the current flowing from a cell is D.C.

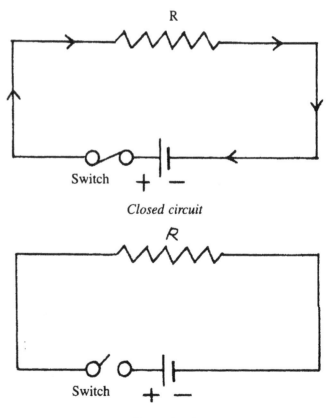

R

Switch + −

Closed circuit

R

Switch + −

Open circuit

415

Alternating Current

An Alternating Current or A.C. is that electric current which changes its direction every half of the time period.

Resistance

All the electric conductors obstruct the flow of electric current. This obstruction is called the conductors' resistance. The better a conductor, the lesser is its resistance. There are two types of electric resistance: (i) Internal Resistance; and (ii) External Resistance. The resistance between the electrodes and chemicals inside the cell is called internal resistance, and the resistance of the outer circuit is called external resistance.

The unit of measurement of resistance is ohm.

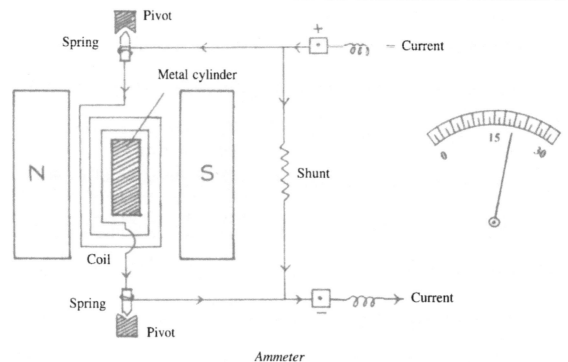

Ammeter

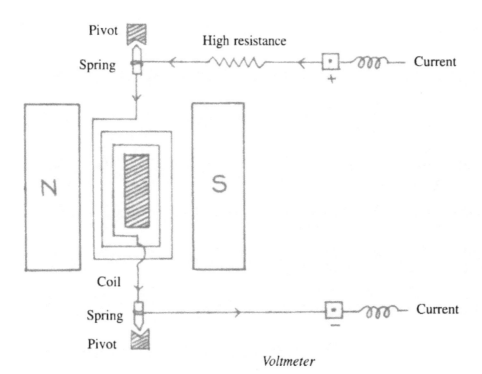

Voltmeter

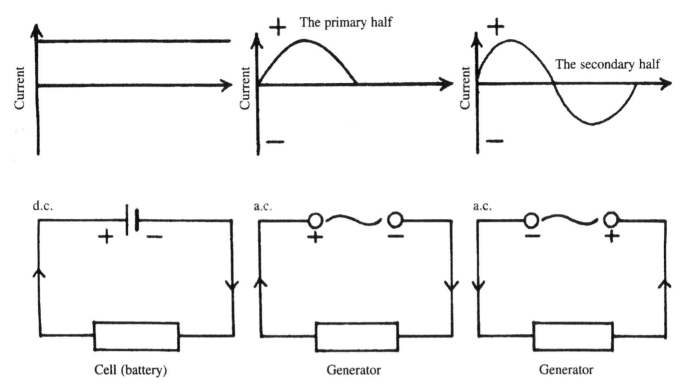

In direct current, the direction of the current does not change. In alternating current, the direction of the current gets reversed every half-time period.

Measurement of Electric Current

The international unit for measuring the electric current is ampere. It is measured by an ampere meter or ammeter. An ammeter is actually a low resistance galvanometer connected in a parallel sequence. An ammeter has a coil. The coil joins the two poles of horse-shoe magnet. During flow of electric current through a circuit connected with an ammeter, the coil deviates in direct proportion to electric current passing through it. The deviation in the coil indicates the measurement of the current. This is measured with an indicator which moves on the circular scale of the ammeter. The circular scale is generally calibrated in ampere or milliampere.

Unit of Potential Difference

The standard unit for measuring the potential difference is volt. It is measured by a voltmeter. It also measures current like an ammeter. Its circular scale is calibrated in volts.

ELECTRIC GENERATOR AND ELECTRIC MOTOR

A generator transforms mechanical energy into electric energy. It is used to generate electricity on a large scale. In a generator, an armature is revolved in a magnetic field to generate electric current. There are two types of generators: (i) A.C. Generator, and (ii) D.C. Generator. An A.C. generator produces alternating current and a D.C. generator generates direct current. Generator is also called Dynamo.

A.C. Generator

An A.C. generator has four parts: (i) Armature; (ii) Field magnet; (iii) Slip rings; and (iv) Brushes. Armature is a rectangular block of coils, made of an insulated copper wire wound on a laminated core of soft iron. Field magnet is a powerful permanent magnet or an electric magnet. Armature revolves between the poles of the magnet. Slip rings are two metal rings. Each ring is permanently attached with an end of armature wire. A pair of carbon rods is called brush. The brushes are connected to slip rings with the help of springs. A terminal each is fixed on the two brushes which help transmission of electric current to outer (external) circuit.

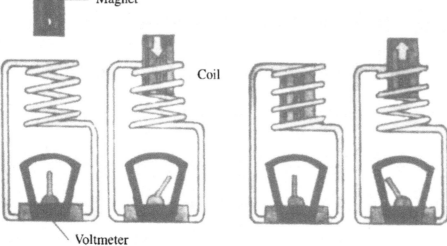

Magnet

Coil

Voltmeter

Faraday (above) saw that if a stationary magnet is placed near a coil, no current is produced. But as soon as the magnet is moved to and fro, current gets produced. The generator is based on this principle of the creation of electromagnetic energy.

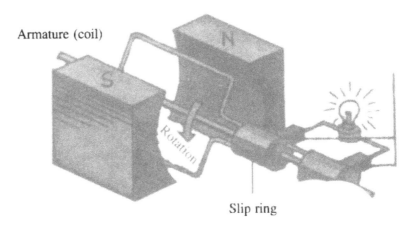

A.C. Generator

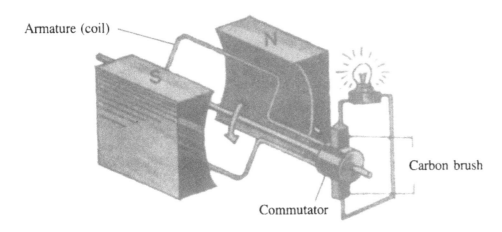

D.C. Generator

When a conductor is revolved in a magnetic field, the magnetic flux of the field registers a change. This change causes current to flow through the closed circuit of the conductor. This effect is called electromagnetic induction. A generator functions on this effect. When a turbine makes an armature revolve between the poles of a magnet, the alternating current is produced due to the electromagnetic induction. The brushes transmit this current to the external circuit. The turbine is moved either by water falling from a height, or steam.

D.C. Generator

A D.C. generator is just like an A.C. generator, except that it has a commutator in place of the slip rings. It also functions on the basis of electromagnetic induction. The commutator, however, has no permanent brush. In a D.C. generator, the direction of electric current changes after every half circuit.

Electric Motor

An electric motor transforms electric energy into mechanical energy. If an electric current is passed through a motor, it will rotate like an electric fan. The functioning of an electric motor is just opposite to the working of a generator or dynamo.

There are two types of motors: (i) A.C. Motor; and (ii) D.C. Motor. An A.C. motor is operated by alternating current and a D.C. motor is run by direct current.

419

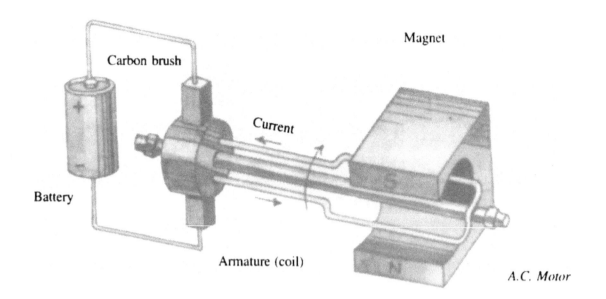

Carbon brush

Magnet

Current

Battery

Armature (coil)

A.C. Motor

An A.C. motor is just like an A.C. generator. It is, however, connected with the source of alternating current instead of a load. Similarly, a D.C. motor is like a D.C. generator. But it has a battery in place of a load.

If an A.C. motor is connected to a D.C. source, and a D.C. motor is connected to an A.C. source, there will be no motion in the motor. In other words, it will not function.

■■

TRANSFORMER

Transformer is a device used to increase or decrease the alternating electric potential difference. It also functions on the principle of electromagnetic induction.

There are two types of transformers: (i) Step-up transformer; and (ii) Step-down transformer. A step-up transformer converts alternating current of a low potential difference into alternating current of a high potential difference. On the contrary, a step-down transformer converts alternating current of a high potential difference into alternating current of a low potential difference.

In an ordinary transformer, electric resistant copper wires are wound on the laminated core of soft iron. These are called coils. A transformer has at least two coils. The coil which receives alternating current is called the primary coil. And, the coil which gives out alternating current is called the secondary coil. In a step-up transformer, the number of windings on the secondary coil is more than that on the primary coil. The wires of the primary coil are thick, while those of the secondary coil are thin. Just opposite to a step-up transformer, in a step-down transformer, the primary coil has a more number of windings than the secondary coil. Moreover, the wires of the primary coil are thin and those of the secondary coil are thick. In a step-up transformer, the electric potential difference of the output increases, but the electric current decreases. In a step-down transformer, however, the electric potential difference of the output decreases while the electric current increases.

A transformer transmits electricity from power-generating stations to far-off places. It is also used in various electric machines where a potential difference more or less than the mains line is required. It is also used in voltage stabilizers.

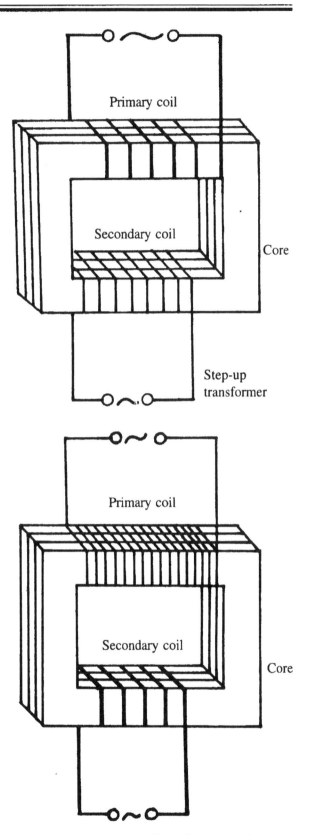

Primary coil

Secondary coil

Core

Step-up transformer

Primary coil

Secondary coil

Core

Step-down transformer

MAGNETS AND MAGNETIC FIELD

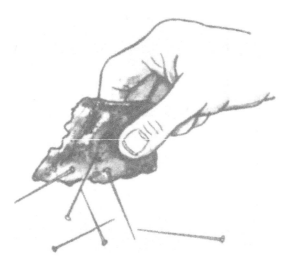

Mineral rock magnetite has the quality of attracting small pieces of iron towards it.

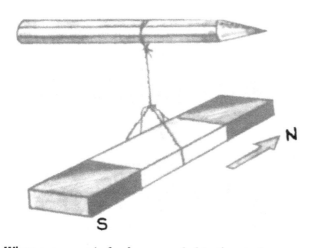

When a magnet is freely suspended in the air, it comes to rest with one pole pointing in the north direction and the other pole pointing in the south direction.

Around 800 B.C., the people of Asia Minor found the lodestone in Magnesia. The stone had peculiar properties. It could attract small pieces of iron. Hung freely, it invariably stopped pointing in the north-south direction. After the place of its findings, Magnesia, the stone came to be known as magnet. The alternating property of the magnet was called 'magnetism'. Actually, the lodestone is magnetic—an iron ore. It is a ferrous oxide.

There are two types of magnets: (i) Natural magnet; and (ii) Artificial magnet. The magnet in which the property of magnetism is present naturally is called natural magnet. When magnetism is artificially induced into a substance, the magnet so is called an artificial magnet. The artificial magnet can also be made from steel. A magnet can be made in any shape. But, bar magnet, horse-shoe magnet and needle magnet are common.

Magnetic Poles

When a magnet is brought closer to iron filings, its ends attract most of the filings. The middle part of the magnet attracts the least amount of filings. The two ends of a magnet, on which stick the maximum amount of iron filings, are called its 'poles'. The middle part of a magnet is called the 'neutral zone'. When a magnet is suspended freely, its ends point towards north and south directions. The end pointing towards the north direction is called north pole, and the end

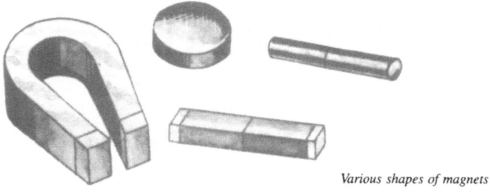

Various shapes of magnets

pointing towards the south direction is called south pole. The north pole of a magnet attracts to the south pole of another magnet, but repels its north pole. Thus, the similar poles of two magnets repel, while their opposite poles attract each other.

Magnetic Field

Magnetic field is that region of space around a magnet where the magnetic force is effective.

How to find the magnetic field? To find the magnetic field, place a bar magnet on a sheet of cardboard. Spread iron filings around it. Tap the cardboard with your fingertip. The filings get arranged in a particular pattern. The reason: when iron filings are placed in the magnetic field, they become magnetised. The magnetised filings then get arranged in peculiar lines according to the direction of the magnetic field. These lines

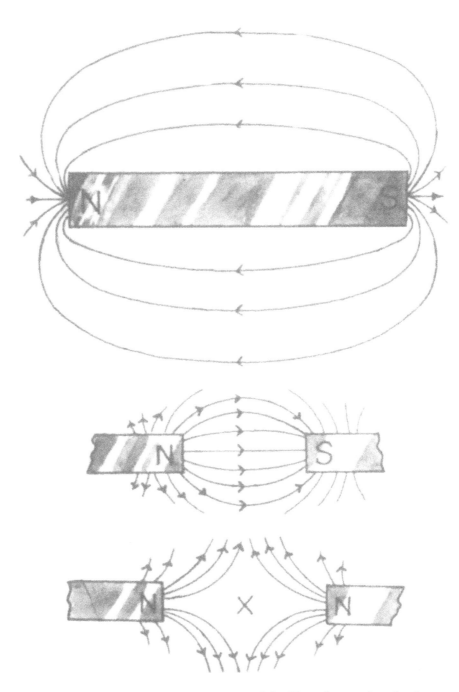

The opposite poles of magnets attract, and the like poles repel each other.

423

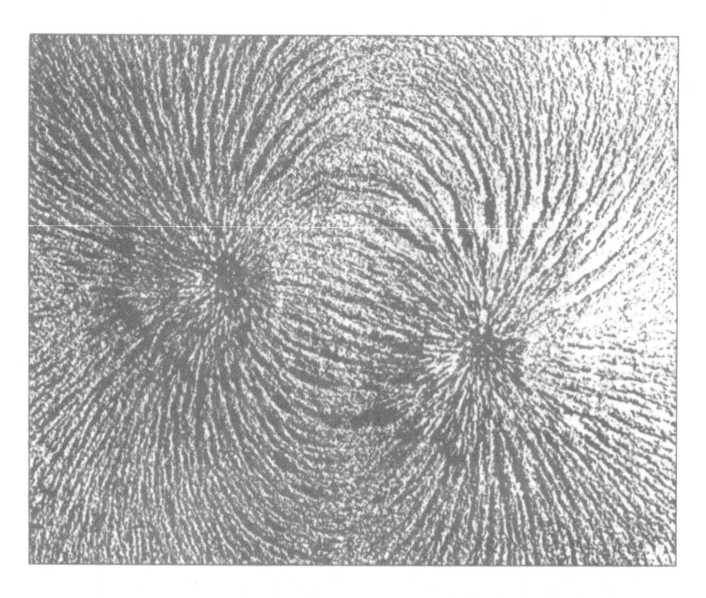

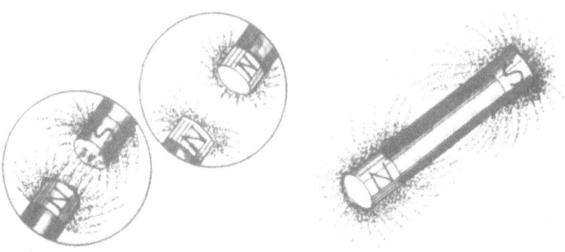

Magnetic lines of force

are called 'the magnetic lines of force'. These lines are imaginary. They go from north pole to south pole of a magnet. These lines also indicate the direction of the force to be exerted on an object when placed within this field. The region, where the lines of force are more dense, has high magnetic intensity and, the region, where the lines of force are less dense, has low magnetic intensity. The direction of the lines of force is from north pole to south pole outside the magnet. The magnetic lines of force never cross one another, because a magnetic field cannot have two directions at one point.

In ancient times, the sailors used magnets in compasses to find out direction. A compass has a small needle-shaped magnet. A compass, however, cannot tell the exact direction. Reason: the magnetic needle does not point towards the geographical North Pole. It points towards the magnetic North Pole, which is 1,600 km. away from the actual North Pole. The geographical South Pole is about 2,400 km. away from the magnetic South Pole. At some places, the iron ores or magnetic matters under the earth also affect a compass. So, the needle cannot provide reliable information about the direction.

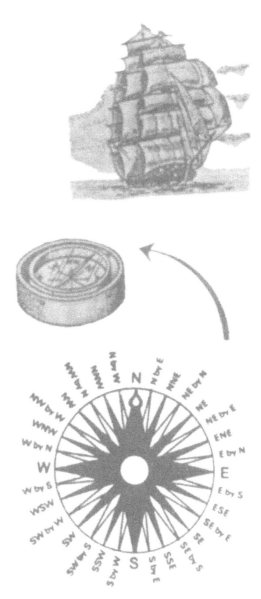

In ancient times, sailors used the magnetic compass to find out the directions.

425

FERRO-MAGNETIC SUBSTANCES

Although every substance in the world has magnetic property, there are certain substances which are attracted by magnets and can be highly magnetised. Such substances are—iron, nickel, cobalt and their alloys. These are called ferro-magnetic substances.

The magnetisation and demagnetisation of ferro-magnetic substances can be explained through the Domain theory.

According to the Domain theory, a ferro-magnetic substance is made up of thousands of tiny domains. In both magnetised and unmagnetised substances, each domain is magnetically saturated, which means all its atoms are magnetised in one direction. In an unmagnetised state, the domains are so unevenly distributed that one domain destroys the effect of another domain. Thus, the magnetic effect of the substance is completely destroyed.

When an intensive magnetic force is applied to these substances, the domains arrange themselves in the direction of the force. In such a situation, even after the removal of the magnetic force, the magnetism of the substance remains.

When a ferro-magnetic substance is heated up to a certain temperature, its domain structure gets disturbed due to thermal agitation, and the substance becomes demagnetised. Falling from a height or hammering may also destroy the magnetism of a substance.

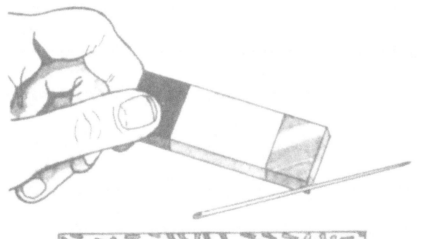

Demagnetised bar

Magnetised bar

ELECTROMAGNET

When an insulated copper wire is wound on a rod of soft iron, and an electric current is passed through it, the iron rod becomes a magnet. One end of this magnet points towards the North Pole and the other one towards the South Pole. Thus, the iron rod is temporarily magnetised. But with the stoppage of electric current, the rod loses its magnetism. When this experiment is repeated with a steel rod, some magnetism is retained by it even after the current is stopped. Therefore, the soft iron does not have the capability of retaining its magnetism, while the steel does have. This property of the magnetic substances is called 'retentivity'. An electromagnet can be made more powerful by increasing the number of the copper wire windings on the rod, and enhancing the value of the electric current passing through it.

Electromagnets are largely used in electric bells, telephone receivers, loudspeakers, etc. Electromagnetic cranes are used to lift iron blocks. Electromagnets are also utilised to make iron ore concentrate. They are also used in the wheels of the high-speed trains.

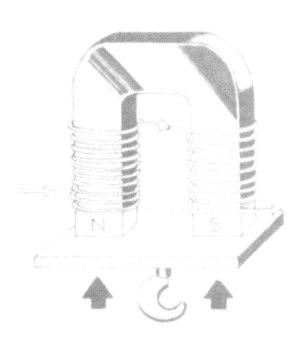

Electromagnet

Electromagnets are used to lift up heavy pieces of iron

THE EARTH'S MAGNETISM

Our planet behaves as if it had a huge magnet within it. A freely suspended magnet invariably stops in the north-south direction, which proves that a magnetic force is working on it due to the earth's magnet. The north pole of the freely suspended magnet points towards the North Pole of the earth; and its south pole points towards the South Pole of the earth. It proves that the North Pole of the earth's bar magnet is towards the earth's south; and the south pole is towards its north.

Scientific studies have shown that the magnetic axis and the geographical axis meet at an angle of 17°.

The earth's magnet-like behaviour has been explained through various principles. But no principle can stand the test of scruptiny regarding all the facts. In fact, there is still no principle that could be acceptable to all. According to one principle, the gases in the atmosphere get ionised by the cosmic and radioactive rays from the sun. As the earth moves on its axis, powerful electric currents flow in the atmosphere around it. Due to these currents, the ferrous substances under the earth get magnetised. That is why the earth has become a magnet.

Another principle claims that magnetic substances—iron and nickel, for example—are present under the earth in a molten state. As the earth rotates, convection currents are generated in these fluids, which produce the effect of a self-excited dynamo. The earth has become a magnet due to these currents.

The poles of the earth's magnet change a little position-wise from time to time. This leads to a change in the angle between the magnetic axis and the geographical axis.

North Pole

South Pole

The earth behaves like a huge magnet

MINERALS & METALS

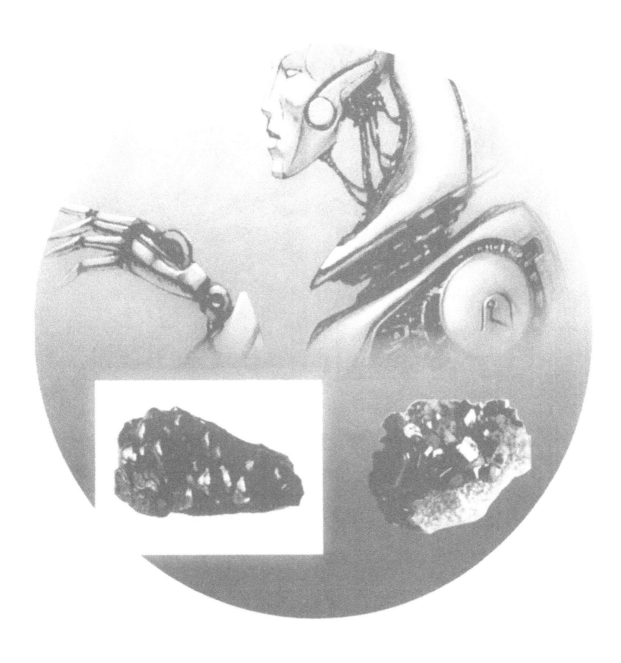

MINERALS

Some metals like gold, silver, copper, platinum etc. are found in nature in free state but most metals occur in combined state as minerals. Minerals are crystalline elements or compounds of elements that occur naturally in the earth. If a mineral contains sufficient quantity of a metal which can be extracted profitably from it, it is known as the ore of the metal. For example, galena is the ore of lead, bauxite that of aluminium and cinnabar is an ore of mercury. The earth's crust contains about 8% of aluminium, 5% of iron and 4% of calcium.

Ores are of two types – native ores and compound ores. Native ores contain elements in free state but with some impurities. In compound ores, elements remain combined with other elements. All minerals are non-living i.e. inorganic in nature. Coal, oil, gas etc. are all organic substances, that is why they are not put in the category of minerals. Emeralds, aquamarine, corundum, topaz, agate, jasper etc. are some well known precious minerals.

Geologists have discovered about 3,000 minerals, of which only about 100 are common and useful. Our atmosphere is rich in oxygen. Seawater contains chlorine and bromine is obtained from the sea grass. The earth's crust contains sulphur and silicon. All these elements are quite reactive, that is why most of the elements are found as oxides, carbonates, chlorides, sulphides, sulphates and sillicates. Some of the important ores are:

Natural ores	(Cu, Ag, Au, Hg, As, Sb, Bi, S)
Oxide ores	(Fe, Al, Mn, Sn)
Sulphide ores	(Zn, Cd, Hg, Cu, Pb, Ag)
Carbonate ores	(Fe, Pb, Zn, Mg, Ca, Ba)
Halide ores	(Na, K, Mg, Ca, Ag)
Sulphate ores	(Ca, Sr, Ba, Mg, Pb)
Silicate ores	(Li, Na, K, Mg, Ca, Al)
Phosphate ores	(Ca, Li, Fe, Mn)

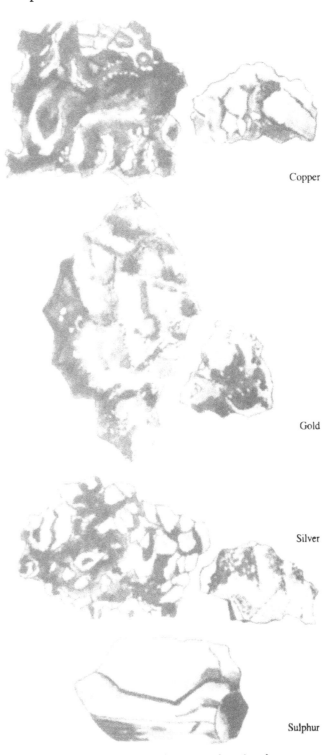

Copper

Gold

Silver

Sulphur

Copper, Gold, Silver and Sulphur are found in free state

431

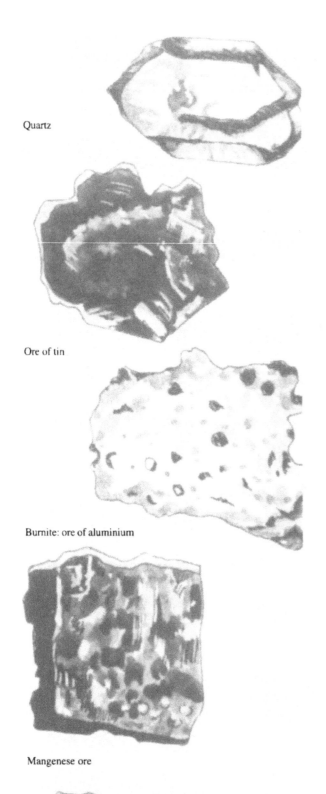

Quartz

Ore of tin

Burnite: ore of aluminium

Mangenese ore

Cinnabar: ore of mercury

The earth's core is very rich in iron and nickel but the outer part of the core contains silicates also. Mantle is rich in pyraxin and olivine. The earth crust contains the maximum quantity of silicate minerals. Felspar, quartz and mica are silicates and commonly constitute the rocks. The earth's crust contains about 62% of silica, 16% of alumina, 6% of iron oxide, 3% of manganese oxide, 6% of lime, 3% of sodium oxide and 3% of potassium oxide. The quantity of metallic elements in the earth's crust like copper, lead, zinc, silver etc. is not very large. These metals are confined only to some particular regions.

The outer layer of the earth is called lithosphere which is mainly made up of rocks. Most of the rocks are made of different minerals. The sedimentary rocks made due to the erosion of igneous and metamorphic rocks contain compounds of gold, tin, platinum etc. Gold in USA and tin in Malaysia are obtained only from such compounds. Haemetite, the ore of iron, is also obtained from the sedimentary rocks.

Usually, the metals are found as minerals in igneous rocks. In rock cracks and cavities, melted magma contains chromium, nickel, copper, silver, gold, lead, zinc etc. in solid form. Uranium and thorium are found in granite rocks or in sedimentary rocks formed from granite rocks.

Most minerals are crystalline in nature. Minerals are identified by virtue of their lustre, hardness, colour, streak, spectrum etc. The scientists who study minerals are known as mineralogists.

METALS

Elements such as sodium, gold, silver, zinc etc. fall in the category of metals. Most metals are heavy, shiny and solid. Mercury is the only metal which exists in liquid state. Metals are malleable and ductile i.e. by beating they can be converted into sheets and wires can be drawn from them.

Till today, we know about 80 metals. Most metals, are found in the earth's crust. Some metals occur in free state but most of them are in a combined state as compounds. A mineral from which a metal can be extracted profitably is called its ore. Metal ores are obtained from mines. The mixture of one metal with the other metal or non-metal is called its alloy. For example, mixture of copper and zinc is known as brass.

Similarly bronze is an alloy of copper and tin while steel is an alloy of iron and carbon.

All metals are good conductors of heat and electricity. Metals are sonorous i.e. they produce sound when beaten by something. Most metals can be melted and cast into the desired shape and size.

Gold, silver and copper are good conductors of heat and electricity. The above metals are also called **Coinage Metals** because during the ancient times, these metals were used to prepare coins. Gold, silver and platinum are expensive metals. In the present days, these metals are mostly used to make ornaments. Copper and aluminium are used to make electric wires. Iron and aluminium are used for the manufacture of machines, trains, engines, railway tracks, motorcars, buses and aeroplanes.

Metals have proved to be a great help in the development of human civilisation

METALLURGY

These ancient daggers are obtained from the graveyard of Tutenkhan Pheroh of Egypt of about 15 Century B.C.

A silver-cauldron of Celtic Iron-Age

Metallurgy is the branch of chemistry in which we study about the different types of metals and their extraction from their ores. Metallurgy is of two types: Extractive metallurgy and Physical metallurgy. Several methods are used to extract metals from their ores. In physical metallurgy, we study the structure and properties of metals. This also includes the study of alloys.

Metallurgists study the surface flaws with the help of microscopes. The analysis of surface flaws tells how metals can be protected from rust and corrosion. Heat treatment, tempering, anodising, annealing, electroplating, galvanizing, carburizing etc. also fall under the category of metallurgy. Hammering, casting, rolling, extrusion etc. are the branches of physical metallurgy. Nowadays, composite materials are also being studied under metallurgy. Composite materials are made by reinforcing fibre glass or plastic fibres with metals.

Metallurgical Methods

1. **Concentration of Ore:** Concentration of ore means removal of unwanted materials from the ore and thereby increasing the percentage of metal in the ore. Ores are concentrated by several methods such as by hand-picking process, gravity separation method, oil floatation method.

2. **Extraction of Metal:** After the concentration of the ore, the metal is extracted by three methods, namely (a) calcination; (b) roasting; and (c) smelting.

3. **Purification of Metal:** The metals obtained by the above three methods are not pure. They contain impurities like silica, phosphorus, carbon etc. Impure metals are purified by the following methods:

Distillation Process: Metals like zinc, cadmium, mercury etc. which can be easily volatilised are purified by the distillation method.

Metallurgy

Melting Method: Metals like tin which are easily melted are purified by the melting process. Impure metal is melted in a slopy furnace. The pure metal comes down from the slope and impurities remain on the slope.

Electrolytic Process: In this method, an electrolytic solution of the metal is taken. When electric current is passed, the metalic ions from Anode come to the Cathode and get deposited on it as pure metal. Anode goes on dissolving and impurities settle down in the solution as anode mud and pure metal goes on depositing on the cathode. Copper, tin, gold, silver, zinc etc. are purified by this method. Besides these methods, there are oxidation processes and poling methods which are also used to purify metals.

Furnaces: Metals are extracted by heating in different types of furnaces. The following furnaces are used in metallurgical processes.

A gold-cup of 500 B.C.

435

Kiln: This is the oldest form of furnace. In a kiln, material and fuel are heated in air to such a temperature that the material does not melt but waste gases come out of the chimney. These furnaces were used to make metals soft and bake bricks and clay utensils. Nowadays, gas or electricity is used as fuel in kilns.

Blast Furnaces: The outer surface of this furnace is made of iron while the inner surface consists of thermally insulated bricks. The ore along with coke and limestone is fed from the top into the furnace. From the lower part, air is sent into the furnace through nozzles. Waste gases come out through the chimney. The ore melts because of the high temperature and goes into the hearth. Here metal and impurities are separated out. Blast furnaces are mainly used for extraction of iron.

Reverberatory Furnaces: In this furnace, one part is meant for the fuel while the other part is a hearth where the ore is heated up. Air is forced into the furnace through a hole. Hot gases come out from the chimney. These furnaces are used for both oxidation and reduction. These furnaces are mainly used for calcination and roasting.

Muffle Furnaces: In this type of furnaces, hot gases heat the ore from all sides. These furnaces are used in zinc extraction and assaying and annealing of silver and gold.

A helmet made with an alloy of copper, iron and gold

An axe of Celtic-iron age

A bronze ware of 700 kg. was made in China in 14-11 Century B.C.

436

Electric Furnaces: Electric furnaces are mainly used to create high temperatures. Ore is oxidized in these furnaces. These furnaces are of three types: Induction furnace, Resistance furnace and Arc furnace. Induction furnaces generate heat by electromagnetic induction, resistance furnaces generate heat by electric current and arc furnaces generate heat by producing an arc with high voltages.

Besides the above mentioned furnaces, there are the very simple Open-hearth-furnaces, the Regenerative furnaces and the Bessener processes that are used for the extraction of metals from their ores.

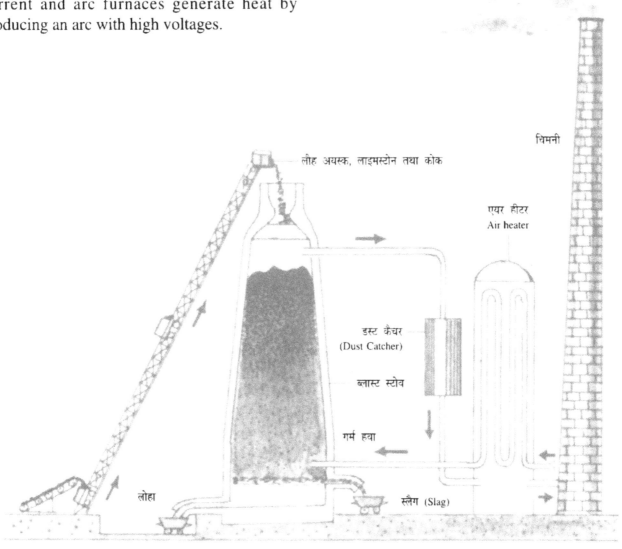

A blast furnace with a chimney

437

GOLD

Latin Name	:	Aurum
Symbol	:	Au
Atomic Number	:	77
Atomic Mass	:	196.97
Colour	:	Yellow, Shiny
Melting Point	:	1063°C
Boiling Point	:	2660°C
Density	:	19.3 times heavier than water

Gold is an attractive metal. It was even known in the neolithic age. According to the fossils of older civilisations, perhaps gold was the first metal discovered by man. The gold ornaments obtained from the graveyard of Tutenkhan Pheroh of Egypt are considered to be made about 1400 B.C. Queen Asariya got a

A throne of gold found from the graveyard of Tutenkhan

gold statue of Semiramis made which was 12 metres high and weighed 30 tons. The gold statue of goddess Rhea weighed 250 tons. The whole Inca civilisation declined due to its people's greed for gold. Francisco Pizarro plundered many of their gold artifacts and got them converted into gold bricks. Explorers undertook many risks in the search of this valuable metal.

After Egypt, Mesopotamia, India and China obtained the knowledge of gold in the 10th century B.C. Alchemists in the middle ages made many efforts for making gold from cheaper metals but they did not succeed. Robert Boyle proved for the first time that gold is an element and it cannot be prepared from other substances. But the scientists of our time can convert lead into gold by the bombardment of neutrons on lead.

The biggest nugget of gold was found in Australia. It weighed 214.32 kg. From this nugget, 70.92 kg of pure gold was obtained.

Coins of Electrum (alloy of gold and silver) were introduced in Lydia (Asia Minor) about 690-650 B.C. Mughal Emperor Shahjahan introduced a gold coin in 1654 which even today is considered the costliest in moulded coins. The biggest stock of gold coins was excavated in 1814 in Brescello near Italy. These were buried in about 37 B.C. These coins contained about 80,000 Aureous coins of Roman Empire.

Gold is found everywhere in the world in very deep mines. It often occurs free in stream deposits, quartz veins or with pyrites. It is also found in the sea water. About 5 tons of sea water contain one milligram of gold. Gold occurs in large quantities in South Africa, Russia, Canada and the United States of America. In India, gold mines are located in Karnataka.

In the beginning, gold was mainly used in jewellery and statue making because of its non-tarnishing nature and brightness. Pure gold is

Gold was the reason of the decline of ancient Inca civilisation

A Death-mask made of gold in Celtic-iron age

called 24 carat gold. If an alloy is 22 carat, it means 22 parts out of 24 are gold. The rest is other metals. 22 carat gold contains 87.7% of pure gold.

Gold is least affected by air, water and sun. It does not dissolve in any acid except aquaregia. Aquaregia is a mixture of concentrated nitric and sulphuric acids. Gold is very heavy but a very soft metal. It is highly malleable and ductile. Gold leaf may be as thin as ten-thousandth part of a millimeter.

Gold is not only used in statue and jewellery making but also in industries. Gold is a very good conductor of electricity. Therefore, it is used to make small components in delicate electrical circuits. Gold-plated items do not corrode easily. Gold is also used in the electronic appliances such as in transistors, diodes and microchips. Gold is used in religious places such as temples, churches and mosques to build statues, ornaments, etc. Gold is also used in satellites. It is used as currency in International trade. Every country has its gold reserves in the form of heavy brick-shaped blocks called *ingots* and the prosperity (wealth) of a nation is determined by the amount of gold reserves it possesses.

439

SILVER

Latin Name	:	Argentum
Symbol	:	Ag
Atomic Number	:	47
Atomic Mass	:	107.87
Colour	:	Shiny White
Melting Point	:	961°C
Boiling Point	:	2180°C
Density	:	10.5 times heavier than water

Pure silver is shiny white. It is a rare and costly metal. Silver is known to man from ancient times. All chemists used a half moon for symbolizing silver. The remains of Sumerian civilisation show that these people knew the art of extracting this metal from lead. Egyptians also knew about silver. Silver ornaments and utensils have been found in the excavations of Harappa and Mohenjo-daro. Argentina has been named after argentum (Latin name for silver) because this country had large deposits of silver. About 327 B.C., when Alexander the Great invaded India, his soldiers suffered from gastrointestinal disease while the high officials of his army had no impact of this disease. The cause of this incident could be traced only after a period of 2,000 years. It was concluded by the scientists that soldiers used tin cups for drinking water while high officials used silver utensils. Silver has a property of purifying water. It can destroy 650 types of disease-causing bacteria. Ten-millionth part of one gram of silver is sufficient to purify one litre of water. This was the cause which protected the high officials of Greek army from gastrointestinal disease. Cyrus, the king of Persia in 500 B.C. used silver utensils for drinking water during his journeys. These were called sacred silver vessels.

Silver is the best conductor of heat and electricity

Silver wares found in France dating 3 millennium B.C.

In olden days, silver was used in the form of currency coins. Romans introduced silver coins in about 269 B.C.

Silver occurs in nature in sufficient quantity both in free as well as combined state. In 1860, one nugget of silver was found in Spain which weighed about 8 tons.

The main ores of silver are:

- Silver glance or argentite — Ag_2S
- Horn silver or chlorargyrite — $Agcl$
- Ruby silver or pyrargyrite — Ag_3SbS_2
- Silver Copper glance or Stromeyrite — $(Ag, Cu)_2S$
- Stephonite — Ag_5SbS_4
- Proustite — $Ag_3 As S_2$

Maximum quantity of silver is obtained in Mexico. North and South America, Australia, Canada, Japan etc. also produce a good amount of silver from their mines. Silver is prepared by a number of different processes such as the cyanide process, the **Desilverisation of lead** and **Electrolysis processes**. The Cyanide process is the most popular omong these. The silver obtained from this process is purified by the *Electrolytic process.*

Though silver is less reactive, it tarnishes when it remains in contact with air for long. It makes different compounds and salts with different acids.

Silver is used in making jewellery, utensils, statues, coins etc.

Silver is a very good conductor of heat and electricity. It is widely used to make electric contacts in electronic equipments. Silver reacts with nitric acid. Its compounds are sensitive to light, that is why silver chloride, silver bromide and silver iodide are used in photography. Silver is also used in food industries. Other metals can be electroplated with silver. Silver coated brass utensils are used in five-star hotels to serve the food. Silver *ingots* are used as currency in International Trade.

441

IRON

Latin Name	:	Ferrum
Symbol	:	Fe
Atomic Number	:	26
Atomic Mass	:	55.847
Colour	:	Greyish-Black
Melting Point	:	1539°C
Boiling Point	:	2800°C
Density	:	7.9 times heavier than water

Iron is known to the man from prehistoric times. In our times, it is a cheap metal but about 5000 years ago, Egyptians had to part with a lot of gold to obtain a very small quantity of iron. In those days, iron was more valuable than gold. Studies of the fossils reveal that in about 1500 B.C., people of South-East Asia extracted iron by heating an ore. The iron obtained by man for the first time was from the meteorites. Indians also knew how to make good quality iron in ancient times. Ashoka Pillar near Qutub Minar in Delhi is an example of this. It is standing there for about 1,500 years but has not rusted so far. The gift given to Alexander the Great by Porus was 15 kg of good quality iron. Damishk was famous for high quality iron. Swords made in Damishk were used all over the world.

The first piece of Iron, obtained by man was from Meteorite

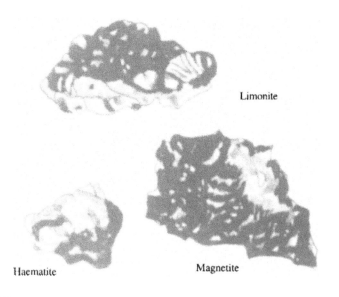

Limonite

Haematite

Magnetite

442

Iron ore extraction from a mine

Iron is obtained from ores. The main ores of iron are:

- Oxide ores: Magnetite Fe_3O_4, Red Haematite Fe_2O_3, Grey Haematite $2Fe_2O_3 \cdot 3H_2O$.
- Siderite or $FeCO_3$.
- Sulphide ores: Iron Pyrites—FeS_2.

Iron may be considered the basis of the modern machine age. Countries which produce large quantities of iron and steel are the richest and most powerful in the world.

About 90% of iron is produced in 11 countries of the world. These are Russia, Brazil, Australia, Canada, USA, India, China, South Africa, Sweden, Liberia and Venezuela. Iron mines of India are found in Bihar, Orissa, M.P., Karnataka, Maharashtra, Tamil Nadu and Goa.

Big Blast furnaces are used to extract Iron. In these furnaces, the ore is heated with coke and limestone. This gives sufficiently pure Iron.

King Soloman was told by the royal mason that the construction of temple was made possible after the making of spade and pick-axe by a blacksmith

Forms of Iron

Iron is obtained in four forms—Cast iron, Steel, Wrought iron and Pigiron. These forms have different percentage of carbon in iron. Cast iron is considered the best among all. Cast iron has 0.2% of Carbon in it. Cast iron is again of two types – White Cast Iron and Grey Cast Iron. White Cast Iron contains Carbon in combined state While Grey Cast Iron contains carbon in the form of Graphite crystals. Stainless steel contain 73% of Iron, 18% of Chromiun, 8% of Nickel, and 10% of Carbon. Inva is another alloy contain Iron and Nickel. Alnico is a mixture of Iron, Aluminium, Nickel and Cobalt.

Two types of compounds are formed by iron ferric and ferrous. In ferrous compounds, the valency of iron is 2 while in ferric it is 3.

Iron compounds are mainly used in making dyes and inks. Our blood also contains iron. Iron is a magnetic metal. It is used in machines. Iron oxide is used in making magnetic tapes. Iron is used in the construction of buildings, bridges, iron structures and machines on large scale.

The use of Iron excelerated the growth of civilization

COPPER

Latin Name	:	Cuprum
Symbol	:	Cu
Atomic Number	:	29
Atomic Mass	:	63.55
Colour	:	Reddish orange
Melting Point	:	1083°C
Boiling Point	:	2567°C
Density	:	8.93 times heavier than water

Man has been using copper for more than 10,000 years. Copper is a native metal obtained by man in the beginning along with silver and gold. The time between 5000 and 3000 B.C. is known as copper age. In this period, tools and weapons were made of copper. It is thought that in the construction of the Pyramids of Egypt, copper tools and appliances were used. Copper appliances have been found in the excavations of Harappa. The armours of Sumerian soldiers were made of copper. Copper utensils, ornaments and statues have been found in the excavations of ancient places.

During the Roman Empire, Cyprus Island was the main place of occurrence for copper. On the basis of Cyprus, it has been named as Cuprum. The English word copper has also originated from it.

The ancient man obtained tin ore along with copper from which tin metal was extracted. Several alloys of copper were made. Bronze also is an alloy of zinc and copper.

Copper is found in large quantities in USA, Russia, Canada, Sweden, Zambia, Chile, Germany, France etc. In our country, copper is found only in Bihar, Rajasthan and Andhra Pradesh in small quantities. About seven million tons of copper is extracted in the world every year. Copper is extracted from its ores by several methods. It is purified by the process of

Copper was being used even 10,000 years ago

electrolysis. The following are the main ores of copper:

Sulphide ores: Copper Pyrites $CuFeS_2$, Bornaite Cu_3FeS_2, Copper glance Cu_2S

Oxide ores: Malachite $CuCo_3$ Cu (OH)2 Azurite $2CuCo_3$. Cu (OH)$_2$

Copper is generally obtained from Copper Pyrites (CuS_2). These are usually purified in Bessemer convertors. Later, by the Electrolytic process pure Copper is obtained. We prepare Brass, Bronze, Bellmetal, Gunmetal, German Silver, etc. from Copper.

After silver, copper is the best conductor of electricity. About two-third of copper produced in the world is used in electrical industry. Rest one-third is used in making alloys, utensils and bullets.

Copper becomes red-brown by the reaction of air. Copper compounds are useful for the living world. They are used as insecticides. Some copper compounds are used in paint-industry also.

MANGANESE

Latin Name	:	Manganese
Symbol	:	Mn
Atomic Number	:	25
Atomic Mass	:	54.938
Colour	:	Grey
Melting Point	:	1244°C
Boiling Point	:	1962°C
Density	:	7.2 times heavier than water

Manganese was discovered in 1774 by the Swiss Chemist, Johann Gahn. Most manganese is obtained from the mineral pyrolusite or manganese dioxide. Pyrolusite is mixed with powdered aluminium and heated. This sets the manganese free.

Manganese is found in several minerals such as pyrolusite, manganite, braunite, hausmannite etc. It is also found in plants, human and animal bodies in minor quantity.

Ores of manganese are found in Russia, India, Brazil, Ghana and South Africa.

Manganese is added to steel to harden it and to make it more resistant to wear and tear. It is also added to bronze.

Manganese dioxide is used in dry cells, ceramics and dyes. It is also used in the industry as a catalyst and an oxidizing agent.

Manganese sulphate is used in paints, varnishes and fertilizers. Potassium permanganate is a powerful oxidizing agent and is used as a disinfectant.

Manganese is used to kill the germs in the water. It is generally known as red medicine used in ponds and wells. A solution of very little quantity of Manganese is used to wash vegetables.

"Don't think me alone. Manganese is with me. You can't break me"

TIN

Latin Name	:	Stannum
Symbol	:	Sn
Atomic Number	:	50
Atomic Mass	:	118.69
Colour	:	Silver white
Melting Point	:	233°C
Boiling Point	:	X
Density	:	7.3 times heavier than water

Tin as a metal was known to man even about 6,000 years ago. A ring and a bottle of tin made in about 1580-1350 B.C. have been found from a graveyard of Egypt. These things are considered to be the oldest items made of tin. One flower pot of tin was also found in Egypt made about 1,200 years ago. Romans used tin in making mirrors.

Cassiterite is the main ore of tin which is found in maximum quantity in Malaysia. Bolivia, Indonesia, Belyin Kango, Thailand and Nigeria also have large deposits of tin ore. Half of the tin produced in the world is used by the United States of America alone.

Tin is a silvery white soft metal and it does not rust. It is mainly used in making cans. A fine coating of tin is done on iron cans so that food materials may not decay.

Tin is also used in making alloys. Soldering alloy is made by mixing tin and lead and is used in soldering electronic components. Bronze is also an alloy of tin. Stannic chloride, a compound of tin, is used in dyes and tooth-pastes. Tin is also used in electroplating. Tin compounds are sprayed on glass sheets to make them electrically conducting. When tin is heated at 1500°C to 1600°C, it starts burning with a bright flame.

Tin is mainly used in making containers

447

MERCURY

Latin Name	:	Hydrargyrum
Symbol	:	Hg
Atomic Number	:	80
Atomic Mass	:	200.59
Colour	:	Bright white liquid
Melting Point	:	–39ºC
Boiling Point	:	359ºC
Density	:	13.5 times heavier than water

Mercury is the only metallic element which remains in liquid state at room temperature. Ancient Romans named it as God Mercury.

Mercury is used in making thermometers

Mercury was a very important metal for alchemists. They tried to make gold from cheaper metals using mercury. Mercury has been found in the Egyptians pyramids built in around 1500 B.C.

Mercury occurs as mercuric sulphide in the red-coloured mineral, cinnabar. It is obtained by heating cinnabar and condensing the vapours given off in the process of heating. The biggest deposits of this mineral are found in Almaden (Spain).

Mercury expands on heating and contracts on cooling. It remains in liquid state from –38.87°C to +356.58°C. At –39°C it freezes. Mercury can dissolve some metals in itself. These are called mercury amalgams. Amalgams are used to coat gold on metals and for filling the cavities of teeth, making mirrors and in the laboratories. Mercury is also used in making thermometers, barometers, manometers, diffusion pumps etc. Mercury vapours are used in mercury vapour lamps and quartz mercury lamps. It is also used in batteries, switches and electrolytic cells. Some compounds of mercury are used in medicines also. Mercury and its compounds are very poisonous in nature. They can cause mental illness. Therefore, they should be handled with great care.

The thermometer is filled with mercury and is used to measure the temperature of the body. Its range varies from 40°C to 350°C. The mercury in Fortin Barometer helps to measure the pressure of the air. Hence mercury is a very useful element.

ALUMINIUM

Name	:	Aluminium
Symbol	:	Al
Atomic Number	:	13
Atomic Mass	:	26.90
Colour	:	Silvery white
Melting Point	:	660°C
Boiling Point	:	2467°C
Density	:	2.7 times heavier than water

Aluminium is widely used in making aircrafts and people now call it the 'flying metal'.

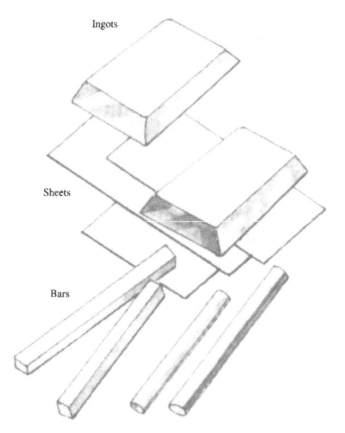

Ingots

Sheets

Bars

Aluminium sheets and bars are extensively used in industries

Aluminium is found in nature in a combined state. That is why man could know about this metal quite late. It was discovered by Friedrich Wohler, a German chemist, in 1827. It was obtained for the first time by heating aluminium chloride with potassium. For about 60 years, it remained an expensive metal. Then in 1886, a cheaper way of producing aluminium was discovered. It is now a widely-used metal.

Earth's crust contains about 8% of aluminium in the form of several different minerals. The following are the main ores of aluminium:

- **Oxides:** Bauxite $Al_2O_3.2H_2O$
- **Silicates:** Felspar $K_2O.Al_2O_6. 6SiO_2$
- **Mica:** $K_2O.3Al_2O_3.6SiO_2, 2H_2O$
- **KGolin:** $Al_2O_3. 2SiO_2. 2H_2O$
- **Fluorides:** Cryolite: $3NaF.AlF_3$
- **Sulphates:** Alunite: K_2SO_4 $Al_2(SO_4)_3 4Al(OH)_3$
- **Phosphates:** Turquoise: $AlPO_4.Al(OH)_2 H_2O.$

Aluminium mainly comes from the mineral bauxite. Firstly, alumina is obtained from the bauxite by washing and refining and then aluminium is obtained by electrolytic process.

Aluminium can be obtained from the Bayer's process, the Hall-Heroult Process of Electrolysis and Serpeck's Process. The alloys or mixed metals of Aluminium is used to prepare utensils and pressure cookers of high quality.

Aluminium is also used to prepare many scientific instruments.

Today, aluminium is so widely used that it is called as the champion of metals. It is a very light but strong metal. It does not get rusted. Its alloy, duralumin, is used in aeroplane industry. Electric wires are made of aluminium. Aluminium foils are used for wrapping foods, candy and cigarettes. Aluminium is also used for making door- and window-frames and handles.

450

LITHIUM

Name	:	Lithium
Symbol	:	Li
Atomic Number	:	3
Atomic Mass	:	6.941
Colour	:	Silvery white
Melting Point	:	180°C
Boiling Point	:	1347°C
Density	:	Lightest metal (0.53)

Lithium was discovered by the Swedish chemist, Johan Arfvedson in 1817. It was extracted from the mineral petalite. After a period of 38 years, lithium production started on a large scale by electrolytic process.

The percentage of lithium in the earth's crust is very small (0.0065%). There are 20 minerals which contain lithium. Lepidolite, spodumene and amblygonite are its main ores. Granite also contains lithium. One cubic kilometer of granite can yield about 1,12,000 tons of lithium.

Lithium is a silvery white soft metal. This can be categorised under the group of Alkali metals. This is the lightest metal. It is highly reactive with hydrogen. 250 gms of lithium hydrite contains 700 litres of hydrogen. It is a highly reactive element and burns in air with a white flame.

Lithium is used in making alloys. It is also used as rocket fuel. Rocket nozzle and combustion chamber are coated with lithium so that they may not melt because of high temperature. Lithium increases the power of electrical batteries three times. Lithium is also used in nuclear reactors. Glass containing lithium is more resistant to heat. That is why T.V. tubes are made of this type of glass only. Lithium is also used in enamels, paints, porcelain and textile industry. Lithium carbonate is a very effective drug used to treat mental depression.

Lithium is also used as Rocket-fuel

BERYLLIUM

Name	:	Beryllium (on the basis of beryl mineral)
Symbol	:	Be
Atomic Number	:	4
Atomic Mass	:	9.012
Colour	:	White grey
Melting Point	:	1280°C
Boiling Point	:	2970°C
Density	:	1.8 times heavier than water

In 1798, a French chemist, Nicholas Louis Vauquelin discovered a metallic element which he named glucinum. It was renamed later as Beryllium by Claproth. The pure beryllium was discovered in 1828 by Friedrich Wohler of Germany and A.A. Bussy of France. This metal is found in many minerals in small quantities but its main ores are beryl and emerald. Beryl is found in nature in the form of large crystals. Beryllium is next to lithium in lightness but it is a very strong metal. It is stronger than steel. It can bear a load six times heavier than aluminium. High thermal conductivity, high heat capacity and high heat resistance are its other specific characteristics. Because of these properties, it is known as the metal of space age. It is used for making many components of aeroplanes, rockets, satellites and spacecrafts. Beryllium-copper alloy is used for making many components of spacecrafts because it is light in weight and strong. Beryllium makes several compounds which are poisonous in nature. That is why they are used with great care.

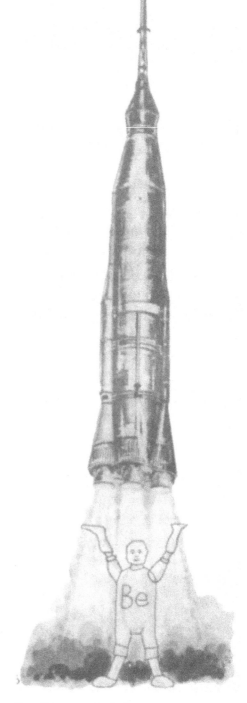

Beryllium is a light but very strong metal. Therefore, it is called as the metal of space age

LEAD

Latin Name	:	Plumbum
Symbol	:	Pb
Atomic Number	:	82
Atomic Mass	:	207.19
Colour	:	Grey
Melting Point	:	327.3°C
Boiling Point	:	1750°C
Density	:	11.3 times heavier than water

Lead is a dense, soft, grey metallic element. A lead statue of ancient Egypt has been found, which is supposed to be 6,000 years old. Lead items have also been found in the Pyramids. These things prove that lead was discovered by Egyptians.

About 3,000 years ago, lead was being used on a large scale by the Romans. In Rome, pipes, utensils, ornaments etc. were made of lead. On one hand, the Roman civilisation flourished because of lead but on the other hand, lead became the cause of its decline also. By the constant use, lead enters the body as a slow poison. This caused many different types of mental disorders in Romans. Nowadays tetraethyl lead is used in petrol which causes air pollution but from few years, lead free petrol is being sold and is in great demand. Lead is not only harmful to man but also to animals and plants.

Lead is mainly obtained from Galena (lead sulphide), Cerussite (lead carbonate) and Massicot (lead oxide). Australia and Russia are its main producing countries. Lead is obtained by heating ores in big furnaces. Lead is primarily obtained from Galena or Lead sulphide which is purified by the process of Electrolysis.

Lead is mainly used in batteries, type metal, solder alloy, etc. It is used in glass and paint industry and also as a shield in nuclear power reactors because it can absorb radiations. It is also used as X-rays shield.

Lead forms many useful compounds. Lead oxide is used in batteries while lead hydroxide is used in paints. Lead compounds are highly poisonous. Lead workers must be protected from it and its compounds.

Australia and Russia have the vast deposits of lead

MAGNESIUM

Early Name	:	Megnium
Symbol	:	Mg
Atomic Number	:	12
Atomic Mass	:	24.305
Colour	:	Silvery white
Melting Point	:	649.3°C
Boiling Point	:	1090°C
Density	:	1.7 times heavier than water

Magnesium is a soft, silvery metallic element. It was discovered by the British chemist, Sir Humphry Davy in 1808. It was named magnium by him, which was known as magnesium later on.

Magnesium is found in nature in abundance. It is found in the sea and lake water. There are about 200 minerals in which magnesium is present. Dolomite and magnesite are its main ores. Most magnesium is obtained by the electrolytic process of molten magnesium chloride. This element is also found in plants and animals. The green colour of trees is due to chlorophyll which contains traces of magnesium.

Mangnesium is a very reactive metal. It burns in air with a dazzling white flame. Because of this, magnesium is used in flares, flash bulbs in photography and fireworks. Magnesium is also used in several alloys. Its alloys are used in making parts and components of aeroplanes, rockets, and spacecrafts. Magnesium compounds are used in dyes, cloth, rubber, cement industries etc.

Magnesium oxide is used to line furnaces for melting metals. Magnesium hydroxide is used to relieve indigestion. Magnesium sulphate, which is commonly known as Epsom salt, is also used in medicines.

Magnesium crystals are used in fireworks

PLATINUM

Early Name	:	Platino-del Pinto
Symbol	:	Pt
Atomic Number	:	78
Atomic Mass	:	195.09
Colour	:	Silvery white
Melting Point	:	1772°C
Boiling Point	:	3800°C
Density	:	21.5 times heavier than water

Man came to know about Platinum only in the 16th century though ancient Aztecs knew how to make platinum mirrors. These mirrors were made by flattening and polishing platinum sheets. It is still a mystery how these ancient Aztecs could create so high a temperature as to make sheets of platinum metal. The Aztecs king, Montezuma presented several platinum mirrors to the king of Spain.

People of Spain were very greedy for gold. They came to know that the sand of some rivers of America contains gold, which can be separated by filtering. When the Spanish army filtered the sand of the Columbian river, Platino-del Pinto, they obtained silvery white metal along with gold particles. They considered this white metal useless and threw it away. It was called platino which meant poor quality silver. They did not know that this was the platinum metal which would later become costlier even than gold.

Platinum was discovered in 1748 by Don Antonio del Ulloa and W. Brownrigg. In 1803-04 some other metals were detected in platinum. In 1803, British chemist, William Hyde Wollaston separated out Rhodium and Paladium. In 1804, Smithson Tennant discovered Osmium and Iridium. Ruthenium was discovered in 1845 by the Russian scientist, Karl Karlavich Klaus.

Spanish people considered Platinum useless and threw it away

In 1803-04, several metals were detected alloyed in Platinum

Platinum exists in nature in small quantities. It is found in the maximum quantity in Canada and Russia. Platinum is also found in South Africa and South America. It is found in the form of nuggets.

Platinum is white in colour and is a very bright and costly metal. It is several times costlier than gold. It does not get rusted. It is not effected by acids also.

Platinum is used in jewellery. It is also used in aeroplanes, radio, television, missiles, jet engines, and electronic instruments. Platinum has its own importance in chemical laboratories. Platinum electrodes are used in several chemical processes. Alloys of platinum and silver are used as tooth-filling cements.

Platinum resistance thermometers/are prepared in scientific laboratories. Now a days, platinum is used to make diamond jewellery. In fact, platinum is one of the costliest metals of today.

URANIUM

Name	: Uranium (named on the basis of Uranus planet)
Symbol	: U
Atomic Number	: 92
Atomic Mass	: 238.03
Colour	: Silvery white
Isotopes of Uranium	: 238U, 232U and 235U
Melting Point	: 1132°C
Boiling Point	: 3818°C and 235U
Density	: 19 times heavier than water

Uranium is obtained from the ore-pitch blende or unaninite

Uranium is a radio-active metal. Before its use in atomic energy, it was just an ordinary metal. It was commonly used in Europe for dying glass and china clay wares. But today, Uranium has become a very important and costly metal. The cost of 4.5 kilograms of uranium is about 2,00,000 dollars.

Uranium was first isolated from Pitch Blende in 1789 by Martin Klaproth. It was named as Uranium on the basis of Uranus planet. Pure Uranium was obtained in 1841 by a French chemist, Eugene Peligot.

Uranium exists everywhere in the earth's crust. Its main ore is Pitch Blende or uraninite. This ore is mainly found in Africa, Kango and Canada. This is also found in Australia, America, East Africa and Russia. It is a silvery white metallic element. Its compounds exhibit fluorescence when illuminated by light. It is a radioactive element and forms a series of fission products. This element turns into lead in millions of years by the process of radioactive decay. In this process, radioactive rays are emitted from it. Natural Uranium is a mixture of two main isotopes – Uranium 238 and Uranium 235.

Uranium 235 is used as nuclear fuel in fission reactions. It is mainly used in nuclear

A nuclear bomb explosion

The 4.5 kilogram of Uranium costs more than 2,00,000 dollars

reactors and atomic weapons. Uranium 238 is used in nuclear breeder reactors. The heat liberated by it is used for boiling water, which, in turn, runs the turbines and produces electricity

Uranium is a very important fuel. Atomic power plants make use of Uranium for producing electricity and saving coal and petrol. One gram of Uranium can give heat energy equivalent to 12,250 tons of coal. Only enriched Uranium is used in Nuclear reactors. The process of Uranium enrichment is known only to a few nations of the world. Uranium has revolutionised the whole world. Uranium is used to prepare Atom bombs. Uranium has three isotopes namely:

238U	Half life span or period 4.5×10^9 years.
232U	1.39×10^{10} years.
235U	7.1×10^8 years.

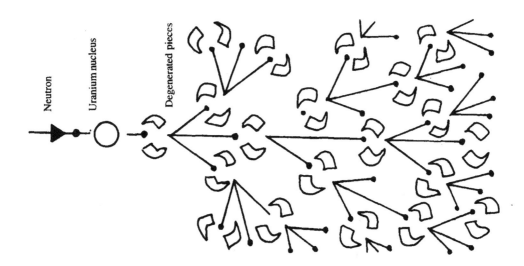

Nuclear fission

458

SCIENTISTS AND INVENTIONS

It is very difficult to imagine the state of early man on earth when he had no home, no tools or weapons and no clothes. He had not learned to grow food for himself. He had to depend solely on his skill as a hunter and on gathering such edible fruits and nuts as he could find. In his search for food, man was different from the animals around him.

Man, as compared to other living beings, is endowed with a more active brain. It is only his brain which has made him far more progressive and advanced than any other animal.

Probably the first invention of man was a primitive tool made from a split stone which served a number of purposes. From this simple tool man developed the hand axe, knife and many other tools.

After this, roughly 500,000 years ago, man discovered how to make fire. Man had started using the skins of animals as clothes and living in caves. By 10,000 B.C. he had learned how to grow crops. Some 5,000 years ago, the wheel was invented which revolutionised transport. All these inventions resulted in the development of human civilisation.

The word 'Science' has been derived from Latin word 'Scientia' which means knowledge. Science is one of the most important subjects of study of our time.

Man used natural colours in his cave-paintings about 2,000 years ago

461

The theories given by Pythagorus, Aristotle, Socrates, Plato, Archimedes, etc. have made significant contributions to scientific developments.

Alchemy – the ancient science of Egypt – contributed greatly in the development of Chemistry till the end of the 16th century. Charak, Sushrut of India, Jabir Ibn Hayyan, Al-Razi and Ibn-Sina, etc. were the famous scientists of that time.

The age of science really started in the middle of the 17th century. Robert Boyle for the first time introduced the method of experiments and its importance in the field of science. During this period, several institutions of science and scientists started working in England, France and Germany. Many scientists like Galileo, Newton, etc. brought forth many new principles and theories in the different fields of science.

Sometimes inventions occur all by accident. While working on cathode rays, Wilhelm Roentgen of Germany accidentally discovered the X-rays in 1895. The famous antibiotic Penicillin too was accidentally discovered by Sir Alexander Fleming. But such accidents occur rarely, and it needs an extra-intelligent brain to turn these chance-achievements into scientific discoveries. However, most of the inventions and discoveries are the result of a systematic research work. Sometimes necessity compels the scientists to discover new things. German scientists had to develop the technology of rockets and missiles to destroy England and the U.S. scientists invented atom bomb to defeat Japan in the Second World War. Radars and Sonars were necessary for self-defence. All these inventions are very useful for us in the present age.

Man succeeded in reaching the space only because of science. Computer has brought revolutionary changes in the field of information and telecommunication technology.

There is a great deal of difference between the words 'invention' and 'discovery'. A discovery is made of a thing that is already present in nature but is not known to man. An invention is a man-made device, for example, fire was discovered while match-box was invented.

Although science is beneficial for mankind, at times it can be harmful as well. Most of the scientists, prior to the 20th century, had not had proper schooling, but nowadays, we have well qualified engineers and scientists working together in well equipped laboratories.

From the very ancient times, the discoveries and inventions are being made in the world and this process will go on for ever.

Aristotle (384-322 B.C.)

GALILEO GALILEI (1564–1642)

Galileo Galilei is one of the greatest scientists of the 16th and 17th centuries. He, with the help of his self-made telescope, noticed the mountains and valleys on the moon, the satellites of Jupiter and the rings of Saturn for the first time. He also proved that the milky way is a cluster of millions of stars.

Galileo was born on February 15, 1564 at Pisa (Italy). At the age of 23, he was appointed as a lecturer of mathematics. In 1581, he invented pendulum. Later his son developed wall clocks with the help of this pendulum. The development of barometer and steam engine was also an outcome of his inventions.

During this time the earth was considered to be the centre of the universe. Galileo, however, proved that earth is not the centre of the universe but like other planets it also revolves round the sun. Galileo was the first to prove Aristotle's statement wrong that if two balls of different masses are allowed to fall simultaneously from the same height, the heavy body will hit the ground first. For this he selected the leaning tower of Pisa and two metal balls weighing 100 pounds and 1 pound respectively. These balls were allowed to fall freely at the same time from the roof of the leaning tower of Pisa. Both the balls hit the ground at the same time, proving a myth wrong.

Old Galileo was imprisoned for blasphemy. After his release from the jail he became blind and died on January 8, 1642 in Italy. He

Galileo Galilei (1564-1642)

collected his thoughts and theories in his famous book *Dialogues Concerning the Two Principal Systems of the World*. The great scientist Galileo also paved the way of space research for us.

463

SIR ISAAC NEWTON (1642–1727)

Newton is famous for his law of Gravitation which he propounded after seeing an apple falling from the tree to the ground. Sir Isaac Newton was born on December 25, 1642 at Woolsthorpe, Lincolnshire. Even after 350 years, he is still considered as the Father of Physics.

As a child, Newton was not a very intelligent student. He did his graduation in 1665 from Cambridge University. In 1669, Newton was appointed as the professor of Mathematics in Trinity College. In 1672, Newton was elected the Fellow of the Royal Society of London. In 1689, Newton was elected the Member of Parliament and in 1703, he became the President of the Royal Society. Thereafter he was re-elected for the same post every year until his death. In 1705, he was knighted by Queen Anne in a special ceremony at Cambridge.

Newton propounded the Law of Gravitation and his three laws of motion which are still relevant and taught to the students of Physics. In 1687, he published a book entitled *The Mathematical Principles of Natural Philosophy*, in which the law of gravitation and laws of motion were explained. Newton showed that the white sunlight in fact is composed of seven colours namely, violet, indigo, blue, green, yellow, orange and red. His researches

Sir Isaac Newton (1642-1727)

concerning light has been published in the book *Opticks*. He also invented the mathematical method known as Calculus.

Even in old age, Newton worked in the field of Astronomy. At the age of 85, he died on March 20, 1727. Even today Newton is considered as one of the greatest scientists of the world.

SIR HUMPHRY DAVY (1778–1829)

Sir Humphry Davy is well known in the world for his invention of safety lamp for coal miners. He also discovered nitrous oxide, a well known anaesthetic gas known as laughing gas. He obtained this gas by heating ammonium nitrate. Apart from this he also developed the methods of analysing the compounds of nitrogen and oxygen and obtaining potassium nitroso sulphate. But his most important contribution was the use of electricity in the experiments of chemistry. He obtained several new elements by the method of electrolysis.

Sir Humphry Davy was born in Penzance (England) on December 17, 1778. He began his career with a medical practitioner but soon he acquired a good knowledge of natural sciences. On the basis of his knowledge, he was able to get a job in the Royal Institution. Here, he used to deliver popular lectures on the various topics of science. His fame spread far and wide and he was appointed a lecturer in the Royal Institution. He was made a knight in 1812 and later became Baronet. On May 29, 1829, he died at Geneva. He obtained sodium, potassium and some alkaline earth metals through electrolysis. He proved that chlorine is an element.

In 1813, Faraday worked with him as his assistant and accompanied Davy on a tour of Europe during which they studied iodine and proved that diamond is an allotrope of carbon. Development of electric arc was also one of his inventions. He also established the catalytic action of red hot platinum on the mixture of inflammable gases. We cannot forget Davy's contributions in the field of science.

Sir Humphry Davy (1778-1829)

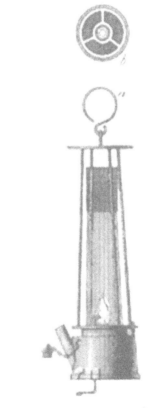

The safety lamp invented by Davy

MICHAEL FARADAY (1791–1867)

Michael Faraday is known as the father of Electromagnetic Induction. In 1831, he made the first dynamo to produce electricity. As a result of this invention, thousands of power stations are producing electricity all over the world. Without the invention of dynamo, production of electric power could not have been a reality. Faraday also made significant contributions in the field of electro-Chemistry.

Michael Faraday, a British scientist, was born on September 22, 1791 at Newwington (England). His father was a blacksmith. At the age of 13, Faraday had to work as an apprentice in a book binding shop. He developed an interest in science after he attended some lectures of Sir Humphry Davy. He requested Davy to accept him as his assistant. In 1813, he began to work as Davy's assistant. In 1824, he became the member of the Royal Institution. Soon, he became the Director of the laboratory of the Royal Institution. In 1823, he was appointed as lecturer of Chemistry in the Royal Institution. In 1839, he fell seriously ill. After that his memory became quite weak. In 1861, after his retirement from the Royal Institution, he moved to a house in Hampton Court that was offered to him by Queen Victoria. On August 25, 1867, he breathed his last.

Most of the contributions made by Faraday are related with Physics. In Chemistry, his most important contribution is the 'laws of electrolysis'. He discovered paramagnetism and diamagnetism. He also proved that by high pressure, gases can be converted into liquid state. He also synthesized some chlorides of carbon. Two electrical units are named after him. One is Faraday unit that is used for measuring the quantity of electricity and the other is Farad unit, used to measure the capacity of a capacitor.

Michael Faraday (1791-1867)

CHARLES ROBERT DARWIN (1809–1882)

Charles Robert Darwin was the British Naturalist who is best known for his theory of evolution. He proved that all living things on earth have descended from the common ancestors who existed millions of years ago. All living beings – plants and animals – have evolved in an orderly way and continue to change even today.

Charles Darwin was born on February 12, 1809 at Shrewsbury. As a child, he was very much interested in studying and collecting insects and minerals. At the age of 16, he was sent to Edinburg University to study medicine but he had no interest in medicine. After that he was sent to Cambridge to study theology. At Cambridge, the professors of Botany and Geology became his good friends.

In 1831, a scientific expedition was being sent to sail around the world in HMBS Beagle under the command of Captain Fitz Roy. A naturalist was needed in this expedition for which Darwin's name was recommended by his two friends. He went on the voyage for five years and made collections of the bones of extinct animals and plants' remains. He surveyed the rain forests of Brazil. He found the fossil of an extinct giant animal Megatherium near Bahia Blanca. The strange marine iguanas, tortoises and the finches on the Galapagos Islands in the Pacific puzzled him at first because similar, yet quite different forms of the same animals appeared on separate islands.

Darwin studied the iguanas and tortoises in the Galapagos islands

Charles Robert Darwin (1809-1882)

During the same period, another naturalist, Alfred Russel Wallace also drew similar conclusions about evolution. In 1858, both Darwin and Wallace presented a joint paper. In 1859, the famous book of Darwin, *The Origin of Species* by Natural Selection was published. The first edition of the book was sold out on the day of its publication itself.

In 1868, his second book entitled *The Variation of Animals and Plants Under Domestication* was published. His other famous books are – *Insectivorous Plants, The Power of Movement in Plants* and *Descent of Man*.

Darwin died on April 19, 1882 at the age of 74 and was buried in Westminster Abbey near the tomb of Sir Isaac Newton.

FRIEDRICH WOHLER (1800–1882)

The German scientist Friedrich Wohler discovered that it was possible to synthesize an organic compound urea from an inorganic chemical. It was one of his key contributions to revolutionise the concepts of organic and inorganic Chemistry.

Friedrich Wohler was born at Aschersheim near Frankfurt. He studied Chemistry under the guidance of Leopold Gmelin and Berzelius. In 1825, he became a lecturer at Berlin Technical school and then in 1831 in Casscel. In 1836, he became a lecturer of Chemistry at University of Gottingen. Here he assisted Liebig in many programmes. This outstanding scientist obtained cyanogen iodide in his student life itself. He demonstrated Pheroah's serpent action by burning mercuric thio-cyanide. He synthesized silver isocyanide and gave its chemical formula.

Most of the work of Wohler was carried out at Berlin. In 1827, he obtained aluminium metal by the reaction of potassium and aluminium chloride. This was a great achievement in the field of Chemistry. In 1828, he did some experiments with hydrocyanic acid and obtained the crystals of urea. Synthesis of urea in the laboratory was one of the greatest achievements because scientists like Berzelius had a concept that such compounds can only be obtained from nature and their synthesis in the laboratories is not possible. But Wohler, who was the student of Berzelius, could make it in the laboratory. In this way, he gave birth to Synthetic Organic Chemistry.

Friedrich Wohler (1800-1882)

In 1832, while he was working with Liebig, he extracted Bezoic acid from bitter almonds. He also carried out research work on quinone, hydroquinone and quinhydrone. He also worked on Alkaloids. He obtained phosphorus from bone ash and acetylene from calcium carbide. He also worked on hydrides of boron and silicon and on their rare compounds. This great chemist died in 1882.

GREGOR JOHANN MENDEL (1822–1884)

Gregor Johann Mendel, the son of a farmer, was born on July 22, 1822 at Heinzendorf, Austria. He was a naturalist and studied science at the University of Vienna. After his education, he returned to his monastery at Bruno in 1847 and taught natural sciences in the school there.

For conducting experiments in genetics, Mendel grew pea plants in his monastery garden and did cross pollination between the different varieties. From 1856 to 1864 he conducted various experiments on pea plants on genetics and drew many conclusions regarding how one characteristic is related to the other. These conclusions were published in the magazine of Natural History Society. No scientist paid any attention to the research work done by Mendel for a period of 34 years. His work could be valued only around the end of 19th century when similar conclusions were drawn by the scientists of Holland, Germany and Austria.

After 16 years of his death on January 6, 1884, the laws of genetics propounded by Mendel could be understood by the scientists. According to him, some characteristics of the parents are inherited by the coming generations. These inherited characteristics are called hereditary characters. After this, Mendel became famous as father of genetics. He gave three laws – law of dominance, law of segregation and law of independent assortment, of genetics which are taught to the students even today.

Gregor Johann Mendel (1822-1884)

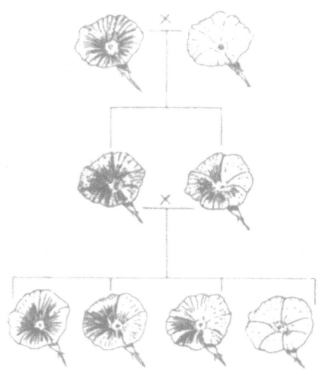

He conducted various experiments on pea-plants

470

LOUIS PASTEUR (1822–1895)

Both science and mankind are indebted to the French chemist Louis Pasteur for his contribution in the development of science of Microbiology. He discovered the process of pasteurization to stop fermentation of milk and butter. He is well known for discovering the vaccine against the rabies.

Pasteur was born in Dole, France in 1822. He was the son of a poor tanner. After primary education, he went to Paris where he heard the lectures of Bellard and Duma. He became the assistant of Bellard and devoted himself to research work. His interest bent towards the study of crystals. In 1848, he was appointed as the lecturer of Physics at Lycee. In 1852, he became the lecturer of Chemistry at Strasberg. In 1857, he was appointed the Director of Ecole Normale, Paris. Here he worked on Fermentation by which originated a new branch of science called Microbiology. He invented rabies

Louis Pasteur (1822-1895)

vaccination and experimented it on a nine-year-old boy whom a mad dog bit 14 times.

In 1888, the Pasteur institute was established and he remained its Head until his death.

FRIEDRICH AUGUST KEKULE (1829–1896)

The German chemist, Friedrich August is well known all over the world for his famous discovery of the unique structure of the benzene molecule. Born in Darmstadt, Kekule was interested in the architecture and he planned a career as an architect. But his contact with Liebig changed his mind towards Chemistry.

After receiving his doctorate in 1852, he started his studies at Paris. In 1856, he was appointed as the lecturer of Chemistry at Heidelberg University. In 1858, he became the professor of Chemistry in Belgium at Ghent University. He moved to Bonn in 1865.

There is a famous story about the invention of the structure of benzene. It is said that on one night, in 1865, Kekule had a dream in which he saw the chains of the atoms of Carbon join together like snakes with each other's tails in their mouth. From this dream, he presented the ring structure of benzene.

On the basis of tetravalency of Carbon, he explained the Carbon bonds in molecules. He discovered that the atoms of Carbon can join onto each other in three different ways – in open chains, closed chains and ring chains. This discovery became so important that it has created more than 700,000 Carbon compounds to the present day.

Kekule in the 19th century brought revolutionary changes in writing the structural formulae of Carbon compounds. He called benzene and its compounds with a different name of aromatic compounds. This implies to all ring compounds. Hofman called other compounds as aliphatic compounds. The compounds of hydrogen and carbon are named as saturated and unsaturated hydrocarbons. Kekule, by adding ane, ene, and ine to the name of hydrocarbons classified them as alkanes, alkenes and alkines.

This revolutionary scientist of Chemistry died on July 13, 1896. His contributions in the field of science can never be forgotten.

Friedrich August Kekule (1829-1896)

  ou

THOMAS ALVA EDISON (1847-1931)

Thomas Alva Edison, during his life time took out more than 1,000 patents for his inventions. In the history of science perhaps there is no other scientist who has been credited with so many inventions and discoveries. Edison was born in Milan, USA on Feb. 11, 1847. When he was only 10 years old, he set up his first laboratory in the basement of his home. During 1860, he worked in the telegraph office of USA and Canada. In 1876, he sold telegraph printer he had invented. With this money, he could set up a good laboratory for his work at Menlo Park, New Jersey. In this laboratory he invented the Carbon-resistance transmitter and the phonograph. This was the first phonograph with which human voice could be recorded and reproduced.

In 1879, he invented the electric bulb. He also invented thermionic emission which is known as Edison effect today. In 1880, he lighted Menlo Park with 100 electric bulbs. This was a small demonstration of his invention. New York was the first city of the world where electric lights were used. Edison laid a cable of 75 km length and lighted more than one thousand houses on the night of September 4, 1882.

In 1887, he made a new laboratory in West Orange. Here in 1891, he invented Kinetograph. It was the first movie camera. In this camera, Eastman film was used. After developing the film, people used to see it with kinetoscope.

Edison was known as the wizard of technical age. He continued his scientific research until his last breath on October 18, 1931.

Thomas Alva Edison (1847-1931) invented more than 1,000 things

ALBERT EINSTEIN (1879-1955)

Albert Einstein is known as the father of modern Physics. He was born on March 14, 1879 in Ulm, Germany. He became a Swiss citizen in 1901. In 1933, he moved to USA due to the fear of brutality of Hitler.

Right from his childhood, Einstein was interested in Science and Mathematics. His father was not very rich to bear the expenses of his education, so he had to earn along with his studies. When he was 15, his family moved to Italy. From Italy, he was sent to Switzerland for education. A rich relative helped him in getting higher education. Inspite of being highly intelligent, he could not get a job.

In Jurich University, his brilliance in Physics and Mathematics came to the surface. In 1900, he completed his education and became a citizen of Switzerland. After his education, he joined the Swiss patent office as a clerk. In his leisure, he used to solve the complex problems of Mathematics.

While working in the patent office, he published a paper which changed the concept of scientists about physical quantities. This research paper was on the Theory of Relativity. This paper brought worldwide fame for Einstein. In this paper he also gave the formula of $E = mc^2$ (E = energy, m = mass and c = velocity of light) regarding the conversion of mass into energy. This formula later became the basis of atom bomb which destroyed Hiroshima and Nagasaki in 1945. He felt deep sorrow at the destruction of these two cities by atomic explosions. After this, he spent his whole life on the peaceful uses of atomic energy.

Einstein gave the quantum theory of photoelectric effect for which he was awarded the Nobel Prize in 1921.

After retiring from Princeton University in 1845, Einstein continued his research on the unified field theory. He could not complete this work in his life time. On April 18, 1955, this great scientist died at Princeton Hospital. He died while sleeping. After his death, his brain was preserved in Princeton Hospital. To honour him, an element 'Einsteinium' has been named after him.

Theory of Relativity is Einstein's major contribution to Science

ENRICO FERMI (1901-1954)

Born in Italy on September 29, 1901, Enrico Fermi was the first scientist to discover chain reactions. He made the first nuclear reactor of the world and marked the beginning of nuclear physics. In fact, he is known as the 'father of nuclear physics'.

In 1922, he obtained his Ph.D. degree in the field of X-rays from the University of Pisa. In 1927, he was appointed as a lecturer of Physics in the University of Rome. He was elected a member of Italian Academy in 1929. In 1933, he discovered a fundamental particle called neutrino for which he was awarded the Nobel Prize in Physics in 1938.

In those days, Mussolini's dictatorship caused a chaos in Italy. Fermi's wife was a Jew, so it directly affected him. Fortunately, Fermi was invited to give a lecture at the University of Columbia. Fermi, along with his family, went to USA and never came back to Italy. In 1939, he was appointed as professor of Physics in the University of Columbia and became a US citizen in 1944.

At the University of Columbia, he started working on nuclear chain reactions. After a devoted hard work, he succeeded in making the first nuclear reactor in 1942. Fermi and his associates worked on the Manhattan project for making the atom bomb. In 1945, this team developed the atom bomb.

After World War II, Fermi joined the University of Chicago where the Institute for Nuclear Studies was named after him. He died on November 28, 1954 at the age of 53. In his honour, an element 'Fermium' has been named after him.

Enrico Fermi (1901-1954)

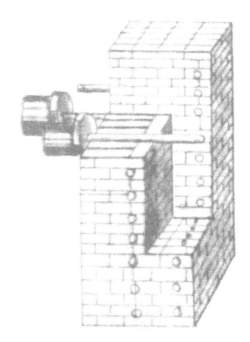

Fermi made the first nuclear reactor

J. ROBERT OPPENHEIMER (1904-1967)

J. Robert Oppenheimer was a successful administrator, a well known physicist, mathematician, researcher and an educationist. He is known to the whole world as the first director of the Manhattan project for developing the first atom bomb. A team of scientists started working together to develop the first atom bomb in 1943 at Alamos and successfully completed and tested it on July 16, 1945 at 5.30 a.m. On August 6, 1945, the first atom bomb was dropped at Hiroshima, Japan which killed 80,000 people and injured about 70,000. On August 9, 1945, the second bomb was dropped at Nagasaki, another city of Japan, which killed 40,000 people and injured 25,000. These inhuman acts of devastation gave a big jolt to Oppenheimer and he resigned from his post.

Oppenheimer was born in New York on April 22, 1904. His parents were rich Jews of Germany who had settled in the USA.

In 1925, he graduated in Physics from the Harvard University. After this, he joined Gavendish laboratory, England and worked with Ernest Rutherford.

In 1929, he came to the University of California and was appointed Associate Professor in 1931. Here he did many experiments in nuclear physics.

In 1941, President Roosevelt sanctioned the Manhattan project for making the atom bomb. A team of scientists was selected for this work and Oppenheimer was appointed the chief director of the project.

After World War II, Oppenheimer resigned from his post but again in 1947, he was made the chairman of the US Atomic Energy Commission. In 1963, he was honoured by the Fermi Award for the Atomic Energy Commission. Oppenheimer died on February 18, 1967 at Princeton.

J. Robert Oppenheimer (1904-1967)

DR. HOMI JAHANGIR BHABHA (1909-1966)

The credit of starting the nuclear energy programme in India goes to Dr. Homi Jahangir Bhabha. After constituting the Atomic Energy Commission of India in 1948, Dr. Bhabha was appointed its first chairman. Indian scientists worked for the development of atomic energy under the able guidance of Bhabha. He was not only a well known scientist but also a successful administrator. He is known as Oppenheimer of India.

In 1956, the first atomic reactor, Apsara was installed under his supervision. Two other nuclear reactors, Cirus and Zerlina were also installed under his guidance. It was the result of Bhabha's efforts that the first nuclear power plant for the production of electricity came into existence at Tarapur. Two years later, a plutonium plant was made and on May 18, 1974, an underground atomic explosion test at Pokharan was conducted. This was the result of the efforts made by Dr. Bhabha.

Dr. Homi Bhabha was born on October 30, 1909 in Bombay (now Mumbai), in a wealthy Parsi family. His primary education was completed in Bombay (now Mumbai). After graduation, he was sent to the Cambridge University for higher education. From Cambridge, he obtained an engineering degree

Dr Homi Jahangir Bhabha (1909-1966)

in 1930 and a Ph.D. in 1934. Here he worked with a world renowned scientist, Niels Bohr. He also worked with Enrico Fermi and Pauli.

Dr. Bhabha worked a lot on cosmic rays. In 1945, he established the Tata Institute of Fundamental Research and became its director. He was the chairman of the first United Nations' Conference on Peaceful Uses of Atomic Energy held at Geneva in 1955. Dr. Bhabha died in an air crash on January 24, 1966 when he was on his way to attend an International conference.

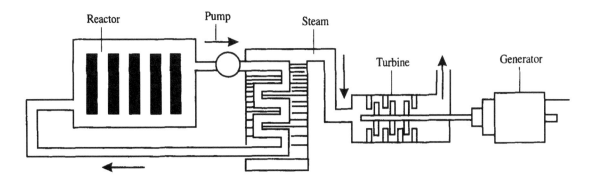

GREAT INVENTIONS

The development of human civilisation is a result of inventions made by men. Metal smelting started about 4000 B.C. which was one of the greatest ancient inventions. The wheel and plough came to be known around 3000 B.C. The speed of inventions was slow upto 14th century but from the 17th century, science developed rapidly. Some great inventions are listed below:

1450   Printing Press, Johannes Gutenberg (Germany)

1590   Compound Microscope, Zacharias Janssen (Netherlands)

1593   Thermometer, Galileo Galilei, (Italy)

1608/09   Refracting T,elescope, Hans Lippershey (Netherlands); Galileo Galilei (Italy)

1668   Reflecting Telescope, Isaac Newton (Britain)

1698   Steam pump, Thomas Savery (Britain)

Hot air balloon made by Montgolfier Brothers

1712	Beam Engine, Thomas Newcomen (Britain)	1800	Lathe, Henry Maudslay (Britain)
1733	Flying Shuttle, John Kay (Britain)	1804	Steam Locomotive, Richard Trevithick (Britain)
1767	Spinning Jenny, James Hargreaves (Britain)	1815	Saftey lamp, Humphry Davy (Britain)
1783	Hot Air Balloon, Montgolfier Brothers (France)	1815	Stethoscope, Rene T.H. Laenec (France)
1784	Improved Steam Engine, James Watt (Britain)	1836	Revolver, Samuel Colt (USA)
1785	Power Loom, Edmund Cartwright (Britain)	1837	Telegraph, William Cooke and Charles Wheatstone (Britain); Samuel Morse (USA)
1792	Cotton Gin, Eli Whitney (USA)	1839	Steam Hammer, James Nasmyth (Britain)
1800	Electric Battery, Allessandro Volta (Italy)	1845	Sewing Machine, Elias Howe (USA)

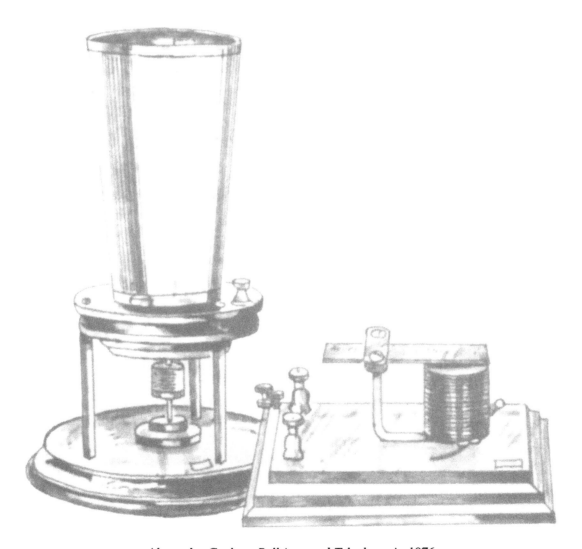

Alexander Graham Bell invented Telephone in 1876

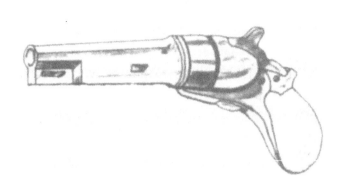

Samuel Colt made the first revolver

1867	Dynamite, Alfred Nobel (Sweden)
1872	Typewriter, Christopher L. Scholes (USA)
1876	Telephone, Alexander Graham Bell (USA)
1877	Phonograph or Gramophone, Thomas Alva Edison (USA)
1878	Cathode Ray Tube, William Crookes (Britain)
1878/79	Electric bulb, Joseph Swan (Britain) and Thomas Alva Edison (USA)
1880	Machine Gun, Hiram Stevens Maxim (USA)
1884	Steam Turbine, Charles Algemon Parsons (Britain)
1885	Petrol Engine, Karl Benz and Gottlieb Daimler (Germany)
1888	Pneumatic Tyre, John Boyd Dunlop (Britain)

1893	Diesel Engine, Rudolf Diesel (Germany)
1895	Radio, Guglielmo Marconi (Italy)
1903	Aircraft, Wilbur and Orvile Wright (USA)
1926	Television, John Logie Baird (Britain) and Vladimir Zworykin (USA)
1930	Cyclotron, Ernest Lawrence (USA)
1937	Jet Engine, Frank Whittle (Britain)
1944	Digital Computer, Howard Aiken (USA)
1947	Polaroid Camera, Edwin H. Land (USA)
1948	Transistor, William Shockley, John Bardeen, W.H. Brattain (USA)
1955	Hovercraft, Christopher Cockerell (Britain)

Alfred Nobel invented the Dynamite

1971	Micro Processor, Intel Corporation (USA)
1972	Commercial Video Games, Pocket Calculator and Home Video System.
1975	Body Scanners.
1977	Apple II, Personal Computer.
1978	First Test Tube Baby.
1980	Compact Disk Audio System.
1985	Compact Disk Memory System.
1986	Challenger Space Shuttle.
1987	Docklands Light Railway (Britain)
1990	High-Definition TV System.

Edison invented Gramophone in 1877

SCIENCE AND TECHNOLOGY

Measurement of Time: Sundial and water clocks were the first instruments discovered to measure time in about 1500 B.C. In the sundial, the sun casts the shadow on the digits and gives the indication of time. In the water clock, water from one container falls drop by drop into another container. The container has an hour scale by which time is measured.

The first mechanical wall clock was developed in 1088 in China. It was 10 meters high and was run by water power. The first mechanical clock in Europe was made in 1200. The first clock was made in Spain in 1276. The oldest mechanical clock which is still working is in the Salisbury Cathedral. It was made in 1368.

Glass Manufacturing: Glass was first made by melting soda and sand in about 3000 B.C. in Syria and its neighbouring countries. Glass blowing was also invented in Syria in about 100 B.C.

First Balance: The first weighing balance was developed in Syria and its neighbouring countries between 4000 and 5000 B.C. This balance was used to weigh gold. Stone weights were used to measure weight. These were cut in the shape of animals.

Gunpowder: Gunpowder was probably first made in China or India by mixing Sulphur, Charcoal and Saltpetre. Around 850 A.D., gunpowder was used by the Chinese for making fireworks and explosives. Gunpowder came in use during the 13th century in Europe. The credit of inventing gunpowder in Europe goes to an English monk named Roger Bacon.

Invention of Spectacles: Impaired sight has always been a challenge to the scientists. As long ago as the year 1000, an Arab scientist Alhazen demonstrated the action of image formation by

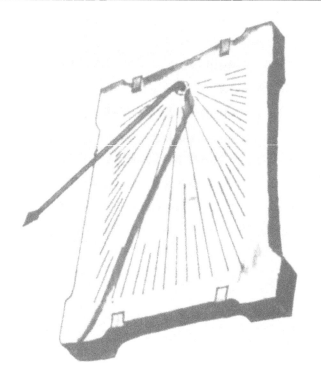

The Sundial

Glass was first made in Syria by melting soda and sand in about 3000 B.C.

482

lenses. He also told that people who had weak eyesight could see properly with the help of lenses. During the year 1200, Roger Bacon made a pair of rudimentary spectacles. By 1430, Italians developed spectacles to view the distant objects clearly. Bifocal lenses were invented in the 18th century by American statesman, Benjamin Franklin.

Production of Petrol: Oil was extracted by drilling in the USA for the first time in 1841 and the oil well was made in 1859. Petrol was obtained from the crude oil in 1864. Petrol had not much utility before the invention of motor-car. After the invention of the motor-car in 1883, the first petrol station was opened in France in 1895. The first petrol refinery was developed in 1860 in the USA. In 1870, Standard Oil Company, the biggest in the world of that time, was established. In 1890, the high quality petrol production started.

Steam Engine: The first successful steam engine was made in 1712 by the British engineer, Thomas Newcomen. It was used to draw water out of the mines. James Watt modified the Newcomen engine. In 1765, he made a new type of steam engine which was more powerful and fast. Steam engines of Watt were used for the first time in 1785 to run cotton mills. These steam engines proved very useful in the industrial development. In the 19th century, the steam engines came to be widely used in road and water vehicles. In 1803, the locomotive was invented in Britain. The first successful rail engine was made by George Stephenson in 1814.

Spinning Machine: Spinning machine was invented in 1700 in Britain. Before this, yarn was made either by hand or by Charkha. The first spinning machine was Spinning Jenny invented by James Hargreaves in 1764. It was a hand-operated machine. This could spin very thin yarn. Another spinning machine was Arkwright's Water Frame which was made in 1769. Samuel Crompton combined both these machines and made a new machine named spinning mule which marked the beginning of the textile industry.

The first automatic spinning machine was made in 1801 in France. This was known as

The first oil well was constructed in 1839

483

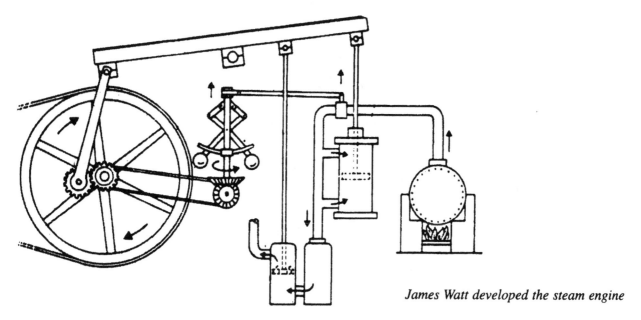

James Watt developed the steam engine

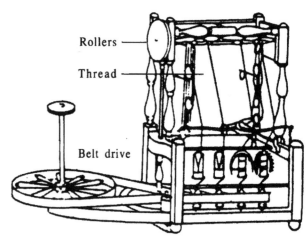

Rollers

Thread

Belt drive

Arkwright's water-frame machine

Loom and was capable of creating designs of silk clothes. It was invented by Joseph Marie Jacquard. It used a set of punched cards for creating new patterns. Nowadays computers have replaced punched cards.

Sewing Machine: The first sewing machine was made in 1830 in France by Barthelemy Thimmonier. This machine was able to put 200 stitches in one minute. The first successful

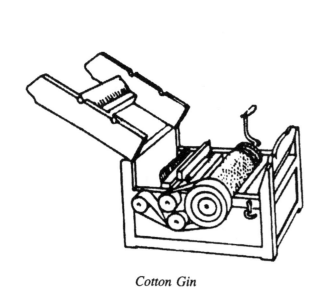

Cotton Gin

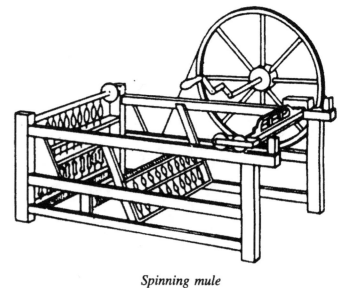

Spinning mule

sewing machine was made by Elias Howe of USA in 1845. Isaac Singer of USA gave the modern shape to the sewing machines in 1851.

Harvesting Machines: Harvesting machines are of two types: one is used for cutting the crops and the other for separating grains and fodder. The first threshing machine was developed in Britain in 1786 by Andrew Meikle. The first reaping machine was also invented in Britain in 1826 by Patrick Bell. The first harvesting machine was made in USA in 1831 by Cyrus McCormick. The maximum development work on harvesting machines has been done in the USA. Today these machines are being widely used all over the world.

Invention of Electric Motor: The first electric motor was invented by Michael Faraday in 1821. This was only an experimental motor. The first successful dynamo was made in Belgium in 1870 by Zenobe Theophile Gramme. After this, the first practical electric motor was made by him in 1873. The AC motor was invented by Nikola Tesla in 1888 in the United States of America.

Electric Light: Electric light for the first time was produced by an electric arc by Humphry Davy in Britain in 1802. Its light was very intense, so it was not good for domestic purposes. J.W. Starr and Joseph Swan tried to make an electric bulb but they did not succeed. The first successful electric bulb was made by the famous inventor, Thomas Alva Edison. Wire filament bulbs came into existence only in 1898.

Artificial Dye: The first artificial dye was made in 1856 by the British scientist, William Perkin. Prior to this, all dyes were made from insects and plants. Different artificial dyes have been made only after the invention of Perkin.

Invention of X-rays: X-rays were invented by the German physicist, Wilhelm Roentgen in 1895. These rays were invented accidentally while he was doing some experiments on Cathode rays. Today, X-rays are not only used to locate dislocations and fractures of bones but also in

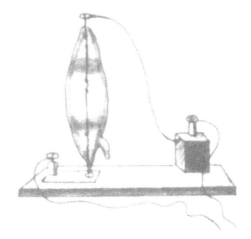

Swan's electric bulb

Electric bulb made by Edison

485

industries. Roentgen named them X-rays because these were not known at that time ('X' means unknown). Roentgen was given the first Nobel Prize in 1901 in Physics for the invention of X-rays.

Invention of Plastic: Plastic was invented for the first time by the British scientist, Alexander Parkes. This was called Parkesine and was made with cellulose and camphor. An

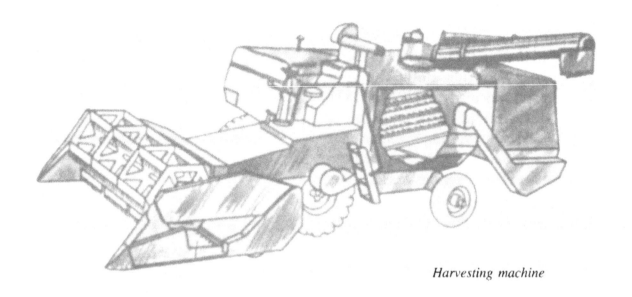

Harvesting machine

William Perkin invented the artificial dyes

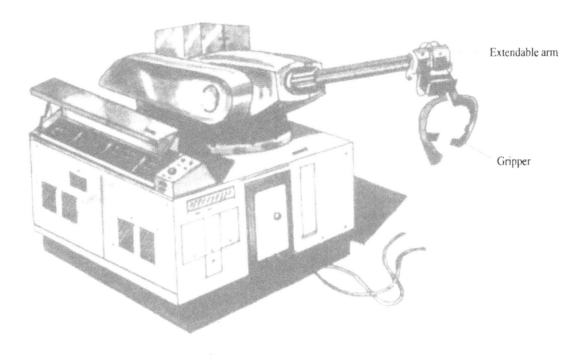

Extendable arm

Gripper

Industrial robots

American inventor, John Hyatt made similar plastic in 1868 which was named as celluloid. The first plastic made of chemicals was Bakelite which was invented by Leo Backeland of Belgium in 1907.

First Telephone: The telephone was invented by Alexander Graham Bell of USA in 1876. The telephone was first used in 1877 in Boston and first public call box was made in Connecticut in 1880. The first automatic telephone exchange was opened in 1892 at La Porte of Indiana. The automatic telephone exchanges came to be used in Europe in 1909.

Beginning of Radio Broadcasting: Radio waves were discovered in 1887 by Heinrich Hertz of Germany. The first signal in Morse Code was transmitted by Guglielmo Marconi of Italy in 1895. The first radio broadcast with music and talks was done by Canadian Reginald Fessenden on December 24, 1906 in USA. The first radio station was established in New York in 1907.

Invention of Television: The first television signal was transmitted by British inventor, John Logie Baird in 1924. This electronic system was different from the one being used today. The electronic television was developed in USA by Philo Farnsworth. Zworykin achieved a great success in 1930 in developing the electronic television.

Nuclear Power: Nuclear energy was first produced in 1942 by the famous Italian scientist, Enrico Fermi in USA. He made the first successful nuclear reactor in Chicago. In this reactor, Uranium was used as fuel. These days such reactors are used for electric power generation.

First Computer: The first computer called Colossus was developed in Britain in 1943. This could very quickly decode the codes of war. The first successful computer was made in USA in 1946. It was called ENIAC. About 19,000 valves and several thousand other electronic components were used in it. It was of the size of a big room.

Maiman invented laser-rays

Invention of Robot: The first robots that could work like human beings were made in Europe around 1700. These were used as toys. Pierre Jacquet-Droz, a Swiss watch-maker, made a writer robot in 1770 which could write any message of 40 letters with its hand. Industrial robots were developed in 1960s. These are being used in factories for operating machines, welding, painting etc.

Invention of Laser: The principle of laser action was given in 1951 by the US scientist, Charles H. Townes. He invented laser in 1953. Laser was made by T.H. Maiman of USA in 1960. It was a ruby laser and it was far more bright than the sun.

■■

LIGHT

Light is a form of energy which makes the objects visible. Plants cannot survive without light and life on earth is not possible without plants. Light is a kind of radiant energy which is not visible like the other forms of energy. We can only see those objects on which the light energy falls.

Light has always been a mystery for man. Different philosophers had different views regarding the visibility of things. About 6 B.C., Pythogoras suggested that light consists of rays, that, acting like feelers, travel in straight lines from the eyes to the object and the object is seen when these rays touch the object. Plato brought forth the idea of divine rays which come out from the eyes and on getting mixed with the sun rays, fall on the objects and make them visible. Aristotle proved the rectilinear propagation of light and gave the laws of reflection.

After this, no significant progress was made in the field of optics for about 1000 years. During 11th century, an Arabian philosopher, Alhazen studied the structure of eyes and explained their functions. He knew the laws of refraction. In 13th century, Roger Becon developed telescopes and microscopes with the help of different combinations of lenses.

During 16th century only, some researches were made in the field of optics but 17th century was certainly an important period for optics. During this period, Newton, Huygens, Romer, etc. made significant inventions in the field of optics. Newton gave Corpuscular theory of light, according to which light travels in the form of corpuscles. Huygens gave wave theory of light and Romer determined the speed of light in 1675.

In 1873, James Clark Maxwell showed that when the electric or magnetic field in a circuit changes, it does not remain confined in the circuit but the vibrations produce a type of waves

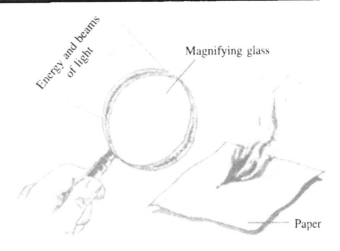

Light is a type of radiant energy. That is why a piece of paper or cloth gets burnt when the sun rays are focused on it through a magnifying glass.

Photosynthesis takes place in the plants and trees through light.

491

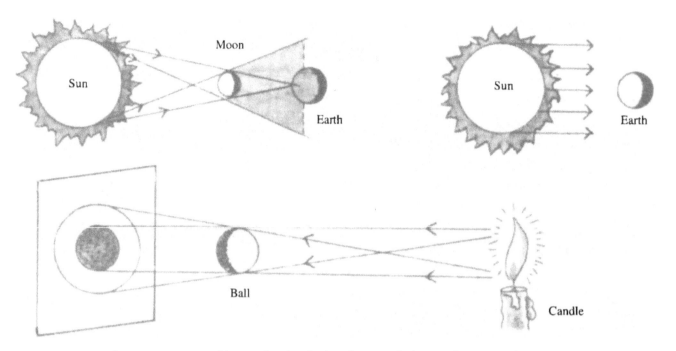

An untransparent object makes its shadow because light travels in straight lines.

which are known as electromagnetic waves. These waves spread in all directions with the speed of light. These waves propagate because of the periodic motion of electric and magnetic fields.

When the frequency of these electromagnetic waves is confined within a limited range, they appear as visible light and the colour of the light is a function of the frequency of the electromagnetic waves. In 1887, Hertz produced the electromagnetic waves in the laboratory and proved that these waves have all the properties of light. In fact, light travels in the form of electromagnetic waves. Electromagnetic waves have the following two properties:

i. These waves can travel in vacuum and their velocity in vacuum is equal to the speed of light.

ii. Electromagnetic waves are transverse in nature.

In addition to the visible light, some waves are not visible to the eyes but their effects can be felt. Gamma rays, X-rays, Ultraviolet rays, Infrared rays and Radio waves are all electromagnetic waves which are not visible to the eyes.

In 1990, Max Planck while working on Black Body radiations found that some of the experimental results cannot be explained with the help of classical theory of light. He gave a new theory of light known as quantum theory. According to quantum theory of radiation, light does not travel as continuous waves but in the form of wave packets. Each unit of wave packets is known as quantum of radiation. Scientists like Einstein and others could realise the importance of quantum theory of light. They also called a quantum of light as photon. Some of the natural phenomenon could not be explained by quantum theory of light. The dual nature of light created a problem to the scientist in the beginning but in 1924, Heisenberg and Schrodinger mathematically solved these problems and dual nature of light was accepted by the scientists. De Broglie, a French scientist also gave the dual nature of matter. Moving electrons travel as particles but waves are always associated with them. These are called matter waves. Scientists of modern times do not ponder much over whether light travels as particles or waves. Nowadays, Quantum concept is used to explain all the phenomena related to light.

492

Radiation :

The form of energy which can be transmitted from one place to another without any material medium is called radiation. The term radiation refers to electromagnetic energy. Electromagnetic waves can travel without any material medium. X-rays, Infrared rays, Radio waves, Photons etc. are all radiations. The range of visible radiations lies between 4000Å and 8000Å. Radiations, whose wavelengths are less than 4000Å or more than 8000Å, are not visible to the eyes because our eyes are not sensitive for such radiations. These are called invisible radiations. Infrared rays, Ultraviolet rays, Gamma rays, X-rays are all invisible radiations.

Speed of Light :

As soon as we switch on the bulb, the light reaches our eyes. The time taken by the light in this case is so little that we cannot measure it, but if light travels a long distance, time taken by it can be measured. Time taken by the light to reach from the moon to the earth is about 1.3 seconds and that from the sun to the earth is 8 minutes 18 seconds. Light from the stars takes very long time to reach the earth's surface.

Though the speed of light is fastest, it travels from one place to another with a definite speed. The space in which light travels is called medium. Light can travel through air, water, glass and even vacuum. The speed of light is maximum in vacuum. In vacuum, it travels with a speed of 2,99,792.5 km/sec.

James Clark Maxwell (1831-79)

Heinrich Hertz (1857-94)

Max Planck (1858-1947)

493

SPECTRUM

In 1666, Sir Isaac Newton let a ray of light, coming from a hole of a window in a room, pass through a prism. The ray of light produced a strip of seven colours on a screen with violet colour on one end and red on the other. This strip contained red, orange, yellow, green, blue, indigo and violet colours. This strip of seven colours was called spectrum of light by Newton. These seven colours in short are expressed as VIBGYOR. By this experiment, Newton proved for the first time that the sunlight which appears to be white is really made up of seven colours. This marked the beginning of the spectroscopy.

We see many spectra in our daily life. The glass candle in a room splits light into different colours. During rainy season, the large number of water droplets hanging in the sky split the sunlight into seven colours and produce rainbow which appears in the opposite direction of the sun. Sometimes we see two rainbows together in which one shows the sequence of colours opposite to the other.

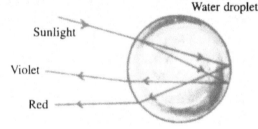

Rainbow appears when the sunlight falls upon the water droplets.

After the study of the spectrum for a long time, scientists found that beside red and violet colours, there exist some other radiations also. These radiations are not visible to the eyes but

Newton's experiments proved to be the beginning of spectroscopy.

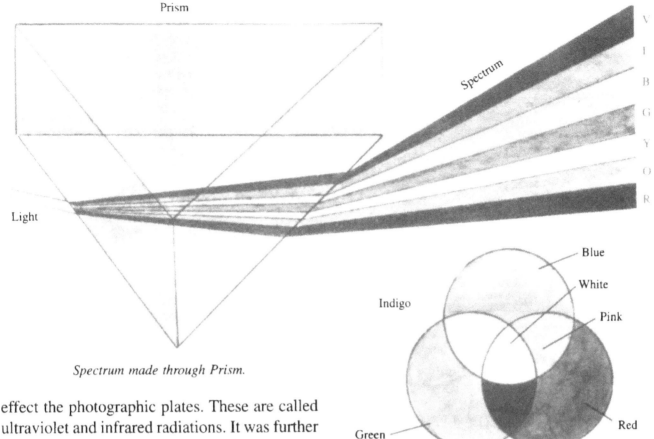

Spectrum made through Prism.

effect the photographic plates. These are called ultraviolet and infrared radiations. It was further investigated that electromagnetic spectrum contains a large number of frequencies extending from gamma rays to radio waves. The spectrum is studied according to frequencies or wavelengths and not by different colours. The visible spectrum as compared to total electromagnetic spectrum is very small. The different wavelengths of visible spectrum are given below:

Wollaston carried out several experiments to obtain pure spectrum. J. Fraunhofer obtained pure spectrum with the help of a prism and made the discovery of plane grating. Fraunhofer made significant contributions in the field of spectroscopy. In 1860, Kirchoff and Bunson studied the spectra of many elements. They proved that the spectrum of every element has some specific lines. They also showed that a

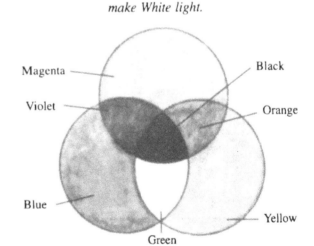

Colour-mixing: Red, Green and Blue together make White light.

Red, Blue and Yellow are the primary colours in paint industry. Mixed together, they make different colours. The colour made by any two of the primary colours is called the complementary colour.

Colour	Violet	Indigo	Blue	Green	Yellow	Orange	Red
Wavelength A°	3900-4460	4460-4650	4650-5000	5000-5700	5700-5900	5900-6200	6200-7600

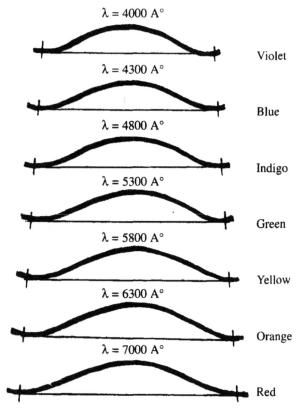

λ = 4000 A° — Violet

λ = 4300 A° — Blue

λ = 4800 A° — Indigo

λ = 5300 A° — Green

λ = 5800 A° — Yellow

λ = 6300 A° — Orange

λ = 7000 A° — Red

Wavelengths of the different parts of light.

Electromagnetic Spectrum

Name	Wavelength (in Angstroms)
Gama Rays	10^{-3} Å to 0.01 Å
X-Rays	0.1 Å to 100 Å
Ultraviolet Rays	100 Å to 4000 Å
Visible	4000 Å to 8000 Å
Infrared	8000 Å to 10^7 Å
Micro Waves	10^7 Å to 10^{11} Å
Radio Waves	10^{11} Å

substance in excited state emits some specific wavelengths and can also absorb the same wavelengths at lower temperatures.

Spectra are of two types, namely, Emission spectrum and Absorption spectrum. Emission spectrum of a substance is obtained by giving energy to its atoms or molecules. This external energy excites the atoms or molecules to higher energy state. When they return to their ground state, the absorbed energy is emitted as the characteristic wavelengths of that substance. The emitted wavelengths may be in visible or invisible range. If the emitted wavelengths are lower than the violet light, they constitute ultraviolet spectrum. On the other hand, if the emitted wavelengths are greater than the red light, they constitute infrared spectrum. Both of these are invisible radiations. All shining objects produce emission spectrum.

If the emission wavelengths are allowed to pass through the vapours of the same substance, the emission lines will be absorbed by the vapours. If the continuous light is allowed to pass through the sodium vapours filled in a tube, you will see two black lines in the yellow region of the spectrum. These are two sodium lines. In sun's emission spectrum also, we see some black lines. These are also absorption lines which were first seen by Fraunhofer and are called Fraunhofer lines.

The emission spectra is also of two types, namely, line emission spectrum and continuous emission spectrum. The absorption spectra is of three types, namely, line absorption spectrum, band absorption spectrum and continuous absorption spectrum. Spectra is studied with the help of prism or grating spectrographs. Nowadays recording spectrographs are used on which spectrum is recorded on the chart paper.

■■

Fraunhofer lines, seen in the spectrum in sunlight.

SOLAR SPECTRUM AND ELECTROMAGNETIC SPECTRUM

Solar spectrum is called the emission spectrum of the sun and it contains many bright and dark lines superposed on the continuous spectrum. The black lines were first observed in 1802 by Wollaston and later studied in great detail by Fraunhofer in 1815. Today these lines are called Fraunhofer lines.

In 1860, Kirchoff and Bunson found out the origin of these lines. According to them, if the emission lines of a substance are passed through the vapours of the same substance, they get absorbed. It is also clear that the elements responsible for the emission of these lines are present in the sun. About 36 elements are present in the sun, which are also found on the earth. Scientists have studied about 20,000 Fraunhofer lines. There are some elements which were first found in the sun. For example, helium was first found in the sun and then Ramsay obtained it

Robert Wilhem Bunson (1811-99).

from clevite on earth. The study of Fraunhofer lines have helped a lot to understand the internal structure of the sun. The sun is composed of gaseous substances whose temperature is not the same everywhere. The core has the highest

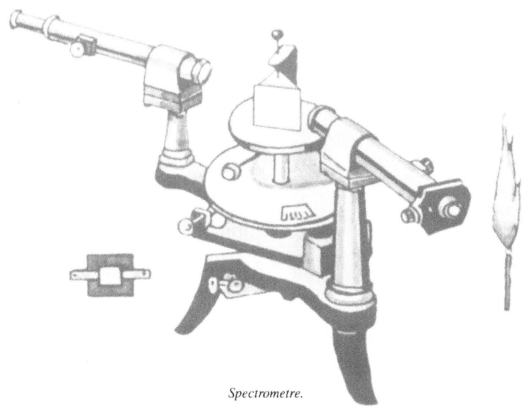

Spectrometre.

497

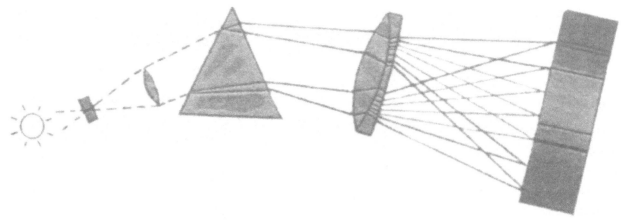

Spectrum of sunlight through spectrometre.

temperature while the outer surface has the lowest temperature. The colours of the stars give us a fair idea of their temperatures. For example, red stars are less hot, yellow more than the red and white stars are the hottest stars.

Electromagnetic spectrum :

Radio waves, infrared waves, ultraviolet rays, visible light, X-rays and gamma rays are all electromagnetic waves. The infrared rays start from the end of the red colour and are extended upto the wavelength of 4×10^{-2} cm. After this, microwaves and radio waves come. The ultraviolet rays have wavelengths below the visible light. The wavelength of ultraviolet rays extends upto 1×10^{-8} m. After this, X-rays spread in the region 1×10^{-8} m to 4×10^{-12} m. At the end

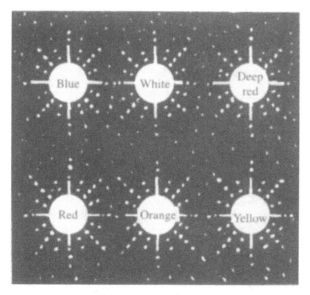

The colours of the stars tell of their temperature.

of the spectrum are gamma rays. No two regions of the spectrum have the dividing line between them. All these rays touch each other. These rays grouped together, are called electromagnetic spectrum.

Ultraviolet spectrum :

Ultraviolet spectrum was discovered in 1801 by Ritter. A little bit of Sun's spectrum gets below the violet colour which is not visible to the eyes but is capable of producing chemical reactions. This region of spectrum extends from 4×10^{-7} m to 1×10^{-8} m.

Ultraviolet rays affect the photographic plates. The sunlight is quite rich with ultraviolet rays but most of it is absorbed by the atmospheric gases. Only a very small part of it reaches the earth. Ultraviolet rays are high energy photons. These rays are produced artificially also and are used to kill bacteria. These rays are also used to treat many diseases. These rays also produce fluorescence and phosphorescence effect when they fall on some substances.

Infrared spectrum :

Infrared spectrum was discovered by William Harshell in 1800. Some part of the solar spectrum beyond red also exists which is not visible to the eyes but produces thermal effects. This extends from 7.8×10^{-7} m to 1×10^{-3} m.

Infrared rays are used in diathermy for treating different ailments. Infrared telescopes are used as night vision devices. These telescopes are used in aeroplanes and tanks.

Radio waves :

Different radio waves are used in modern telecommunication systems such as radio, T.V., telephone, fax, etc. Some important radio frequencies are given below along with their uses:

VLF (Very Low Frequency): Scientists use them for giving special signals.

LF (Low Frequency): AM modulated radio & ship communications.

HF (High Frequency): Ham radio.

VHF (Very High Frequency): FM modulated radio and black and white television.

MF (Medium Frequency): Police radio and MF amplitude modulated radio.

UHF (Ultra High Frequency) : Colour TV.

SHF (Super High Frequency): Space and satellite communications.

The entire electromagnetic spectrum is essential for the human beings.

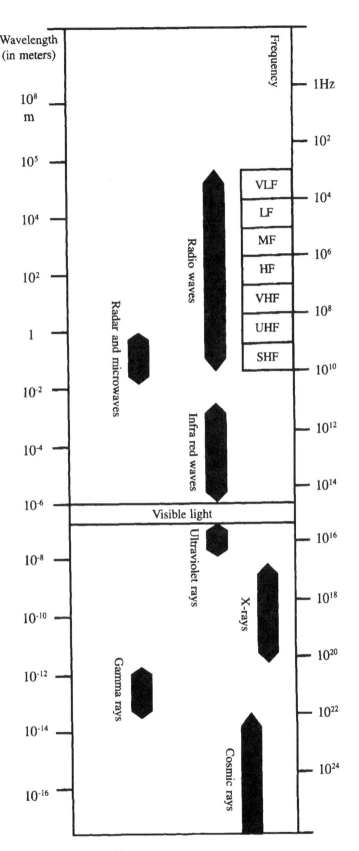

Electromagnetic spectrum.

FLUORESCENCE AND PHOSPHORESCENCE

Some substances, when illuminated by light, emit radiations for a short interval of time. Such substances are called fluorescent materials and this phenomenon is known as fluorescence.

When the light of short wavelengths falls on a fluorescent material, it is absorbed by the material. This energy excites the electrons in the atoms of the substance. The absorbed energy is re-emitted in the form of fluorescent light of larger wavelength and heat. The energy of the emitted light is less than the absorbed light.

Fluorescence :

Fluorescent substances have many uses. Fluorescent tubelights are being used in the houses on a large scale. These tubes are made of glass and their walls are coated with Magnesium tungstate and Beryleum silicate. A tubelight consists of two tungsten electrodes at the two ends. After evacuation, it is filled with some inert gas and a few drops of mercury are put in it. When the current is passed through the tube, electric discharge takes place. Mercury emits ultraviolet light which hits the walls of the tube and gets converted into visible light by the coated material. Fluorescent screens are used in TV picture tubes. Paints and inks used in advertising contain fluorescent substance. When ultraviolet rays are inserted into acidic quinine, this solution shines in the blue colour.

Phosphorescence :

When a substance is illuminated by light and it continues to glow even after the removal of light, such substances are called phosphorescent materials and this phenomenon is called phosphorescence. Phosphorescence can be understood as follows:

When light shines on certain materials, it is absorbed by the electrons. This causes the atoms to go into the excited state. When the atoms come back to their original state, they emit light.

If the light is given out immediately it is called fluorescence but if it is released over a period of time it is called phosphorescence. There are certain substances which glow for hours together after the removal of light.

Some paints and dyes contain phosphorescent materials such as fluoresine, eosine, rhodamine, stilbine etc. These materials are used for making advertising sign boards and road signs. Such paints are also used in the dials of watches and electric switches. They glow in the dark due to the phosphorescence property. Similarly, when ultraviolet light shines on calcium sulphide, it shines for long hours in a dark room. This happens due to phosphorescence.

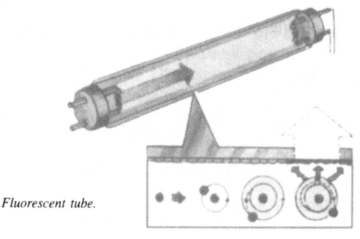

Fluorescent tube.

500

REFRACTION OF LIGHT

When a ray of light goes from one transparent medium into the other, it deviates from its path. This change of direction or deviation is called the refraction of light. If the ray of light enters from the rarer into the denser medium, it bends towards the normal. On the other hand, if the ray enters from denser to rarer medium, it goes away from the normal.

We observe many causes of refraction of light in our daily life. A coin at the bottom of a glass full of water, appears raised because of refraction. In a similar way, a fish appears nearer the surface of the water. Rivers and ponds seem to be shallow because of refraction. A pencil half dipped in a glass full of water appears bent due to refraction. Twinkling of stars is also associated with the refraction of light. The sun at the time of dawn and dusk, though below the horizon, can be seen because of refraction.

There are two rules of refraction:

(i) Angle of incidence, angle of refraction and normal are at one surface only.

A pencil, half-dipped in water, appears bent due to refraction.

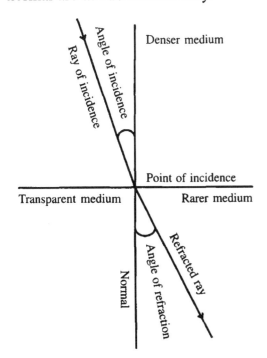

Refraction of light from denser to rarer medium.

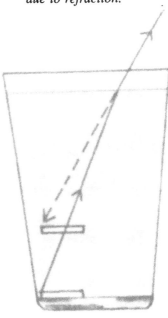

A coin at the bottom of a glass full of water, appears raised due to refraction.

(ii) The measurement of the signs of angle of incidence and angle of refraction is constant. This is known as refractive index.

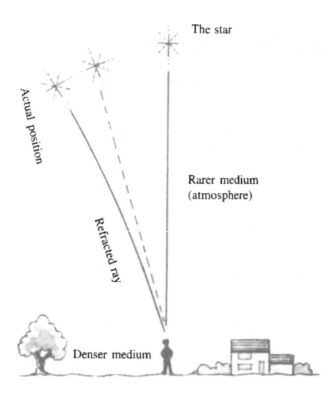

The beams of light deviate from their path due to refraction.

You must have seen twinkling stars in the night. The intensity of their light varies while twinkling. This is caused by the refraction of light. Our atmosphere is never still and calm. Cold and hot winds always keep on blowing in it. This makes the refractive index of air varying. The rays of light coming from a star, when pass through the atmosphere, get deviated from their path. Due to this deviation, light of the stars does not reach our eyes for a very short time. This makes the stars seem twinkling.

Visibility of the sun even when it is below the Horizon :

The sun at the sunrise is visible even when it is below the horizon. This is due to the refraction of light. The air above the surface of the earth becomes rarer with altitude and its refractive index decreases. When the sun is below the horizon at the time of dawn and dusk, its rays travel from rarer to denser medium and due to refraction bend towards the normal. When these rays reach our eyes, the sun appears to be raised with respect to its original position and it is visible to the eyes even when it is below the horizon.

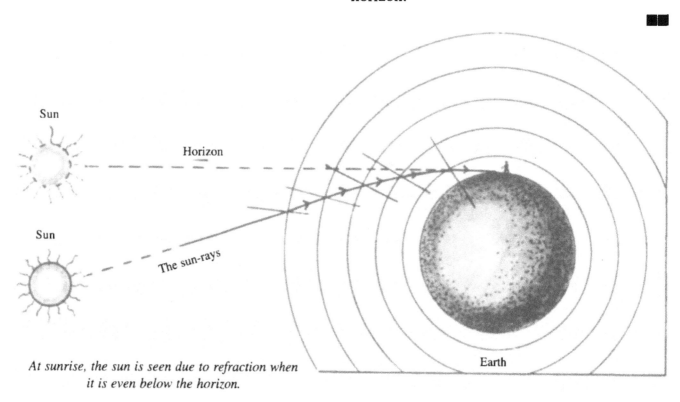

At sunrise, the sun is seen due to refraction when it is even below the horizon.

502

MIRAGE

When a ray of light enters from denser to rarer medium, it goes away from the normal. As the angle of incidence increases, the angle of refraction also increases. For a particular value of angle of incidence, the angle of refraction becomes 90°. The angle of incidence at this value is called critical angle. When the angle of incidence becomes larger than the critical angle, the light does not go into the rarer medium but comes back into the denser medium itself. This is known as total internal reflection of light. There are many phenomena in nature which occur due to total internal reflection of light. Mirage in deserts is also a result of total Internal Reflection.

A mirage familiar to most people is the shimmer on a hot, dry road. The mirage looks like a pool of water.

In deserts, sandy land becomes quite hot due to high temperature. This makes the nearby layers of earth rarer than the cool upper layers. In this condition, light from the top of a tree gets refracted and goes away from the normal. There

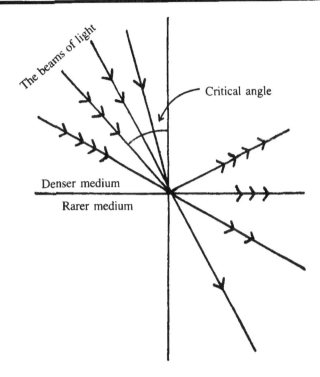

The total internal reflection of light.

reaches a condition, when the angle of incidence becomes more than the critical angle. In this situation, total internal reflection of light takes place. When these rays of light reach the eyes

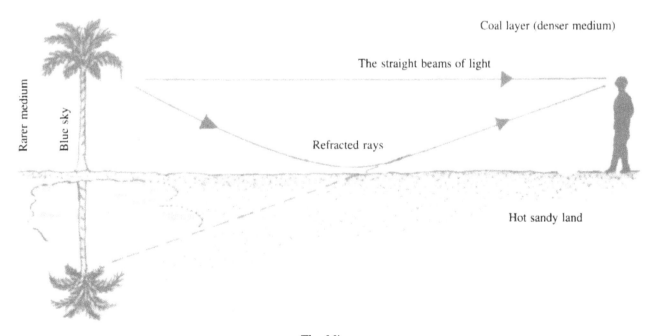

The Mirage.

503

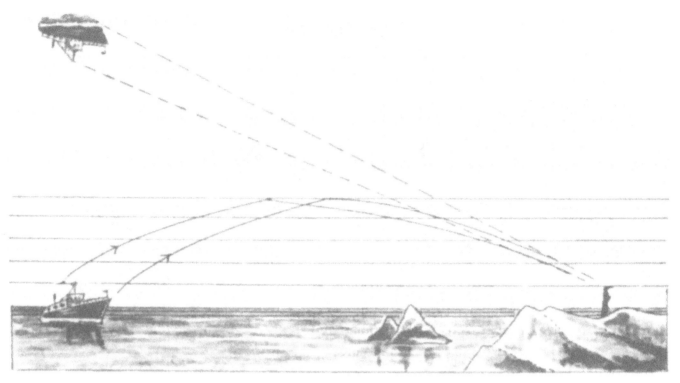

The looming in cold region.

of a traveller, he sees an inverted image of the tree. This image gives an illusion of the presence of water. A similar phenomenon is observed by deer in the deserts. A deer runs for water but in reality there is no water. Deer goes on running from one place to another in search of water and loses its life.

Looming :

As we see the mirage in hot deserts, a similar phenomenon called looming is seen in cold regions. On the South Coasts of Italy, there is a village from where Sicily is seen to be hanging upside down. This is called Fata Morgana.

In cold countries, the air near the earth's surface gets cooler and denser as compared to the upper layers. As we go above the earth's surface, the air becomes rarer and rarer. The rays of light coming from a ship or some other object, travel from denser to rarer medium and get away from the normal. There reaches a condition when angle of incidence becomes larger than the critical angle and total internal reflection of light takes place. This makes the inverted image of the ship hanging in the sky. In looming hills, valleys and snow covered peaks at 120° across the horizon appear inverted in the sky.

There are other examples of total internal reflection also:

(a) The cutting of diamond at different angles produces shining due to total refraction.

(b) Light comes out from a 90°-prism due to total internal refraction.

Two things are necessary to be kept in mind in case of total internal refraction:

(a) The light rays should travel from a denser medium to a rarer medium.

(b) In the denser medium, the measurement of the angle of incidence is more than the critical angle.

LENSES

A lens is a specially-shaped piece of transparent material used to bend rays of light. Lenses are usually of two types – Convex lens and Concave lens. Convex lenses are thick at the centre and thin at the edges while concave lenses are thin at the centre and thick at the edges. These two types of lenses are available in six standard forms, namely, double convex lens, plano convex lens, concavo-convex lens, double concave lens, plano concave lens and convexo-concave lens.

In standard forms also, lenses are of two types – thin lenses and thick lenses. Usually thin lenses are used most. Lenses are mainly spherical and cylindrical.

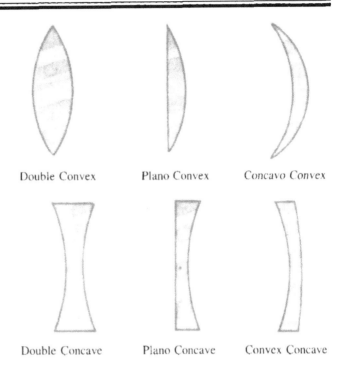

Double Convex   Plano Convex   Concavo Convex

Double Concave   Plano Concave   Convex Concave

The standard form of lenses.

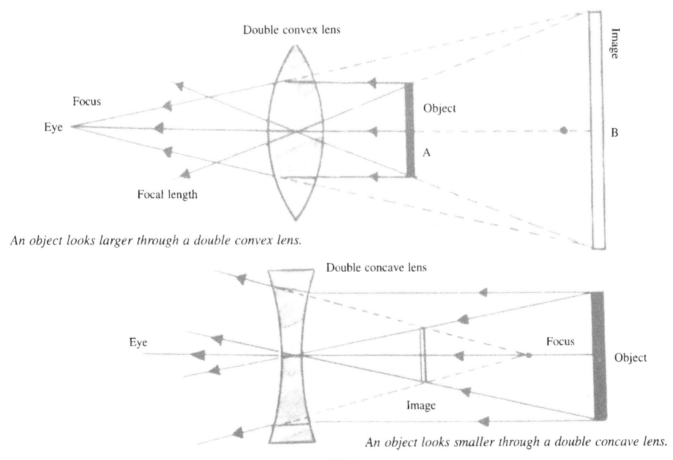

An object looks larger through a double convex lens.

An object looks smaller through a double concave lens.

The line passing through the two centres of curvature of the lens is called principal axis. A point on the principal axis at the centre of the lens is called the optic centre.

When the rays of light pass through a lens, they bend towards or away from the principal axis due to refraction of light. Thus a lens simply bends the rays of light. The larger the bending of light, the higher is the power of the lens. Every lens has a focal point. The point where the rays coming from the infinity get concentrated, is called the focus of the lens. Usually the lenses of small focal lengths bend the light rays more.

Lenses form images by the phenomenon of refraction. Camera lens forms the image on the film just as our eyes form the image on the retina. The images formed by lenses suffer with the following aberrations:

Chromatic Aberration :

The image of a white object formed by a lens is usually coloured and blurred. This defect is called chromatic aberration. This defect is removed by making a combination of two lenses – one of crown glass and the other of flint glass. Such a combination makes an achromatic lens. These lenses are made in such a way that

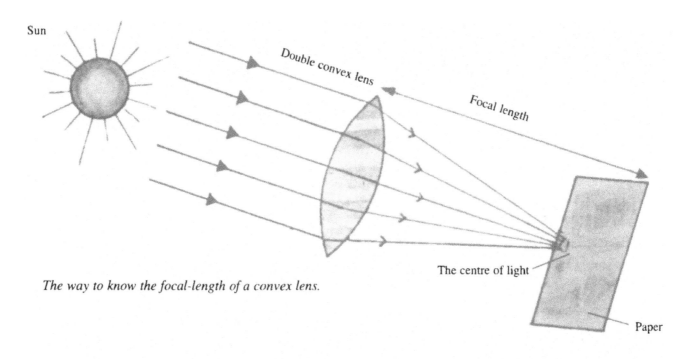

The way to know the focal-length of a convex lens.

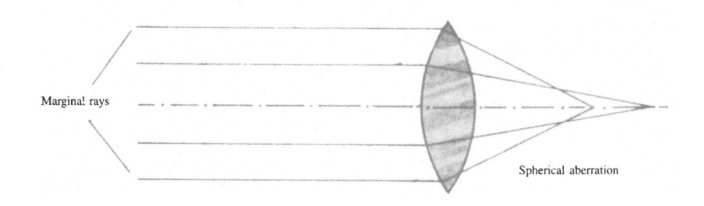

506

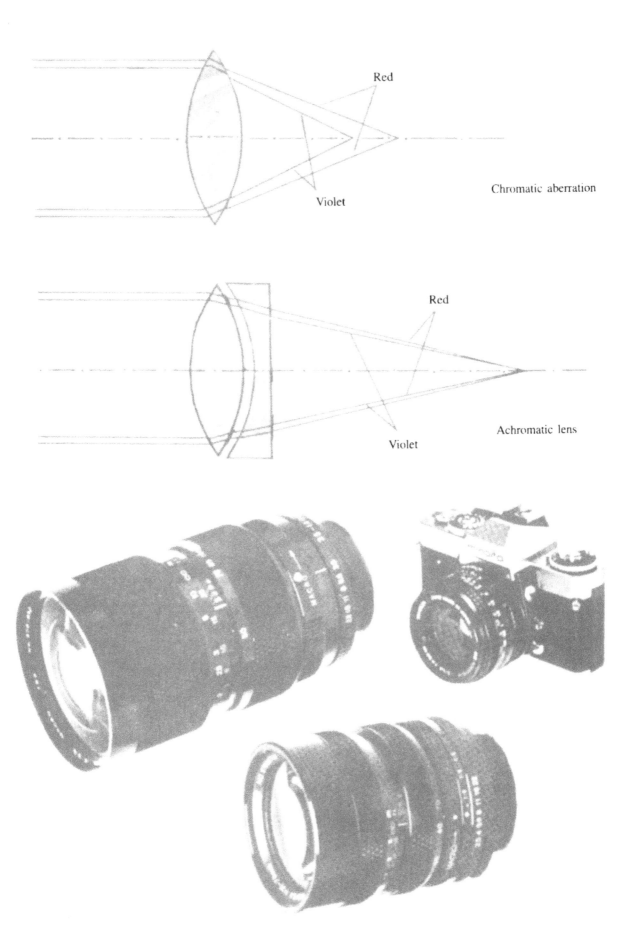

Chromatic aberration

Achromatic lens

High quality lenses are used in the cameras.

507

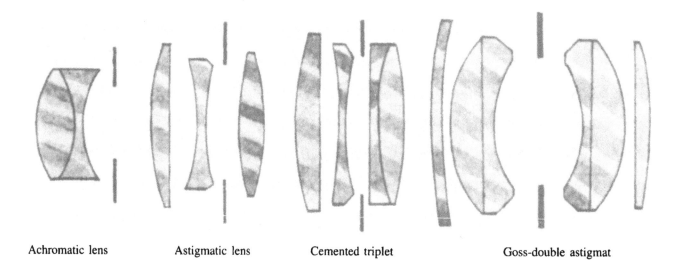

| Achromatic lens | Astigmatic lens | Cemented triplet | Goss-double astigmat |

The primary structure of Photographic lenses.

the deviation produced by one lens is cancelled by the other. The lens made of crown glass is converging in nature while that of flint glass is diverging. The achromat of two thin lenses is only effective for two particular colours and not for all. The image formed by such an achromat contains some percentage of other colours but the presence of other colours does not affect the quality of the image.

In addition to the chromatic aberration, lenses suffer with other defects also, such as, astigmatism, spherical aberration and coma. Lenses having astigmatism produce egg-shaped images. It is very difficult to remove astigmatic defect. Lenses made for objectives of telescopes should be free from spherical aberration, coma and longitudinal chromatic aberrations. Lenses used in photographic cameras should be partially corrected for spherical aberration and coma but should be fully corrected for astigmatism, curvature and distortion.

Today, lenses of various shapes and different focal lengths are being made. Very thin contact lenses which can be put on the pupil of eye are available in the market. Almost all optical instruments use lenses on large scale.

508

REFLECTION OF LIGHT

When light falls on a smooth, polished and bright surface or mirror, most of it bounces back into the same medium. This is called reflection of light. The different images, we see with the help of the plane mirrors, are formed due to the reflection of light.

There are two laws of reflection of light. According to the first law, the angle of incidence is equal to the angle of reflection. According to the second law of reflection, the incident ray, normal and reflected ray, all lie in the same plane.

If an object is moved away from a plane mirror, its image will also move the same distance in the backward direction as that of the object. The size of the image formed by a plane mirror is the same as that of the object but it is laterally inverted. If a person standing before a plane mirror lifts up his right hand, his image will appear lifting up its left hand. The letters in a book appear laterally inverted in the plane mirror. This is called lateral inversion.

When two plane mirrors are bent at an angle, and an object is lying in between them, the image of this object formed by the mirror, acts as an

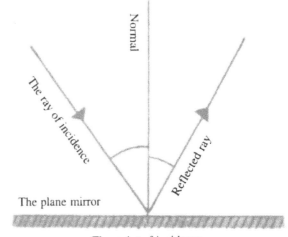

Reflection of light on a plane mirror.

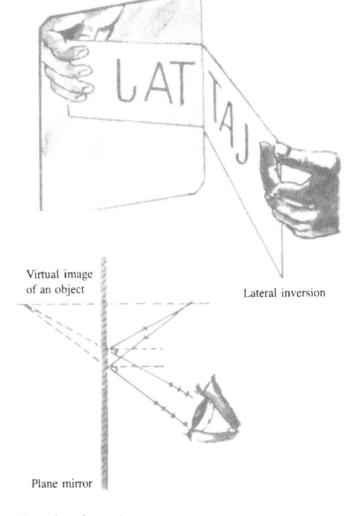

Lateral inversion

Virtual image of an object

Plane mirror

The image of an object through a plane mirror.

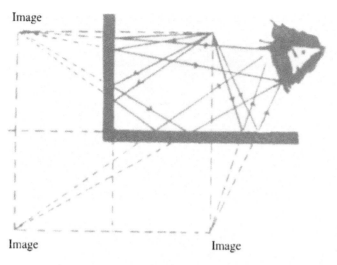

Image

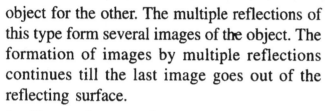

Image   Image

The three images of an object when two mirrors are placed at 90° angle.

object for the other. The multiple reflections of this type form several images of the object. The formation of images by multiple reflections continues till the last image goes out of the reflecting surface.

1. If two mirrors are bent on an angle (suppose θ), then the number of images is as follows:

   If 360/θ is even, then η = (360/θ − 1)

   If 360/θ is odd, then η = (360/θ − 1), if the object is symmetrical or η = (360/θ), if the object is non-symmetrical.

2. If the mirror is turned to θ°, then the refracted ray will turn to 2θ.

3. To see an image of a person, the length of the mirror should be half of the height of that person.

Mirrors are mainly of three types – plane, convex and concave. Plane mirrors are normally used for seeing the face and combing hair. Concave mirrors are used for shaving and examining ears. They are also used in telescopes. Convex mirrors are used in motor vehicles for seeing the rear view. All the three types of mirrors work on the phenomenon of reflection of light. In a plane mirror, the size of the object and the image is the same while in a convex mirror, the size of the image is smaller than the object but in a concave mirror, the size of the image depends on the position of the object i.e. how far the object is situated from the mirror.

SIGHT AND EYE

Eye is an important organ of our body. The rays reflected from an object reach the eyes. An inverted image of that object is formed on the retina of the eyes. The electrical signals from every point of the image reach our brain through the optic nerve. Brain makes the erect image of the object and we see the object in the correct perspective.

The working of our eyes resembles a photographic camera. The size of the eye is about 25 mm. It is covered by a white, opaque and rigid layer from outside.

Eyes can see both the nearby and distant objects. The distant point which eyes can see clearly is called the far point. Usually this point is at infinity. The nearest point from the eyes which can be seen clearly is called nearpoint. For a normal eye this distance is roughly 25 cm. The distance between the nearest and farthest point is known as the range of vision. The range of vision is from 25 cms to infinity.

Due to some reasons, eyes of some people develop the following defects:

(i) Short-sightedness or Myopia

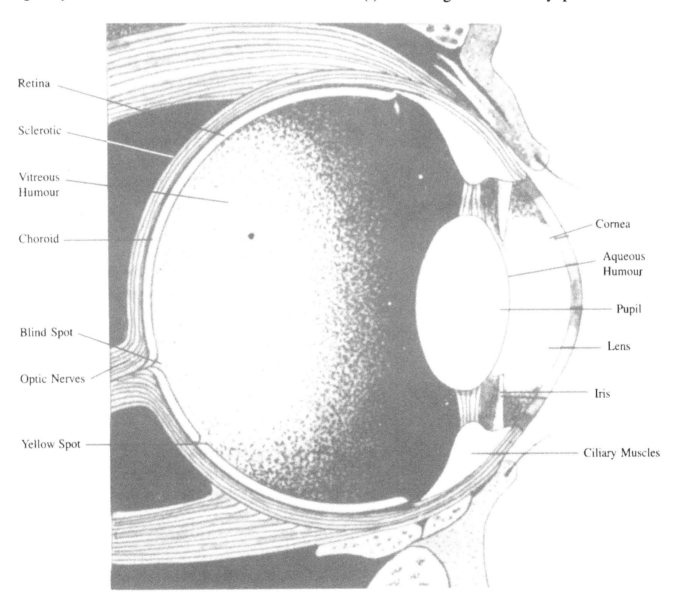

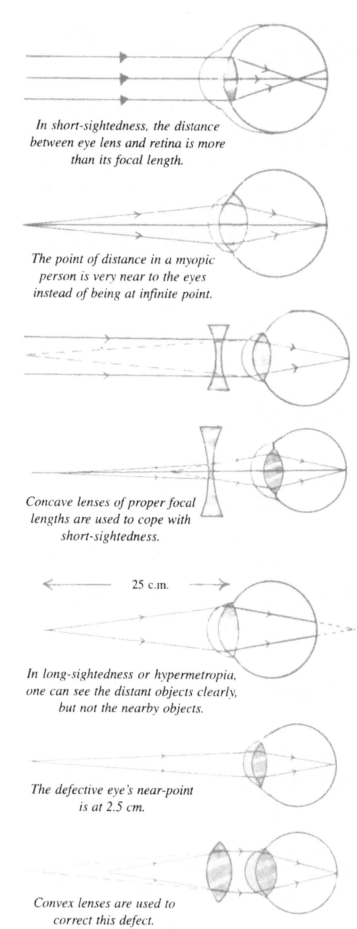

In short-sightedness, the distance between eye lens and retina is more than its focal length.

The point of distance in a myopic person is very near to the eyes instead of being at infinite point.

Concave lenses of proper focal lengths are used to cope with short-sightedness.

← 25 c.m. →

In long-sightedness or hypermetropia, one can see the distant objects clearly, but not the nearby objects.

The defective eye's near-point is at 2.5 cm.

Convex lenses are used to correct this defect.

(ii) Long-sightedness or Hypermetropia
(iii) Astigmatism

Short-sightedness or Myopia :

A person suffering from this defect can see the nearby objects clearly but the distant objects appear blurred. This defect occurs due to the increase in distance between the eye lens and the retina. This causes the image of the object to form before the retina due to which the objects appear blurred. This defect is corrected by the use of spectacles of concave lenses of proper focal lengths. These lenses make the image of the objects at retina. This defect mainly occurs in the eyes of the students.

Long-sightedness or Hypermetropia :

A person suffering from this defect can see the distant objects clearly but the nearby objects appear blurred. The image of the nearby objects is formed beyond the retina. This defect is corrected by using convex lenses of right focal lengths. These lenses make the image on the retina so that the object appears clearly. This defect generally occurs after the age of 40 years.

Astigmatism :

In this defect, the image is formed at different distances from the retina which makes the objects appear blurred. This defect is corrected by using cylindrical lenses of appropriate focal length.

Presbyopia :

A person suffering from this defect can see neither the nearby objects nor the distant objects clearly. This defect is corrected by the use of bifocal lenses. This defect occurs in the eyes of the elderly people.

Colour Blindness :

A person suffering from this defect cannot establish difference between any of the colours. This is a hereditary defect. No cure is there for this disease. This defect is tested on the motorcyclists. No driving license is provided to the people suffering from this disease.

WORLD OF COLOURS

Colour is an aspect of light that enables human beings and some animals to differentiate between objects that are otherwise identical. Eyes tell us about the different colours.

The sunlight, commonly called white light, is made up of seven colours. We can break white light into seven colours by a prism. When white light reaches an object, some wavelengths are reflected and some are absorbed. The wavelengths that are reflected, make up the colours we see. For example, the grass reflects green colour of white light, that is why it appears green.

A white object reflects all the colours, while a black object absorbs partially all the colours.

Red, green and blue are commonly called the primary colours of light. Any two of these colours can easily be mixed to form a third colour. All three can be mixed to form white light. When red and blue overlap, we get magenta, while red and green give yellow. Colour television depends on the principle of additive colour mixing.

Colour concept of paints and dyes is different from that of light. The primary colours for paints and pigments are magenta, yellow and blue. The artist can make any colour using these three primary colours. For example, magenta and yellow give orange colour.

The world of colours is really beautiful but some people are colour blind i.e. they cannot distinguish between the two colours. Colour blind people can see only two colours. This incurable inability often affects people by birth.

MICROSCOPES AND TELESCOPES

Microscope is an optical instrument which is used to see nearby tiny objects enlarged. Telescope is also an optical instrument but it is used to see distant objects distinctly and enlarged.

Microscope :

Microscope is used to see those tiny objects which cannot be seen with naked eyes. It forms the image of the object at a distance of distinct vision or beyond it which can be seen by the eyes very clearly.

Microscopes are of two types – simple microscope and compound microscope. Simple microscope is just a convex lens. When the object is placed between the focus and the optical centre of the lens, an enlarged and erect image of the object is formed. This cannot give larger

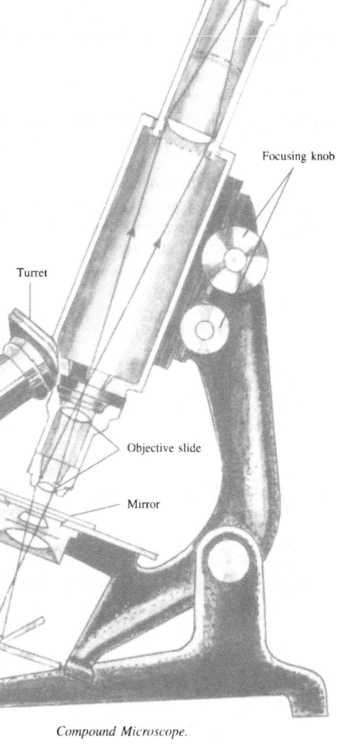

Compound Microscope.

514

Galileo with his telescope.

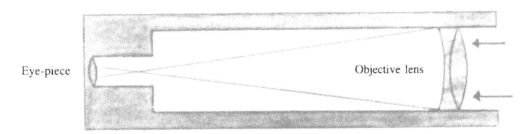

Eye-piece   Objective lens

Ancient astronomical telescope.

magnification, that is why two lenses are used in a microscope. The microscope which used two lenses is called a compound microscope. The lens which remains towards the object is called objective lens, while the one which remains towards the eyes is called eyepiece. Objective lens is smaller in size as compared to eyepiece. Both the lenses are of small focal lengths. Both the lenses are fitted in the metallic tubes. Eyepiece can be moved up and down by a knob. The object which is to be seen is placed before the objective lens at a distance slightly more than its focal length.

A good quality compound microscope can have a magnification upto 2,500 times. For larger magnifications, electron microscopes are used. They can magnify upto 20,00,000 times. In an electron microscope, electron beams are used instead of light beams. It consists of magnetic lenses. The waves associated with the electrons have very small wavelengths, that is why their magnification is very large. Few years back, an electron microscope was developed in Japan, which can magnify upto 10,00,000 times.

Today, scientists have developed microscopes which make use of ultrasonic waves.

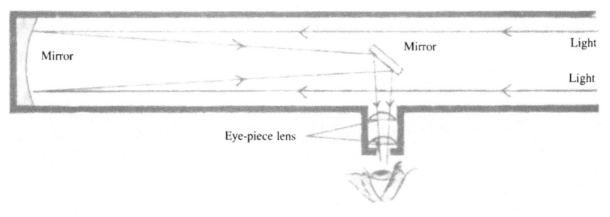

Newton's reflecting telescope.

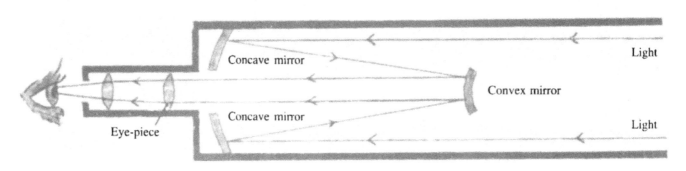

Cassegrainean telescope was developed after Newton's telescope.

Astronomical Telescope :

The telescope was invented in 1608 by a Dutch optician Hans Lipperskey. In 1609, the Italian astronomer Galileo made his first telescope and observed the moons of Jupiter and the rings of Saturn.

Astronomical telescope is used to see the heavenly bodies or the distant objects. This instrument usually consists of two convex lenses fitted in a metallic tube on the same axis. The lens which remains towards the object, is called objective lens while the one which remains towards the eyes is called eyepiece. The aperture and focal length of the objective lens is larger than the eyepiece. The distance between the two lenses is adjusted by a knob. Such a telescope produces the inverted image of the object which can be made erect by using a third lens or a prism.

Reflecting Telescope :

The reflecting telescope was invented by Sir Isaac Newton in 1668. It used a concave mirror instead of a lens. After that N. Cassegrain developed different types of reflecting telescopes.

The reflecting telescope has a large-sized concave mirror. It is not affected by spherical aberration and chromatic aberration. It does not absorb light, that is why its range is very large and the quality of the image is very good.

The largest reflecting telescope is mounted at the Astro Physical Laboratory of Russia. The diameter of its mirrors is 600 cm.

Radio Telescopes :

The radio telescope receives radio waves instead of light waves. A radio telescope has a huge dish-

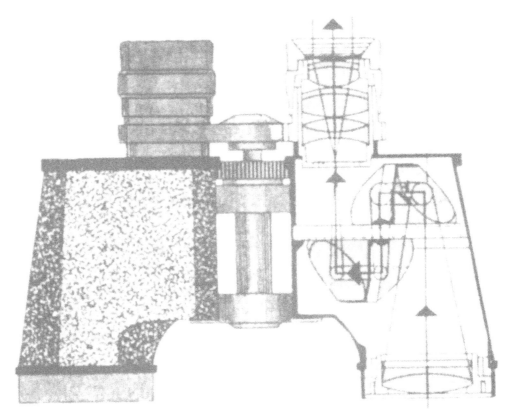

The image seen by binoculars is more clear than those seen by telescope. There are two telescopes with two prisms in each tube of binoculars.

shaped reflector which reflects radio waves coming from heavenly bodies to a detector. This can be rotated in any direction. It is used to study planets and stars. The largest radio telescope, mounted at Arecibo (Puerto Rico) has a dish 305 m across. This can receive and detect the radio waves coming from a distance of 1500 crore light years. Here it is important to note that one light year is the distance covered by light in one year. Its measurement is 9.46×10^{12} kms. France also has a radio telescope almost of the same power.

A few years ago, Hubble telescope was established in the space by a space shuttle. It is ten times more powerful than any other radio telescope situated on the earth.

■■

Radio telescope.

517

LASER

The term 'laser' stands for 'light amplification by stimulated emission of radiation'. Laser beams are electromagnetic waves which are highly directional, coherent, monochromatic and intense. The divergence of these beams is small. The properties of laser light are different than the conventional light. The source which produces laser beams is called laser.

The first laser of the world was developed in 1960 by an American Scientist T.H. Maiman. It was a ruby laser. It made use of a ruby rod and a xenon flash lamp. When the lamp was lighted by the electrical energy, a red beam of light came out from the rod which was brighter than the sun. This beam of light was called laser beam.

The unique properties of laser light have made it highly useful in the field of defence,

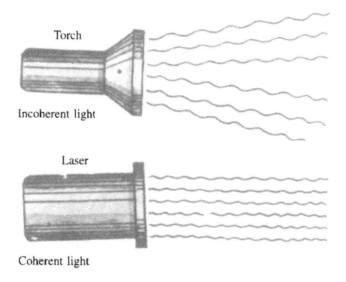

Torch

Incoherent light

Laser

Coherent light

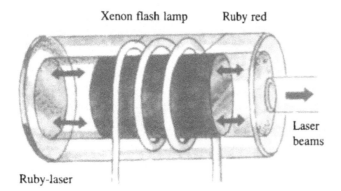

Xenon flash lamp   Ruby red

Laser beams

Ruby-laser

science, industry and medicine. Today scientists have developed many substances which are capable of producing laser beams of different wavelengths. The main lasers are He-Ne laser, Argon laser, Carbon dioxide laser, Neodymium YAG laser and glass laser. Today, we have laser beams which can cover the spectrum range from infrared to ultraviolet region including the visible range.

Laser beams have very low divergence. Even after travelling upto the surface of the moon, ruby beams spread only 3 km across. Using laser beams, the distance of the moon was measured in 1969 with an accuracy of 6 inches. Laser based instruments have been developed for defence purposes which are being used for missile guidance, pinpoint dropping of bombs and range finding. Such laser-based weapons were used on large scale in both the gulf wars.

Laser is now being used on large scale in the field of science, technology, medicine, defence and communications. Laser beams are being used for drilling the diamonds, metal cutting, welding etc. These beams have proved highly useful in eye surgery for curing cataract and retinal defects. With laser beams, bloodless surgery has become a reality. These beams have proved very useful in space research. These are also being used in Holography, Photography, radio, television, computer, printing etc. In fact, laser beams are being used almost in every field of life. Laser beams have brought revolution in curing the tissue functioning of the human body. These beams have proved very fruitful in curing many diseases. These are also used for defence purposes in the battlefields.